Library of Congress Cataloging-in-Publication Data

Artificial recharge of ground water

Bibliography: p.
Includes index.
1. Water, Underground—United States—Artificial recharge. I. O'Hare, Margaret P. II. University of Oklahoma. Environmental and Ground Water Institute.
TD404.A79 1986 627′.56 86-2990
ISBN 0-87371-050-9

LEWIS PUBLISHERS, INC.
121 South Main Street, P.O. Drawer 519, Chelsea, Michigan 48118

PRINTED IN THE UNITED STATES OF AMERICA

ARTIFICIAL RECHARGE OF GROUND WATER

Status and Potential in the Contiguous United States

Margaret P. O'Hare
Deborah M. Fairchild
Paris A. Hajali
Larry W. Canter

Environmental and Ground Water Institute
University of Oklahoma, Norman, Oklahoma

MARGARET P. O'HARE is a research assistant in the Environmental and Ground Water Institute at the University of Oklahoma. She received a BS in geology with a special interest in ground water from the University of Oklahoma. She is currently a graduate student pursuing a degree in Environmental Science in the ground water quality management program. Her thesis work focuses on the development of artificial recharge site selection criteria.

DEBORAH M. FAIRCHILD is a senior environmental scientist and has been with the Environmental and Ground Water Institute for the past five years. She received a BS in biology and chemistry from the State University of New York at Oswego in 1975. She worked as a chemist for the U.S. Department of Agriculture while obtaining an MS in soil microbiology from West Virginia University in 1979. Since coming to the University of Oklahoma she has worked on many ground water availability and quality studies focused on petroleum production and agriculture.

In addition, she has worked on several environmental impact statements and has served as conference organizer, information and data management specialist, and quality assurance officer for laboratory studies. Her professional activities have resulted in over 30 publications and presentations. In 1985, Ms. Fairchild received the Distinguished Service Award from the University of Oklahoma.

PARIS A. HAJALI, research assistant at the Environmental and Ground Water Institute, received his BS and MS degrees in Civil Engineering from the University of Oklahoma in 1979 and 1981, respectively. He worked for several years as an environmental engineer for consulting firms in Beirut, Lebanon. He also taught hydrology, solid waste and water and waste water treatment and disposal at the American University of Beirut from 1982 to 1984. He is currently completing a doctorate degree in Civil Engineering, with emphasis in ground water management, at the University of Oklahoma.

LARRY W. CANTER, P.E., is the Sun Company Professor of Ground Water Hydrology, and Director, Environmental and Ground Water Institute, at the University of Oklahoma, Norman, Oklahoma. Dr. Canter received his PhD in Environmental Health Engineering from the University of Texas in 1967, MS in Sanitary Engineering from the University of Illinois in 1962, and BE in Civil Engineering from Vanderbilt University in 1961. Before joining the faculty of the University of Oklahoma in 1969, he was on the faculty at Tulane University and was a sanitary engineer in the U.S. Public Health Service. He served as Director of the School of Civil Engineering and Environmental Science at the University of Oklahoma from 1971 to 1979.

Dr. Canter has published several books and has written chapters in other books; he is also the author or co-author of numerous papers and research reports. His research interests include environmental impact assessment and ground water pollution control. In 1982 he received the Outstanding Faculty Achievement in Research Award from the College of Engineering, and in 1983 the Regents' Award for Superior Accomplishment in Research.

Dr. Canter currently serves on the U.S. Army Corps of Engineers Environmental Advisory Board. He has conducted research, presented short courses, or served as advisor to institutions in Mexico, Panama, Colombia, Venezuela, Peru, Scotland, The Netherlands, France, Germany, Italy, Greece, Turkey, Kuwait, Thailand, and the People's Republic of China.

PREFACE

Artificial recharge involves the transport of water via engineered systems from the surface of the earth to underground water-bearing strata, where it may be stored for future use. Direct artificial recharge systems include spreading basins, recharge pits and shafts, and recharge wells. Basins are typically used in areas with shallow unconfined aquifers; wells are used in areas with deep confined aquifers. Artificial recharge can also be accomplished via indirect means, such as enhanced streambed infiltration and the use of conjunctive wells. As more demands are placed on ground water resources in the contiguous United States, artificial recharge may represent an attractive option for replenishing diminishing ground water resources. Technical requisites for successful artificial recharge projects include favorable hydrogeological conditions, and the availability of a sufficient quantity of source water of adequate quality.

This book provides a summary of the status and potential for artificial recharge in the contiguous United States. While the focus is on artificial recharge for enhancing ground water availability for agricultural uses, the book is not limited to this single purpose. The book is organized into five chapters and three appendices. Key chapters are included on technical descriptions of both direct and indirect artificial recharge methods, site selection and evaluation criteria for artificial recharge projects, case studies of projects and methodologies for determining the potential of geographical areas for artificial recharge. The key appendix contains an annotated bibliography of 345 pertinent references identified from computer-based searches of the literature.

The authors wish to express their appreciation to a number of individuals involved in a variety of ways in the development of this book. First, the support of the U.S. Soil Conservation Service (SCS) and the assistance of Clive Walker from the Washington headquarters office is acknowledged. Dennis Erinakes of the SCS Technical Center in Ft. Worth, Texas, provided valuable guidance and suggestions via a series of meetings with the authors.

Several research staff members of the Environmental and Ground Water Institute (EGWI) at the University of Oklahoma participated in information gathering and/or discussions related to the development and organization of this book. Included were Dr. Gary Miller, Assistant Professor of Environmental Science; and Sam Atkinson, Debi Curry, and Sa Ba Lee, all graduate research assistants. The able typing assistance of Mrs. Leslie Rard and Mrs. Mittie Durham is gratefully acknowledged. Finally, and most important, the authors are indebted to Mrs. Wilma Clark of EGWI for her typing skills and sincere commitment to the preparation of this manuscript.

The authors also wish to express their appreciation to the University of Oklahoma College of Engineering for its basic support of faculty and research staff writing endeavors, and to their families for their understanding and patience.

Margaret P. O'Hare — Deborah M. Fairchild
Paris A. Hajali — Larry W. Canter

To Mom, Carolyn, Anne, John, Michael, and Joan

To My Husband, Burt

To Andreas, Jeanette, Danay, and Aphrodite

To Donna, Doug, Steve, and Greg

LIST OF TABLES

LIST OF FIGURES

CONTENTS

CHAPTER 1

INTRODUCTION

In any given area of the United States, the available supply of ground water does not necessarily constitute the usable supply. For example, in many locations a large percentage of the available supply is too saline for typical uses. In other locations, a large portion of the available supply may be consumed by vegetation that is generally considered to be nonbeneficial economically. Even in areas where all the available supply is of usable quality, it may not be present in sufficient quantities to sustain withdrawals for intended uses over extended time periods. However, a number of water-management practices are available for increasing the usable ground water supply. They include (1) artificial recharge, (2) reduction of natural discharge by nonbeneficial vegetation, (3) demineralization, and (4) recycling (Todd, 1983).

Artificial recharge is the process of replenishing ground water through man's activities. It consists of manipulating input to ground water reservoirs in addition to regulating pumping withdrawals. In the broadest sense, artificial recharge includes not only planned replenishment but also additions to the ground water that are incidental to other activities. Ground water is replenished in many places by irrigation in exceedingly permeable soils, in natural recharge areas where the irrigator knows he will add to ground water storage, and in some places by flooding to leach soluble salts from saline or alkaline soils. However, for purposes of this book, these incidental activities will not be considered as planned artificial recharge activities.

CONCEPTS OF ARTIFICIAL RECHARGE

The concept of replacing salty ground water naturally present in aquifers with fresh water, and the concept of replenishing diminishing ground water supplies, has received increasing attention in recent years. Artificial recharge may be accomplished by direct and indirect methods. Direct methods include water spreading by means of ponds, check dams, pits, furrows, or ditches to increase the amount of water infiltrating from the surface into the ground water aquifer. Impounded surface water can infiltrate permeable portions of unconfined aquifers near the land surface. Direct injection of water into ground water can also be accomplished through wells or shafts. Indirect methods of recharge include inducing movement of water from streams or lakes into underground formations, and preventing the natural flow of ground water to the land surface by lowering ground water levels. Chapter 2 contains more detailed descriptions of these artificial recharge methods.

Water pumped down an injection well into a deep saline aquifer does not immediately mix with the native ground water, but instead forces the native water away, creating a bubble or cylindrical-shaped body of injected water around the well. The boundary or interface between the fresh and salty water will be somewhat gradational, owing to mixing, diffusion and dispersion. From an overall viewpoint, however, the injected body of water remains more or less intact. Movement of water in deep saline zones is very slow, thus injected water will not move rapidly away from the injection site.

Retrieval of the total amount of injected water by pumping from an injection well or a separate recovery well is never 100 percent effective, owing to the partial mixing. Experiments have shown recovery rates of 35 to 75 percent, and have demonstrated that the retrieval efficiency becomes better as more injection/retrieval cycles are performed. To improve the recovery efficiency, it is desirable to initially inject more water than needed, so as to create a buffer zone around the injected body. The difference in density between the native saline ground water and the injected fresher water may tend to cause layering of the water in the aquifer, thereby reducing the recovery potential. This effect seems to be more pronounced in zones having high transmissivities.

The capacity of a geological formation to accept injected fluids over long periods of time is surprising. As a single hypothetical example, injection at a rate of 500 gpm (gallons per minute) for 200 years into a zone 100 feet thick and having an overall porosity of 0.15 would create a bubble extending outward only 2.5 miles from the injection well (Palmer, 1983).

ADVANTAGES OF ARTIFICIAL RECHARGE

Ground water is being used increasingly for domestic and industrial water supplies (McGuiness, 1969). Many surface reservoirs cannot meet local water demands, therefore, the potentialities of aquifers as reservoirs to be alternatively filled and emptied has been examined (Kazmann, 1967). For the most part, aquifers have not been considered as reservoirs but as potential areas for "mining". The advantages of aquifers over surface water impoundments as reservoirs for cyclic storage of water are: (1) permanence; (2) no loss of storage capacity due to sedimentation; (3) no loss of water due to evaporation; (4) less vulnerability to destruction or contamination; and (5) the absence of threat to downstream communities (there is no dam to break and no flood to be feared) (Kazmann, 1967).

Ground water in storage can be withdrawn, replaced, or increased if the aquifer is not hydraulically connected to streams or surface-water reservoirs at numerous locations. For example, during a drought or any period in which water demand increases, water supplies can be obtained by pumping wells and mining ground water from storage. Mining eventually lowers the potentiometric surface or water table, possibly over large geographical areas. Surplus water supplies can be added to the ground water in storage by using artificial recharge. Water added to storage eventually raises the potentiometric surface over the water table, possibly over large geographical areas (Todd, 1983). Water commonly used for artificial recharge includes natural stream flow, high flows diverted from a nearby water course, cooling waters, diverted storm drainage water or wastewater treatment plant effluent (treated or untreated) (Pettyjohn, 1974).

Artificial recharge has many potential uses and advantages. It may be used either to replace ground water where water levels have been drawn down or to temporarily increase local ground water storage for later recovery. Artificial recharge can also aid in smoothing out streamflow variations—that is, aquifers can be used to absorb flood runoff, and release the water back to the stream as runoff subsides (Todd, 1983). It can also be used to store cooling water or heat for industrial processes because the water is kept at a relatively constant temperature underground. Artificial recharge is also used as a barrier to reduce salt-water encroachment and has been investigated as a method for preventing subsidence from worsening. Recharged water may also have better water quality than that which was injected because the water is filtered through natural sand and gravel formations. Other advantages of ground water reservoirs over surface reservoirs are reduced costs, larger capacities, and use of large aquifers as distribution systems (Parker, Cohen and Foxworthy, 1967).

The recharge, movement, and discharge of ground water also may facilitate water-management practices. Any technique that increases natural recharge to, or decreases natural discharge from, an aquifer will increase ground water in storage that can be withdrawn by pumping wells. Production wells can be located according to aquifer properties and boundaries and pumped according to calculated time schedules. The resulting withdrawals induce changes in ground water movement, control the distribution of ground water in storage, and affect the nearby streamflow. The resulting changes in ground water and surface water can result in an improved supply that is available at the desired time and location. Aquifers must be managed so that storage will be reduced during peak consumption and replenished by recharge during periods of surplus storage water (Todd, 1983).

HISTORICAL INFORMATION ON ARTIFICIAL RECHARGE

Research studies on artificial recharge can be traced back to the 1880's when surface spreading was attempted in Denver, Colorado (Dvoracek and Peterson, 1971). As early as 1895, floodwaters were spread over the alluvial fan at the mouth of San Antonio Canyon to sustain the many flowing wells in the Upper Santa Ana Valley in southern California. The practice spread to other parts of the valley and was so successful that downstream users obtained an injunction to prohibit artificial recharge from local waters until adjudication of the waters of the river basin had been completed (Toups, 1974).

By the 1930's all the major ground water basins in southern California were in an overdraft condition. Ground water recharge became necessary during the 1950's because ground water levels along the coast had dropped below sea-level and sea water was intruding into the ground water of Los Angeles and Orange Counties. During the late 1950's and 1960's water from the Colorado River was used to recharge several million acre feet (Toups, 1974).

The first reported attempts to recharge the Ogallala aquifer in the southern High Plains of Texas occurred in 1918. These early attempts were primarily aimed at draining the playa lakes so the fertile soil in the lake bottoms could be used for crop production. Little thought was given to aquifer replenishment. The steady increase in ground water withdrawals starting in the mid-1940's, the rapidly accelerating decline of the water table, and the drought in the 1950's stimulated artificial recharge efforts on the southern High Plains of Texas (Dvoracek and Peterson, 1971)

In the 1960's in the Texas southern High Plains, as well yields further declined, farmers and city water superintendents considered the feasibility of salvaging water from thousands of playa lakes. Playa lakes occur in the area at an average density of about one per square mile, and it was thought they could serve as a water source for recharge before the water was lost to evaporation. In 1970, a symposium was held to summarize artificial recharge activities in the Ogallala aquifer. Many different topics were discussed, including: (1) treatment of streambeds by excavation to increase recharge; (2) salvaging playa lake water by recharging through wells; (3) the importance of site selection; and (4) economic feasibility (Sweazy, 1984).

In 1984 a second symposium was held on artificial recharge of the Ogallala aquifer, and secondary recovery was presented as a new concept (Sweazy, 1984). The concept is based on the fact that the pores of a permeable aquifer are capable of retaining water against the pull of gravity. If, however, the force of gravity is augmented by the imposition of a downward pressure gradient from compressed air or other gas, some of the liquid will be released. The principle has been employed for several decades by the petroleum industry. The application of compressed air to the vadose zone above the water table of the Ogallala aquifer was employed at two sites in 1982. Additional information on secondary recovery is presented in Chapter 2.

OBJECTIVE OF BOOK

Artificial recharge is an important issue relating to current and future uses of ground water resources for agricultural and other purposes. Therefore, the objective of this book is to identify potential increased dependable ground water supplies resulting from artificial recharge activities for the base time year of 1985, and for the future reference years 2000 and 2030.

CONTENTS OF BOOK

This book is organized into five chapters and three appendices. Following this introductory chapter, Chapter 2 provides technical descriptions of both direct and indirect artificial recharge methods. The focus of Chapter 3 is on site selection and evaluation criteria for artificial recharge projects. A survey of case histories of artificial recharge projects is presented in Chapter 4. A methodology for determining the potential of geographical areas for artificial recharge is described in Chapter 5. The methodology includes consideration of current ground water availability problems, current and future agricultural water needs, regional hydrogeological factors, and current or planned artificial recharge projects. The geographical area used in the methodology is the aggregated subarea (ASA) of the water resources regions developed by the U.S. Geological Survey. The results from applying the methodology to the contiguous United States are also described in Chapter 5.

Appendix A contains an annotated bibliography of 342 references identified from computer-based searches of the literature. Each included abstract is categorized into one of the following 10 categories: bibliographic materials; techniques of artificial recharge; wastewater recharge; water quality impacts of artificial recharge; modeling of artificial recharge; economic feasibility; site selection and system design; case studies; miscellaneous; and international examples. Appendix B contains a listing of all counties in the continental United States organized by state and aggregated subareas. Finally, Appendix C summarizes data related to needs and potential for artificial recharge in accordance with the aggregated subareas.

SELECTED REFERENCES

Dvoracek, M.J. and Peterson, S.H., "Artificial Recharge in Water Resources Management," Journal of Irrigation and Drainage Division, American Society of Civil Engineers, Vol. 97, No. IR2, 1971, pp. 210-232.

Kazman, R.G., "Perspectives in Artificial Recharge," American Water Resources Association, Proceedings Series #4, 1967, pp. 187-192.

McGuiness, C.L., "Ground Water Research in the United States," Earth-Sciences Review, Vol. III, No. 3, 1969, pp. 181-202.

Palmer, P.L., "Artificial Recharge of Fresh Water into Florida's Saline Aquifers," Proceedings of the National Water Well Association Eastern Regional Conference on Ground Water Management, Orlando, Florida, 1983, pp. 3-16.

Parker, G.G., Cohen, P. and Foxworthy, B.L., "Artificial Recharge and Its Role in Scientific Water Management with Emphasis on Long Island, New York," American Water Resources Association Proceedings Series #4, 1967, pp. 193-213.

Pettyjohn, W.A., "Flow Analysis of Hydraulic Connectors in Artificial Recharge Systems, A Model Study," OWRR A-015-Ohio(1), July 1974, Water Resources Center, Ohio State University, Columbus, Ohio.

Sweazy, R.M., "Scientific and Economic Research in Support of the Investigation of Secondary Recovery of Ground Water," Proceedings of the Ogallala Aquifer Symposium II, Texas Tech University, Lubbock, Texas, June 1984, pp. 528-533.

Todd, D.K. Ground Water Resources of the United States, Premier Press, Berkeley, California, 1983.

Toups, J.M., "Water Quality and Other Aspects of Ground Water Recharge in Southern California," Journal American Water Works Association, Mar. 1974, pp. 149-151.

CHAPTER 2

DESCRIPTION OF ARTIFICIAL RECHARGE METHODS

Artificial recharge refers to the movement of water via man-made systems from the surface of the earth to underground water-bearing strata where it may be stored for future use (Griffis, 1976). The aspect of this definition which will be explored in this chapter is "man-made systems". Many systems have been developed, with Figure 1 containing a flow chart showing the principal methods of artificial recharge (Motts et al., 1983). Recharging reclaimed water, usually sewage and wastewater, and recharging natural water in lakes, ponds, and streams are the two major categories of artificial recharge. The artificial recharge of naturally occurring surface water may be further subdivided into ground water augmentation by indirect methods involving inducement of surface and ground water toward pumping wells, and by direct methods involving discharge of surface water into the subsurface via spreading or recharge wells.

Direct artificial recharge methods to be described in this chapter include spreading basins, playa lakes, recharge pits and shafts, ditches, and recharge wells. Indirect methods to be described include induced infiltration (enhanced streambed infiltration) and conjunctive wells. Also included will be information on general problems resulting from artificial recharge and potential measures to mitigate these problems. Water quality transformations in the subsurface will also be briefly addressed. Unplanned artificial recharge from irrigation, septic tanks and cesspools, and dams will also be mentioned. Finally, secondary recovery of capillary water as an alternative to artificial recharge will be noted.

A total of 69 abstracts on techniques of artificial recharge are included in Appendix A. As shown in Table 1, 12 abstracts contain general information on multiple techniques.

SPREADING BASINS

A frequently used method of artificial recharge involves surface spreading of water in spreading basins or recharge basins. These basins are commonly built by modest excavation in the existing terrain, or by placing low head dams across natural waterways (Pettyjohn, 1981).

In order to accomplish artificial recharge by spreading basins, highly permeable soils and maintenance of a layer of water over the highly permeable soils is a necessity (Dvoracek and Peterson, 1971). When direct recharge is practiced by spreading water over pervious soils in basins and ditches, the amount of water entering the aquifer depends on three factors: (a) the infiltration rate, (b) the percolation rate, and (c) the capacity for horizontal water movement (Huisman and Olsthoorn, 1983). The infiltration rate, also called the entry, intake or acceptance rate, is the rate at which the water is picked up by the soil. At the beginning of spreading operations and assuming a homogeneous aquifer up to ground surface, the infiltration rate is equal to the

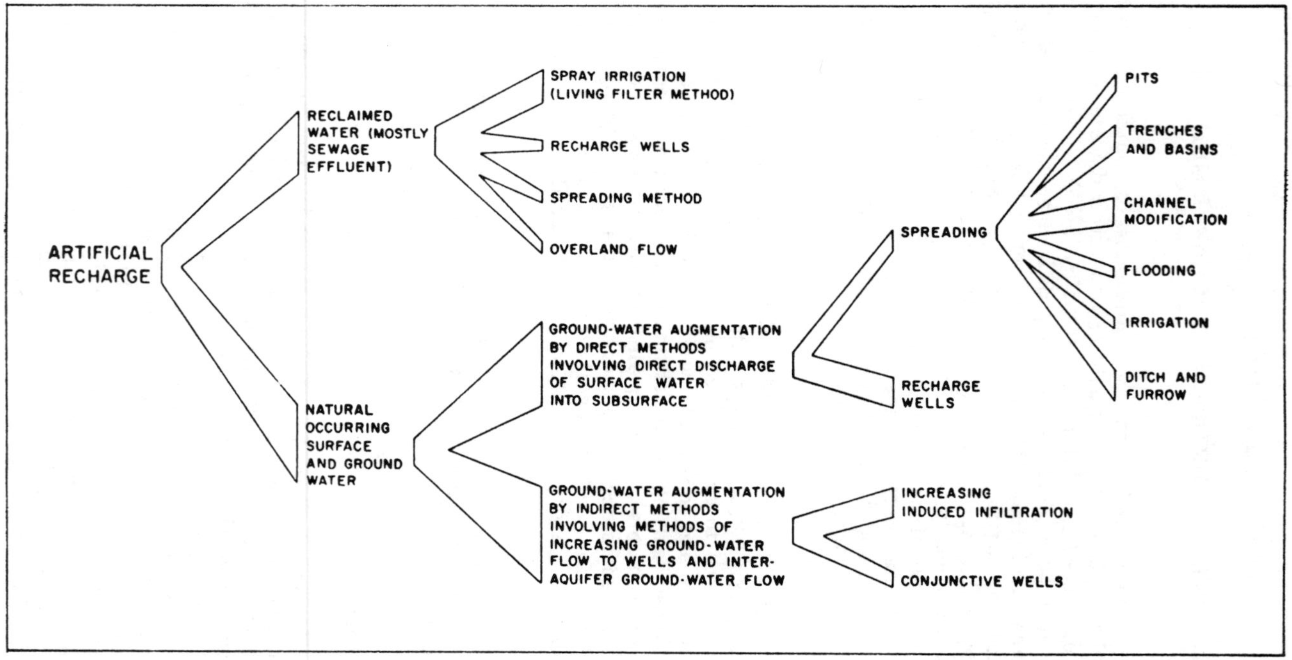

Figure 1: Flow Chart Showing Principal Methods of Artificial Recharge (Motts, et al., 1983)

Table 1: Abstracts in Appendix A Containing General Information on Techniques of Artificial Recharge

Dvoracek and Peterson (1970)
Dvoracek and Peterson (1971a)
Dvoracek and Wheaton (1970)
Dvoracek (1969)
Espina (1980)
Lichtler, Stannard, and Kouma (1980)
Pettyjohn (1981)
Signor (1976)
Sinclair (1977)
Thomas (1978)
Watson and Merritt (1970)
Winqvist and Mareliun (1970)

percolation rate mentioned below. At the point of entry into the aquifer, however, clogging occurs by the deposition of particles carried by the water in suspension or in solution, by algal growth, colloidal swelling and soil dispersion, microbial activity, and so on. This clogging seals pore openings and increases entry resistance. With the same water level in the spreading basin, the entry velocity diminishes and is, after some time, only a fraction of its original value.

The percolation rate is the rate at which the water is able to move downward through the soil. After trapped air has been removed, the percolation rate is a constant, depending on the flow pattern and the coefficient of permeability in the vertical direction. With aeolian formations, the coefficients of permeability are about the same in all directions, but with fluviatile and marine deposits, the vertical coefficient may be a few to many times smaller than the horizontal one. The capacity for horizontal water movement depends on the flow pattern and the coefficient of transmissivity of the aquifer below the ground water table. For a specified case, this capacity is again constant (Huisman and Olsthoorn, 1983).

Spreading basins must be constructed to have an approximately flat bottom which is to be covered evenly by small quantities of water. If the ground selected for spreading operations slopes too much, the amount of earth-moving required for the construction of the basins is, in most cases, the major economic limiting factor. Spreading techniques imply that large surfaces of land are available for the recharge works of Economic and Social Affairs, 1975). However, it should be noted that this method possibly delivers the lowest recharge rate per unit area utilized, and its success depends entirely on the porosity of the spreading area (Dvoracek and Peterson, 1971).

To serve as an illustration, spreading or recharge basins have been used on Long Island since 1935 to control storm runoff at moderate cost in inland areas, especially where ponding might otherwise occur, and to replenish the upper glacial (water table) aquifer by intercepting fresh water that would otherwise be lost through evaporation or diversion to coastal waters through storm sewers (Ku and Simmons, 1983). Most of the recharge basins on Long Island are unlined, open pits ranging in size from 0.1 to 30 acres and averaging 1.5 acres. They are generally about 10 feet deep, but may be as deep as 40 feet. They are fed by storm sewers. Most of the more than 3,000 basins on Long Island are in the central part of the island. About 91 percent of these basins are dry within five days after a one-inch rainfall, owing to the high infiltration rate of 0.2 to 1.0 feet per hour (ft/hr). Those that hold water for longer periods either intersect the regional water table or a perched water table, are excavated in till or moraine of low hydraulic conductivity rather than in outwash deposits, or are clogged with layers of sediment and debris. These basins transmit approximately 300 million gallons per day to the aquifer.

As mentioned earlier, a number of factors affect artificial recharge by spreading basins. Recharge by spreading basins is most effective where there are no impeding layers between the land surface and the aquifer and where clear water is available for recharge; however, a more turbid water can be tolerated than with well recharge (Lichtler, Stannard, and Kouma, 1980). The most common problem in recharging by surface spreading is clogging of the surface material by suspended sediment in the recharge water, by microbial growth, or by both. In coarse-grained material, fine suspended sediment may move 18 inches or more into the soil, making its removal difficult. Cultivation of the land surface usually increases the retention of suspended materials in the upper inch of the soil to more than 90 percent. This facilitates their removal by scraping or scarification. Clogging of the surface by microbial growth can usually be controlled by allowing the basin to dry periodically, which will then allow the organic material to oxidize. This practice requires the basin to be inoperative for varying lengths of time.

A total of 24 abstracts in the techniques category in Appendix A contain information on spreading or recharge basins, including playa lakes; these abstracts are listed in Table 2.

PLAYA LAKES

Playa lakes in the high plains area of Texas, New Mexico and Colorado have been used, or have the potential for use, in artificial recharge projects. Playa, or wet weather, lakes are depressions that collect water after rainfall or periods of snowmelt. Water from playa lakes can be pumped directly to fields for irrigation or used to artificially recharge an aquifer. Heavy clay soils can be broken up and the lake bottom regraded for maximum recharge (Colorado High Plains Technical Coordinating Committee, 1984).

Because of the variability of runoff, the number of playa lakes and the amount of available water changes annually. Some lakes may remain dry for several years. Many playa lakes have tight clay and silt deposits that restrict or even prevent leakage of water from the lake. Nearly all the water is lost to evaporation or to non-beneficial use through vegetation growing in the lake. Many of the lakes are very shallow, so using a centrifugal pump to pump the water directly into a pipeline for irrigation purposes is often not possible. Playa lakes have become a nuisance since extensive irrigation development began in the early 1960s. Some farmers have changed the land surface to eliminate the lakes, decrease their area, or convert them to tailwater sumps.

Table 2: Abstracts in Appendix A Containing Information on Spreading or Recharge Basins, Including Playa Lakes

---

Agricultural Research Service (1978)
Aronovici, Schneider, and Jones (1970)
Aronovici, Schneider, and Jones (1972)
Aronson and Seaburn (1974)
Aronson (1978)
Bianchi, Nightingale, and McCormick (1978)
Bouwer (1970)
Brock (1974)
Gass (1979)
Goss, et al. (1973)
Goss and Jones (1973)
Jones, Goss, and Schneider (1974)
Keys and Brown (1970)
Kincannon and Tiederman (1972)
Nightingale and Bianchi (1977)
Nightingale and Bianchi (1981)
Prill and Aronson (1973)
Ripley and Saleem (1973)
Rubin, Glass and Hunt (1976)
Seaburn and Aronson (1973)
Seaburn and Aronson (1974)
Schneider and Jones (1983)
Signor and Hauser (1968)
Williams and Allman (1969)

In some instances, agricultural practices such as deep chiseling have improved the infiltration capacity so that lakes do not form for extended periods.

In a demonstration project near Lubbock, Texas, playa lakes were modified by excavating concentration pits and using the excavated soil to raise the elevation of some of the previously flooded lands. The benefits from such a modification include: (1) a reduced water surface area and corresponding reductions in evaporation; (2) increased area for farming, thus more revenues to apply to the cost of lake modification; (3) an increase in recharge, because the excavation of the pit was deep enough to expose more permeable soils and increase infiltration from the bottom and sides of the pit; and (4) fewer mosquito problems, because the shallow pond--ideal for mosquito propagation--was destroyed (Colorado High Plains Technical Coordinating Committee, 1984).

RECHARGE PITS AND SHAFTS

From a regional point of view, conditions that permit surface spreading methods for artificial recharge are relatively rare. In the more general case, lenses and layers of earth materials that are only slightly permeable lie between the land surface and the water table. In situations such as these, artificial recharge systems must penetrate the less permeable strata in order to provide direct access to the dewatered aquifer. This may be accomplished with pits or shafts. These structures also permit construction of artificial recharge facilities in relatively small areas (Pettyjohn, 1981).

Recharge pits should be sufficiently deep to penetrate the low permeability surface material (Pettyjohn, 1981). Recharge through constructed pits has given reasonable recharge rates in studies conducted at Texas Tech University (Dvoracek and Peterson, 1971). Rates, expressed as a water level decline, of 1.5 feet per day were experienced. The pits were constructed on the periphery of a playa lake and the pit bottom extended 4 feet to 8 feet below the heavy, very slowly permeable, top soils characteristic in the playa lakes. Side slopes of the pits varied, with the steepest side slopes resulting in the highest rates. However, rates declined as the percentage of the wetted area found in the pit sides declined. Unfiltered runoff waters used in these recharge studies left a thin layer of sediment on the sides and bottoms of the pits which required maintenance in order to sustain the high rates. The steep sides were essentially self cleaning as the sediment layer would dry and curl creating a condition susceptible to removal in the presence of any motion. Winds or other similar motion performed some of the desirable cleaning but more extensive maintenance activities were required for the pit bottoms. Three abstracts in Appendix A focus on recharge pits (Dvoracek and Wheaton, 1969; Ganus, 1970; and McWhorter and Brookman, 1972). In some locations, abandoned gravel pits have been used as artificial recharge basins (Pettyjohn, 1981).

Recharge shafts can be used to transfer surface water through low permeability layers. Ground water recharge through shafts is similar to well recharge except for two features. The shafts possess either circular, rectangular, or square configurations with cross-sectional areas exceeding 3 square feet. The excavation may or may not terminate some distance above the static water level and may be cased, remain uncased, or backfilled with porous material (Dvoracek and Peterson, 1971). Shafts extending below the water table are referred to as hydraulic connectors. This form of recharge allows surface waters to bypass the slowly permeable upper layers of the strata generally existing between the surface and water formation.

Recharge shafts may also be used in conjunction with recharge pits (Pettyjohn, 1981). Both shafts and pits may suffer from decreasing recharge rates with time due to the accumulation of fine-grained materials and to the plugging effect brought about by microbial activity. Rates through recharge pits may be maintained by periodically allowing the facility to become dry or by scraping and removing the accumulated material from the sides and bottom. Shafts are less easy to maintain owing to their smaller diameter and greater depth. In some cases the coarse material used to fill the shaft must be replaced (Pettyjohn, 1981).

DITCHES

Recharge through constructed ditches or trenches may be considered as a modification of recharge pits. A long narrow trench, with its bottom width less than its depth is utilized rather than a large rectangular pit (Dvoracek and Peterson, 1971). The rate of recharge will be dependent upon the porosity of the material into which the trench penetrates. Some data on recharge rates, expressed in liters per second, of ditches in various geological formations are (Department of Economic and Social Affairs, 1975):

Volcanic pumice or porous lava	1,350 l/s/km
Sand dune	43 l/s/km
Sand and gravel beds	840 l/s/km

A ditch system can be designed to best suit the topographic and geologic conditions that exist at a potential artificial recharge site (Pettyjohn, 1981). A layout plan for a ditch and flooding recharge project could include a series of ditches trending slightly down the topographic slope. The ditches could terminate in a collection ditch designed to carry away the water that does not infiltrate in order to avoid ponding and reduce the accumulation of fine material. Appendix A has one abstract related to the use of a ditch or trench (Watson and Whisler, 1977).

RECHARGE WELLS

Recharge or injection wells are used to directly recharge water into deep water-bearing zones. Recharge wells have many similarities to water-supply wells; however, water is pumped into recharge wells rather than being pumped out as in the case of water supply wells. A recharge well can be cased through the material overlying the aquifer and if the earth materials are unconsolidated, a screen can be placed in the well in the zone of injection. The purpose of the screen is to prevent the aquifer from caving into the bore hole and to permit access to the water-bearing materials. In some cases, several recharge wells may be installed in the same bore hole (Pettyjohn, 1981).

Recharge wells are the only feasible method of artificial replenishment in locations where a thick impervious layer exists between the surface of the soil and the aquifer to be replenished (Department of Economic and Social Affairs, 1975). Recharge wells are also advantageous when land is scarce. A relatively high rate of recharge can be attained, with the rate a function of the characteristics of the formation being recharged and well design factors.

A number of factors affect recharge well performance. Most of the problems in recharging through wells, especially in aquifers composed of fine-grained material, involve excessive buildup of water levels in the recharge well because of clogging of the well screen or aquifer. If the recharge water is not low in suspended solids, air, and microorganisms, and is not chemically compatible with the natural ground water and the aquifer material, clogging can cause a recharge operation to be infeasible (Lichtler, Stannard, and Kouma, 1980).

Under ideal conditions a well will accept recharge water at least as readily as it will yield water by pumping. That is, if a well will yield 1,000 gal/min with 30 feet of drawdown, it should accept as recharge 1,000 gal/min with 30 feet of buildup in the well, and a graph of the buildup of water levels in the vicinity of a well when it is recharging should be a mirror image of a graph of the drawdown in the vicinity of the well when it is pumping. Actual conditions, however, are seldom ideal. Factors in nature that alter the ideal conditions in regard to artificial recharge through wells include differences in the physical and chemical quality and temperature between the recharge water and the native water, rearrangement of the gravel-pack material or the aquifer material near the well because of changes in direction of flow, and changes in aquifer response caused by changes in saturated thickness of the aquifer during pumping and recharge operations. Buildup of water levels in recharge wells is usually, but not always, greater than corresponding drawdowns during pumping. Factors that can cause the buildup of water levels in a recharge well to be greater than the corresponding drawdown in a discharging well include the following (Lichtler, Stannard, and Kouma, 1980):

(1) Suspended sediment in the recharge water, including both organic and inorganic matter. This sediment is a major cause of well and aquifer clogging because it may be filtered out of the water and deposited on the well screen, in the pores of the aquifer, or both. The sediment may reduce the area of the well-screen openings and the transmissivity of the aquifer, thereby increasing the hydraulic-head differential necessary to maintain a given recharge rate.

(2) Entrained air in the recharge water, which is another major cause of clogging. The exact manner in which air bubbles become lodged in the interstices of the aquifer is unknown, but the bubbles apparently have the same effect as clay particles or sand grains in that they effectively retard the passage of water. Movement of the bubbles outward from the well or upward to the water table may be prevented by one or both of the following: simple blocking of the air bubbles by sand grains; and the Jamin effect. The Jamin effect results from the fact that a capillary tube which contains restrictions and is filled with a chain of alternate air bubbles and water is capable of sustaining a finite pressure gradient without allowing fluid movement. A sand-and-gravel aquifer probably acts as a series of interconnected capillary tubes containing many restrictions and the forces created by the Jamin effect prevent the movement of the entrained air after it has been introduced into the aquifer. Once air has been entrained in a sand-and-gravel aquifer, it is difficult to remove with simple pumping of the well. Special redevelopment procedures involving the use of a wetting agent such as hexametaphosphate and surging and pumping are necessary to remove the bubbles before the specific capacity of an air-clogged well can be restored. Gas may be liberated from solution if the recharge water is warmed by contact with the native ground water, resulting in clogging similar to that caused by air entrainment. If the two waters are approximately the same temperature or if the recharge water is warmer than the ground water, this is usually not a problem.

(3) Microbial growth in a well. Such growth produces slimes or other products that can clog recharge wells and aquifers. The effects of these products are similar to those of suspended sediment in reducing the transmissivity of the aquifer and increasing water-level buildup in the recharge well. Ground water normally is nearly free of microbes and should not clog the well by microbial growth during

recharge unless it is contaminated during transit to the recharge well.

(4) Chemical reactions between the recharge water and the native ground water, the aquifer material, or both. Such reactions can cause precipitation of insoluble matter that can clog the screen, the pores of the aquifer, or both. The effects of the precipitate are similar to those of suspended sediment in clogging the well and aquifer, reducing the transmissivity of the aquifer and thereby causing greater than normal buildup of water levels in the recharge well. The chemical reactions causing precipitation commonly are complex and temperature sensitive. A knowledge of the dissolved chemical constituents of the native and recharge water and their chemical equilibria, the water temperatures, and any changes in the composition of the waters with time are necessary to accurately predict potential problems from chemical precipitation.

(5) Ionic reactions that result in dispersion of clay particles and swelling of colloids in a sand-and-gravel aquifer. Such reactions can occur when the recharge water is of different chemical composition than the native ground water. When freshwater is recharged into an aquifer containing saline water the ion concentration of the native water is decreased, reducing the electrostatic attraction between the clay particles and permitting them to disperse and form a nearly impermeable barrier near the well bore.

(6) Iron precipitation. Such precipitation can occur in a well if iron-rich water is aerated or exposed to other oxidizing conditions. Soluble ferrous iron is oxidized to the insoluble ferric form. The resulting iron compounds clog the aquifer much the same as suspended particles do.

(7) Biochemical changes in the recharge water and ground water involving iron-reducing bacteria or sulfate-reducing organisms. Such changes can cause clogging under certain conditions. If the recharge water and the recharge system are kept free of bacteria, this should not be a problem.

(8) Differences in temperature between recharge and aquifer water. When water that is cooler than the native ground water is injected into an aquifer, the recharge water has a greater viscosity than the native water. This greater viscosity requires a greater buildup in the recharge well to move cool water at the same rate as warmer water.

Factors that can cause the buildup of water levels in the recharge well to be less than the corresponding drawdown in a discharging well include the following (Lichtler, Stannard, and Kouma, 1980):

(1) Recharge water that is warmer than the native ground water and, therefore, less viscous.

(2) Increase in the saturated thickness and transmissivity of the aquifer due to the higher water levels that result when a water-table aquifer is recharged.

(3) Recharge water that is unsaturated with respect to calcium carbonate. Such water can dissolve parts of a carbonate aquifer or calcium carbonate in a sand-and-gravel aquifer, thereby increasing the size of the pores and channels and increasing the transmissivity and storage capacity of the aquifer. Other chemical reactions also may increase the transmissivity and storage capacity of an aquifer. Solution effects from such chemical reactions usually occur slowly and normally are not noticeable during tests of a few days' duration; however, during a period of years they could become significant.

A total of 24 abstracts in the techniques category in Appendix A contain information on injection or recharge wells; these abstracts are listed in Table 3.

ENHANCED STREAMBED INFILTRATION

A common method of induced recharge consists of setting a gallery or a line of wells at a short distance--say 50 m--parallel to the bank of a river or lake. In an original situation there may be unimpeded outflow of ground water to the river (Huisman and Olsthoorn, 1983). When small amounts of ground water are withdrawn from the gallery parallel to the river, the ground water discharge into the river will decrease, while the water recovered by the gallery consists wholly of natural ground water, originating from rainfall. Each ground water withdrawal, however, is accompanied by a drawdown of the ground water table and when for larger recovery rates this drawdown tends to lower the ground water table near the shore-line to below the water level in the river, surface water from this river will be induced to enter the aquifer and to flow to the gallery. With high rates of withdrawal, the water recovered consists of a small part of natural ground water and a large part of artificial groundwater, originating from the river. These same concepts are also applicable to individual wells located alongside bodies of water (Pettyjohn, 1981). Where saturated streamside deposits are thin, horizontal wells may be more feasible and appropriate than vertical wells. Laterals which radiate from a central collector well can be injected so that they lie close to the bottom of the aquifer and allow the maximum amount of drawdown (Pettyjohn, 1981). One abstract on enhanced streambed infiltration is included in the techniques section of Appendix A (Walton, Hills, and Grundeen, 1967).

Direct streambed infiltration can be enhanced by certain channel modifications. Other than channel dredging, straightening and widening, low head check dams are the most popular method for increasing recharge from existing streams or ephemeral channels (Pettyjohn, 1981).

Enhanced infiltration (induced infiltration) depends, to a large extent, on the quantity of water that can be diverted from a stream or lake. The controlling factor, other than a dependable source of surface water of acceptable quality, is the permeability of the streamside deposits and the sediment on the bottom of the stream. In areas where the stream is separated from the aquifer by materials of low permeability, leakage from the stream may be so small that the system is not feasible. The permeability of a stream bottom fluctuates throughout the year. During periods of low flow, mud and organic matter may accumulate in the channel forming a bed of low permeability that reduces infiltration. The channel is scoured during high flows. The rate of infiltration can vary by 50 percent or more between periods of high and low flows. In many places, and particularly during droughts or

Table 3: Abstracts in Appendix A Containing Information on Injection or Recharge Wells

Baffa (1970)
Bassett, et al. (1981)
Bianchi, Nightingale, and McCormick (1978)
Brown and Signor (1973)
Brown and Silvey (1977)
Bush (1979)
Cluff, et al. (1969)
Conselman, Johnson, and Crawford (1972)
Dvoracek and Peterson (1971b)
Hauser and Lotspeich (1967)
Johnson and Sniegocki (1967)
Johnson (1970)
Keys and Brown (1973)
Kimbler and Whitehead (1977)
Krone (1971)
McCormack (1975)
Pettyjohn (1974)
Rebhun and Hauser (1967)
Schneider, Jones, and Signor (1971)
Scott (1969)
Smith (1972)
Smith (1980)
Sniegocki and Brown (1971)
Vecchioli (1972)

extended dry periods, stream discharge is not sufficient to erode the fine materials from the channel bottom (Pettyjohn, 1981).

Another important consideration in enhanced infiltration is the chemical quality of the surface water source. Many rivers carry appreciable amounts of suspended matter (and other clogging agencies), which are filtered out when the water enters the aquifer. After some time, the bed of the river will be covered by a less pervious layer, a so-called filter skin, with a large resistance against the inland flow of river water. With a constant rate of withdrawal, the groundwater table will drop, thus impeding the processes of recovery and doing damage to other users of ground water (Huisman and Olsthoorn, 1983). In the past, this was not a serious problem as most rivers in their natural state do not have a stable bed. Sand and gravel deposited in periods of low to normal flow and picked up again when the river is in high flow. However, dams have been built in many rivers and their tributaries, to store water from flood flows for subsequent use in dry periods. High river discharges and their scouring effect no longer occur and--as with lakes--a persistent clogging of the river bed will take place. Artificial cleaning of the river or lake bed next to the recovery works may be necessary. Dredging could be used.

CONJUNCTIVE WELLS

A conjunctive well is one that is screened in both a shallow unconfined and deeper artesian aquifer. When ground water is pumped from the deeper aquifer and its potentiometric surface is lowered below the shallow-water table, water from the shallow aquifer drains directly through the well into the deeper aquifer (Motts et al., 1983). Water augmentation by conjunctive wells has the advantage of utilizing sediment-free ground water which greatly reduces the damage of clogging well screens. Other benefits from conjunctive wells are (1) to reduce the amount of evapotranspiration water loss from the shallow water table where it is close to the land surface, and (2) to reduce flooding effects in some places. Environmental effects from the conjunctive well methods must be carefully studied to assure that unwanted dewatering of wetlands or reduction of base flow will not occur. Also the possibility of negative effects of coagulation due to the mixing of chemically different ground waters should be carefully investigated. Four abstracts in the techniques section of Appendix A deal with inter-aquifer pumping and conjunctive wells (Deutsch, 1967; Knochenmus, 1975; Tibbals and Frazee, 1976; and Watkins, 1977).

GENERAL PROBLEMS AND MITIGATION MEASURES

Problems which arise as a result of recharge projects are mainly related to the quality of raw waters that are available for recharge, and which generally require some sort of treatment before being used in recharge installations. They are also related to the changes in the soil structure and the biological phenomena which take place when infiltration begins, to the changes brought to the environmental conditions, and to land-ownership and legal aspects (Department of Economic and Social Affairs, 1975).

A major requirement for waters that are to be used in recharge projects is that they be silt-free. Silt may be defined as the content of undissolved solid matter, usually measured in mg/l, which settles in stagnant water with velocities which do not exceed 0.1 m/hr. This definition comprises a large variety of materials, such as clay particles, organic matter and fine particles of calcite. The silt content of a river depends upon the type of soils in the area of run-off, the vegetative cover of this area, its topographic slopes and the meteorological characteristics prevailing in its catchments and, especially, intensity of rainfall. Average silt contents of from 3,000 to 8,000 mg/l are not uncommon in rivers during flood periods. As soon as the water becomes stagnant after being collected in a pond; or, if the velocity of flood is much reduced, the silt load gradually settles out. The speed of the settling process is governed by the size and specific weight of the suspended particles and by electro-chemical forces acting between the particles and the water which, in their turn, are strongly influenced by the chemical composition of the water. Therefore, there is no clear correlation between initial turbidity and the speed of the settling process (Department of Economic and Social Affairs, 1975).

Retention reservoirs may be used to reduce the silt load to a level acceptable for spreading operations. Costs of maintenance and operation are determined mainly by the cost of periodically removing the silt collected. Chemical flocculants are sometimes used to speed up the settling process or to obtain a cleaner water. Doses of alum, in the range of 100-150 mg/l, may reduce to only a few hours the time required to obtain a water with only about 300 mg/l suspended matter. Recent experiments have shown that specific polymer compounds, known under various trade names, sometimes achieve the same effect with doses of only 0.3-1.5 mg/l.

Suspended matter may clog the soil in two different ways: first, near the surface the interstices of the soil may be filled up and a layer of mud may be deposited on the surface; on the other hand, suspended particles may penetrate deeper into the soil and accumulate there. A layer of mud is formed on the surface by particles, the settling velocity of which exceeds infiltration velocities. Smaller suspended particles are filtered out in the uppermost layer of the soil. The filtration process is governed not only by mechanical factors, but it seems to be strongly influenced by electro-chemical surface forces. Still finer particles, especially very fine grains of montmorillonite clay, are carried further into the soil. Observations in spreading-grounds, composed of medium-grained dune sands, showed that these particles become lodged at from 10 to 20 m below the surface, and some of these particles are carried even deeper (Department of Economic and Social Affairs, 1975).

Methods to remedy or minimize the clogging effect by suspended matter can be classified into four broad groups (Department of Economic and Social Affairs, 1975):

(a) periodical removing of the mud-cake and discing or scraping of the surface layer;

(b) installation of a filter on the surface, the permeability of which is lower than that of the natural strata (the filter must, of course, be removed and renewed periodically);

(c) addition of organic matter or chemicals to the uppermost layer; and

(d) cultivation of certain plant-covers, notably certain kinds of grass.

Scraping of the surface layer is effective only in rather coarse-grained spreading grounds. In soils composed mainly of sand, repeated compaction by heavy machinery may easily nullify any benefit gained from scraping. The installation of filter-layers has been used with good results in gravel recharge pits in Peoria, Illinois. Layers of sand or pea gravel six inches thick were put on the bottom of the infiltration pits as a filtering medium. In order to reduce the necessity of frequently changing the filter, the mud-layer was removed, while still liquid, with the aid of a vacuum cleaner of the type used for cleaning swimming pools. Superficial raking of the dry pea-gravel also helped in restoring infiltration rates. Various chemicals and organic matter have been used to restore infiltration capacities. These have included gypsum, various organic compounds, cotton-gin trash and alfalfa, the latter grown while the pond was still wet, and then spaded under. The growth of a permanent grass-cover has proved to be an effective method for maintaining infiltration capacities, but it is difficult to select a grass which grows under a given climatic and soil condition and is able to withstand alternate periods of flooding and drying.

Clogging by biological activity depends upon the minerological and organic composition of the water and basin floor and upon the grain-size and permeability of the floor (Department of Economic and Social Affairs, 1975). The only feasible method of treatment developed so far consists of thoroughly drying the ground under the basin. Experiences seem to indicate that short periods of operation (about one month), followed by drying, are more effective than prolonged periods of operation, even if they are followed by a prolonged and most thorough period of drying during the hot summer.

Environmental problems may stem from artificial recharge projects (Department of Economics and Social Affairs, 1975). Such projects usually have to be carried out in the vicinity of densely populated and industrial areas, where large quantities of water are needed. The possible close vicinity of spreading basins to population centers often creates various kinds of problems. In particular, it is well known that stagnant water serves as a breeding ground for mosquitoes, flies and a variety of other biological nuisances. The best remedy is to operate parts of spreading basins in sequence, so that water remains in each part for a shorter period than the larvae-stage of the insect's life-cycle. This remedy may, however, lead to unrealistically large land requirements. Another method is to apply insecticides, making certain before hand that they decompose quickly and have no toxic effect on the water. In addition, the installations may become an eyesore unless properly tended, surrounded by trees and more or less concealed. They also may become an accident-hazard, as any body of open water results in trespassing by children and is a target for acts of vandalism.

WATER QUALITY TRANSFORMATIONS

Transformations in the quality of recharged water can occur in the subsurface environment as a result of one or more of the following: biodegradation by and growth of microorganisms; chemical oxidation and reduction; sorption and ion exchange; filtration; chemical precipitation or dilution; and volatilization or photochemical reactions (in spreading basins) (Roberts, 1980). These mechanisms can be considered in terms of two broad classes of potential pollutants--particulates and dissolved organics, with these classes identified by many researchers as deserving of the highest priority attention (Roberts, 1980).

Particulate constituents introduced into the ground by recharge are retained effectively by filtration. This applies to microorganisms as well as other particulate organic materials. In the case of virus, the main method of attachment is by adsorption. The removal of microorganisms including virus during passage of the water through the aquifer is believed to be virtually complete under typical conditions of ground water recharge (Roberts, 1980).

Dissolved organic constituents are affected primarily by biodegradation and adsorption during ground water recharge. Biodegradation offers the potential of permanent conversion of organic substances into harmless products. Biodegradation is likely to occur during ground water recharge to the extent that the organics introduced are at least in part degradable, and other conditions permit. Biodegradation in ground water environments is accomplished for the most part by microorganisms attached to surfaces (attached growth). Hence biodegradation would tend to be enhanced if the aquifer material is finely divided, as sand or silt, therefore having a high specific surface area. Unfortunately, such conditions also are conducive to clogging through bacterial growth. Biodegradation may tend to be less rapid and perhaps less extensive in aquifers consisting of coarse gravels. However, this disadvantage is compensated by greater permeability and reduced clogging of coarser aquifer material (Roberts, 1980).

Adsorption has two roles in water quality transformations in artificial recharge: retardation of solute movement and attenuation of concentration fluctuations. These phenomena are generally beneficial from the viewpoint of water quality. Retardation is most important in the early stages after a change in input concentration (stimulus) has been made. Sorption of uncharged organic species is believed to be influenced largely by the hydrophobic nature of the molecule. Hydrophobic compounds sorb strongly onto (or into) soils, because they are in a sense expelled from aqueous solution into a hydrophobic phase. Hydrophilic substances are not subject to such a chemical potential driving force, and so tend to remain in solution rather than to sorb. Other more specific chemical interactions may play a secondary role. Examples of strongly hydrophobic solutes are the highly chlorinated hydrocarbons such as DDT and PCB compounds. Such substances may be so strongly sorbed as to exhibit aquifer retention times of hundreds or thousands of years under typical recharge conditions (Roberts, 1980).

Attenuation is the result of the interaction of sorption and elution that occurs as a response to fluctuations of input concentration. Briefly, attenuation is a measure of the damping (reduction in amplitude) of fluctuations. The degree of attenuation increases with increasing strength of sorption, increasing distance from the recharge point, and increasing frequency of input fluctuation. Dispersion--a hydrodynamic or hydrogeological phenomenon related to fluid mixing and gradations of aquifer permeability--also contributes to attenuation fluctuations in solute concentration; dispersion is effective for solutes that sorb as well as those that do not (conservative or noninteracting solutes) (Roberts, 1980).

UNPLANNED ARTIFICIAL RECHARGE

While planned artificial recharge has the objective of replenishing the ground water resource, unplanned artificial recharge may occur with no such objective and is incidental to some other human activity, such as irrigation. The amount of water recharged in an unplanned way is vastly greater than the amount deliberately recharged (Department of Economic and Social Affairs, 1975). Irrigation, for example, plays a very significant role in the recharge of ground water. Cesspools, septic tanks, leaky water mains and sewers, industrial waste disposal pits and similar facilities also recharge large amounts of fluids into the underground environment. Some of these operations may significantly affect the composition of ground water and pose important problems related to pollution of water.

Irrigation, particularly during the nongrowing season, may add a significant amount of water to underground storage (Pettyjohn, 1981). An especially attractive prospect is irrigation with wastewater or partly reclaimed wastewater. Here the advantages are twofold: (1) after recharge the water is again available for reuse and (2) the water receives a substantial amount of treatment and renovation as it percolates through the unsaturated zone. These systems must be monitored and managed, however, in such a way that water logging does not occur (Pettyjohn, 1981).

SECONDARY RECOVERY OF CAPILLARY WATER

While not a method of artificial recharge, secondary recovery of capillary water offers promise of enhancing available ground water supplies. The recovery process involves the injection of air into the ground to force aquifer capillary water trapped between rocks of the water-bearing strata to move to the water table (Gass, 1985). Testing to date has been primarily focused on the Ogallala aquifer in the high plains area of Texas. To indicate the potential volume of releasable capillary water, a determination of the amount of this water in storage in the Texas portion of the Ogallala was made (Wyatt, 1984). The capillary water in storage as determined by means of core analysis yielded values ranging from 10 to 40 percent by volume, and averaging 25.9 percent. There being some 3.36 billion acre-feet of volume in the formation between the root zone and the 1980 water table, and assuming a 25 percent moisture content, the total volume of water in the unsaturated zone would be about 840 million acre-feet (Wyatt, 1984). However, not all capillary water can be recovered by air injection.

Field tests have been conducted by the High Plains Underground Water Conservation District (Wyatt, 1984). Three field tests have verified the effectiveness of air drives for secondary recovery of water. An economic analysis of the third test indicated that the estimated 406 acre-feet of increased ground-water storage obtained had cost somewhat less than $50 per acre-foot. A fourth test is investigating (successfully) the possibility of reducing costs by reducing air pressure (Wyatt, 1984). Costs in the range of $20 per acre-foot are considered to be necessary for the practical application of this technology (Gass, 1985).

To illustrate some results of a field test at Wolforth, Texas, holes were drilled to two test levels--one between 30 and 80 feet below the soil surface, and the other between 100 and 150 feet (Gass, 1985). Air was injected in the upper section with a pressure of about 10 pounds per square inch for 60 days. Results were a 20 percent drop in water content in the upper section and an increase of up to 26 percent water by weight in the lower section.

SUMMARY

Man-made systems for moving water from the surface of the earth to underground water-bearing strata include both direct and indirect methods. Direct methods include spreading or recharge basins, playa lakes, recharge pits and shafts, ditches, and recharge or injection wells. Indirect methods include induced infiltration (enhanced streambed infiltration) and conjunctive wells. The most widely used methods include spreading basins and recharge wells. Spreading basins involve the surface spreading of water; they are useful in areas with abundant land availability, highly permeable soils, and shallow unconfined aquifers. Recharge wells require minimal land and are useful for recharging deep, confined aquifers. In addition to hydrogeological factors, the quality of the recharge water can influence the operational efficiencies of recharge techniques. Key concerns are associated with clogging from suspended matter, possible chemical reactions between recharge and aquifer waters, and enhanced microbial growth in the subsurface environment.

SELECTED REFERENCES

Agricultural Research Service, "Percolation Rates with Surface Layers of Gravel or Clay Flocs on Soil Fed Clear and Turbid Waters", Water Resources Research, Vol. 14, No. 3, June 1978, pp. 539-541.

Aronovici, V.S., Schneider, A.D. and Jones, O.R., "Basin Recharge of the Ogallala Aquifer", Journal of the Irrigation and Drainage Division, American Society of Civil Engineers, Vol. 98, No. IR 1, Paper 8760, Mar. 1972, pp. 65-76.

Aronovici, V.S., Schneider, A.D. and Jones, O.R., "Basin Recharging the Ogallala Aquifer through Pleistocene Sediments", The Ogallala Aquifer--A Symposium, Special Report No. 39, 1970, International Center for Arid and Semi-Arid Land Studies, Texas Tech University, Lubbock, Texas, pp. 182-192.

Aronson, D.A. and Seaburn, G.E., "Appraisal of Operating Efficiency of Recharge Basins on Long Island, New York, in 1969", Water Supply Paper 2001-D, 1974, U.S. Geological Survey, Washington, D.C.

Aronson, D.A., "Artificial Recharge on Long Island, New York", Long Island Water Resources Bulletin LIWR-9, 1978, U.S. Geological Survey, Syosset, New York.

Baffa, J.J., "Injection Well Experience at Riverhead, New York", Journal of the American Water Works Association, Vol. 62, No. 1, Jan. 1970, pp. 41-46.

Bassett, R.L., et al., "Preliminary Data from a Series of Artificial Recharge Experiments at Stanton, Texas", Open File Report 81-149, 1981, U.S. Geological Survey, Denver, Colorado.

Bianchi, W.C., Nightingale, H.I. and McCormick, R.L., "A Case History to Evaluate the Performance of Water-Spreading Projects", Journal of the American Water Works Association, Vol. 70, No. 3, Mar. 1978, pp. 176-180.

Bianchi, W.C., Nightingale, J.J. and McCormick, R.L., "Fresno, California, Subsurface Drain Collector-Deep Well Recharge System", Journal of the American Water Works Association, Vol. 70, No. 8, Aug. 1978, pp. 427-435.

Bouwer, H., "Groundwater Recharge Design for Renovating Waste Water", Journal of the Sanitary Engineering Division, American Society of Civil Engineers, Vol. 95, No. SA 1, Paper 7096, Feb. 1970, pp. 59-74.

Brock, R.R., "Hydrodynamics of Artificial Groundwater Recharge", Completion Report, Feb. 1974, School of Engineering, University of California, Irvine, California.

Brown, R.F. and Signor, D.C., "Artificial-Recharge Experiments and Operations on the Southern High Plains of Texas and New Mexico", Water Resources Investigations 10-73, May 1973, U.S. Geological Survey, Austin, Texas.

Brown, D.L. and Silvey, W.D., "Artificial Recharge to a Freshwater-Sensitive Brackish-Water Sand Aquifer, Norfolk, Virginia", Professional Paper 939, 1977, U.S. Geological Survey, Reston, Virginia.

Bush, P.W., "Connector Well Experiment to Recharge the Floridan Aquifer, East Orange County, Florida", Water Resources Investigations 78-73, 1979, U.S. Geological Survey, Tallahassee, Florida.

Cluff, C.B., et al., "Development and Evaluation of Water Harvesting Systems", Report No. OWRR-A-001-ARIZ(6), Dec. 1969, Arizona Water Resources Research Center, University of Arizona, Tucson, Arizona.

Colorado High Plains Technical Coordinating Committee, "Artificial Aquifer Recharge in the Colorado Portion of the Ogallala Aquifer", Draft, Sept., 1984, Denver, Colorado.

Conselman, F.B., Johnson, P. and Crawford, D.A., "Underground Storage of Texas Playa Lake Waters by Injection into the Ogallala Formation under Moderate Pump Pressure", International Symposium on Water Resources Planning, Dec. 1972, Mexico City, Mexico.

Department of Economic and Social Affairs, United Nations, "Groundwater Storage and Artificial Recharge", Natural Resources/Water Series No. 1, ST/ESA/13, 1975, New York, New York.

Deutsch, M., "Artificial Recharge by Induced Interaquifer Leakage", Symposium of Haifa, Artificial Recharge and Management of Aquifers, Mar. 19-26, 1967, Publication No. 72, 1967, International Association of Scientific Hydrology, pp. 159-172.

Dvoracek, M.J. and Peterson, S.H., "Artificial Recharge in Water Resources Management", Report No. OWRR-B-041-TEX(4), 1970, Department of Agricultural Engineering, Texas Tech University, Lubbock, Texas.

Dvoracek, M.J., and Peterson, S.H., "Artificial Recharge in Water Resources Management", Journal of Irrigation and Drainage Division-ASCE, Vol. 97, No. IR2, 1971, pp. 210-232.

Dvoracek, M.J. and Peterson, S.H., "Development of Systems for Ground Water Recharge into the Ogallala Formation", OWRR-B-041-TEX(6), June 1971a, Water Resources Center, Texas Tech University, Lubbock, Texas.

Dvoracek, M.J. and Peterson, S.H., "Recharging the Ogallala Formation Using Shallow Holes", Proceedings of Hydrology and Water Resources in Arizona and the Southwest, Vol. 1, Apr. 22-23, 1971b, Arizona Section-American Water Resources Association and the Hydrology Section-Arizona Academy of Science, Tempe, Arizona, pp. 245-259.

Dvoracek, M.J. and Wheaton, R.Z., "Development of Systems for Ground Water Recharge into the Ogallala Formation", OWRR-B-016-TEX(2), May 1969, Water Resources Center, Texas Tech University, Lubbock, Texas.

Dvoracek, M.J. and Wheaton, R.Z., "Does Artificial Ground Water Recharge Contaminate Our Ground Waters", OWRR-B-041-TEX(3), 1970, Department of Agricultural Engineering, Texas Tech University, Lubbock, Texas.

Dvoracek, M.J., "Use of Wells and Pits to Recharge Underground Formations in Semi-Arid Areas", International Conference on Arid Lands in a Changing World, June 3-13, 1969, June 1969, University of Arizona, Tucson, Arizona.

Espina, J.M., "The Case for Artificial Recharge", The Johnson Drillers Journal, Vol. 52, No. 1, Jan.-Feb. 1980, pp. 7-11.

Ganus, W.J., "Project Aquarius--Geohydrologic Investigation", CONF-700101, Vol. 2, Jan. 1970, National Technical Information Service, U.S. Department of Commerce, Springfield, Virginia, pp. 1174-1183.

Gass, T.E., "Artificial Recharge: Part II", Water Well Journal, Vol. 33, No. 2, Feb. 1979, pp. 32-33.

Gass, T.E., "Air Injection Technique Wrings New Water from Ogallala Aquifer", Ground Water, Vol. 23, No. 1, Jan.-Feb. 1985.

Goss, D.W., et al., "Fate of Suspended Sediment During Basin Recharge", Water Resources Research, Vol. 9, No. 3, June 1973, pp. 668-675.

Goss, D.W. and Jones, O.R., "Movement and Accumulation of Suspended Sediment During Basin Recharge", Underground Waste Management and Artificial Recharge, Vol. 1, 1973, pp. 468-478.

Griffis, C.L., "Artificial Recharge in the Grand Prairie, Arkansas", Bulletin 810, Aug., 1976, Agricultural Experiment Station, University of Arkansas, Fayetteville, Arkansas.

Hauser, V.L. and Lotspeich, F.B., "Artificial Groundwater Recharge through Wells", Journal of Soil and Water Conservation, Vol. 22, No. 1, Jan.-Feb. 1967, pp. 11-15.

Huisman, L., and Olsthoorn, T.N., Artificial Groundwater Recharge, Pitman Advanced Publishing Program, Boston, Massachusetts, 1983.

Johnson, A.I. and Sniegocki, R.T., "Comparison of Laboratory and Field Analyses of Aquifer and Well Characteristics at an Artificial Recharge Well Site", Symposium of Haifa, Artificial Recharge and Management of Aquifers, Mar. 19-26, 1967, Publication No. 72, 1967, International Association of Scientific Hydrology, pp. 182-192.

Johnson, P., "Application of Surface Pressure to Assist Water Recharge into the Ogallala Formation", The Ogallala Aquifer--A Symposium, Special Report No. 39, 1970, International Center for Arid and Semi-Arid Land Studies, Texas Tech University, Lubbock, Texas, pp. 193-204.

Jones, O.R., Goss, D.W. and Schneider, A.D., "Surface Plugging During Basin Recharge of Turbid Water", Transactions of the American Society of Agricultural Engineers, Vol. 17, No. 6, Nov.-Dec. 1974, pp. 1011-1019.

Keys, W.S. and Brown, R.F., "The Use of Well Logging in Recharge Studies of the Ogallala Formation", The Ogallala Aquifer--A Symposium, Special Report No. 39, 1970, International Center for Arid and Semi-Arid Land Studies, Texas Tech University, Lubbock, Texas, pp. 31-48.

Keys, W.S. and Brown, R.F., "Role of Borehole Geophysics in Underground Waste Storage and Artificial Recharge", Underground Water Management and Artificial Recharge, Vol. 1, 1973, pp. 147-191.

Kimbler, O.K. and Whitehead, W.R., "Effect of Viscosity Ratio on the Recovery of Fresh Water Stored in Saline Aquifers", PB-270 780, Apr. 1977, National Technical Information Service, U.S. Department of Commerce, Springfield, Virginia.

Kincannon, D.F. and Tiederman, W.G., "Water Reclamation for Ground Water Recharge", OWRR-A-034-OKLA(1), June 1972, Oklahoma Water Resources Research Institute, Oklahoma State University, Stillwater, Oklahoma.

Knochenmus, D.D., "Hydrologic Concepts of Artificially Recharging the Floridan Aquifer in Eastern Orange County, Florida--A Feasibility Study", Report of Investigations No. 72, 1975, Florida Bureau of Geology, Tallahassee, Florida.

Krone, R.B., "Borehole Recharge: The Compatibility of Recharge Water with the Aquifer", Artificial Groundwater Recharge Conference, University of Reading, England, Sept. 21-24, 1970, Vol. 2, Paper No. 10, June 1971, The Water Research Association, Marlow, England, pp. 263-277.

Ku, H.F.H., and Simmons, D.L., "Use of Recharge Basins for Storm Water Management on Long Island, New York", Proceedings of the NWWA Eastern Regional Conference on Ground Water Management, 1983, Orlando, Florida, pp. 17-43.

Lichtler, W.F., Stannard, D.I., Kouma, E., "Investigation of Artificial Recharge of Aquifers in Nebraska", Water Resources Investigations 80-93, September, 1980, U.S. Geological Survey, Lincoln, Nebraska.

McCormick, R.L., "Filter-Pack Installation and Redevelopment Techniques for Shallow Recharge Shafts", Ground Water, Vol. 13, No. 5, Sept.-Oct. 1975, pp. 400-405.

McWhorter, D.B. and Brookman, J.A., "Pit Recharge Influenced by Subsurface Spreading", Ground Water, Vol. 10, No. 5, Sept.-Oct. 1972, pp. 6-11.

Motts, W.S., et al., "Feasibility of Increasing Water Supplies and Preventing Environmental Damage by Artificial Recharge in Massachusetts", Publication No. 132, Jan. 1983, Water Resources Research Center, University of Massachusetts, Amherst, Massachusetts.

Nightingale, H.L. and Bianchi, W.C., "Artificial Basin Recharge", Water Engineering and Management, Vol. REF, No. HB, 1981, pp. R85-87, R89-92.

Nightingale, H.I. and Bianchi, W.C., "Ground Water Chemical Quality Management by Artificial Recharge", Ground Water, Vol. 15, No. 1, Jan.-Feb. 1977, pp. 15-21.

Pettyjohn, W.A., "Flow Analysis of Hydraulic Connectors in Artificial Recharge Systems, A Model Study", OWRR A-015-OHIO(1), July 1974, Water Resources Center, Ohio State University, Columbus, Ohio.

Pettyjohn, W.A., "Introduction to Artificial Ground Water Recharge", EPA-600/2-81-236, Oct. 1981, Robert S. Kerr Environmental Research Laboratory, U.S. Environmental Protection Agency, Ada, Oklahoma.

Prill, R.C. and Aronson, D.B., "Flow Characteristics of a Subsurface-Controlled Recharge Basin on Long Island, New York", Journal of Research of the U.S. Geological Survey, Vol. 1, No. 6, Nov.-Dec. 1973, pp. 735-744.

Rebhun, M. and Hauser, V.L., "Clarification of Turbid Water with Polyelectrolytes for Recharge through Wells", Artificial Recharge and Management of Aquifers, Symposium of Haifa, Mar. 19-26, 1967, Publication No. 72, 1967, International Association of Scientific Hydrology, pp. 218-228.

Ripley, D.P. and Saleem, Z.A., "Clogging in Simulated Glacial Aquifers Due to Artificial Recharge", Water Resources Research, Vol. 9, No. 4, Aug. 1973, pp. 1047-1057.

Roberts, P.V., "Water Reuse for Groundwater Recharge: An Overview", American Water Works Association Journal, Vol. 72, No. 7, July, 1980, pp. 375-379.

Rubin, H., Glass, J.P. and Hunt, A.A., "Analysis of Storm Water Seepage Basins in Peninsular Florida", Publication No. 39, Sept. 1976, Florida Water Resources Research Center, University of Florida, Gainesville, Florida.

Seaburn, G.E. and Aronson, D.A., "Catalog of Recharge Basins on Long Island, New York, in 1969", Bulletin 70, 1973, New York State Department of Environmental Conservation, Albany, New York.

Seaburn, G.E. and Aronson, D.A., "Influence of Recharge Basins on the Hydrology of Nassau and Suffolk Counties, Long Island, New York", Water Supply Paper 2031, 1974, U.S. Geological Survey, Washington, D.C.

Schneider, A.D., Jones, O.R. and Signor, D.C., "Recharge of Turbid Water to the Ogallala Aquifer Through a Dual-Purpose Well", Miscellaneous Paper MP-1001, Aug. 1971, Texas Agricultural Experiment Station, Texas A&M University, College Station, Texas.

Schneider, A.D. and Jones, O.R. "Basin Recharge of Playa Water", Journal of Irrigation and Drainage Engineering, Vol. 109, No. 3, Sept. 1983, pp. 309-316.

Scott, V.H., "Optimization of Wells and Recharge Facilities in Conjunctive Use of Groundwater Reservoirs", OWRR-B-006-CALIF(1), Oct. 1969, Department of Water Science and Engineering, University of California, Davis, California.

Signor, D.C. and Hauser, V.L., "Artificial Groundwater Recharge Through Basins in the Texas High Plains", Miscellaneous Publication 895, Sept. 1968, Texas Agricultural Experiment Station, Texas A&M University, College Station, Texas.

Signor, D.C., "Recent Advances in Artificial Recharge", Proceedings of the Conference of the Great Plains Resources Council, Apr. 8-9, 1976, Denver, Colorado.

Sinclair, W.C., "Experimental Study of Artificial Recharge Alternatives in Northwest Hillsborough County, Florida", Water Resources Investigations 77-13, 1977, U.S. Geological Survey, Tallahassee, Florida.

Smith, A.J., "Design is the Key to a Good Recharge Well", Public Works, Vol. III, No. 10, Oct. 1980, pp. 66-67.

Smith, C.S., "An Investigation of Filter Media for Recharge Wells in the Grand Prairie Region of Arkansas", MS Thesis, 1972, Department of Agricultural Engineering, University of Arkansas, Fayetteville, Arkansas.

Sniegocki, R.T. and Brown, R.F., "Clogging in Recharge Wells, Causes and Cures", Artificial Groundwater Recharge Conference, University of Reading, England, Sept. 21-24, 1970, Vol. 2, Paper No. 13, June 1971, The Water Research Association, Marlow, England, pp. 337-357.

Thomas, H.E., "Cyclic Storage, Where Are You Now", Ground Water, Vol. 16, No. 1, Jan.-Feb. 1978, pp. 12-17.

Tibbals, C.H. and Frazee, J.M., Jr., "Ground-Water Hydrology of the Cocoa Well-Field Area, Orange County, Florida", Open File Report FL 75-676, Jan. 1976, U.S. Geological Survey, Tallahassee, Florida.

Vecchioli, J., "Experimental Injection of Tertiary-Treated Sewage in a Deep Well at Bay Park, Long Island, New York -- A Summary of Early Results", New England Water Works Association, Vol. 86, No. 2, June 1972, pp. 87-103.

Walton, W.C., Hills, D.L. and Grundeen, G.M., "Recharge from Induced Streambed Infiltration Under Varying Groundwater-Level and Stream-Stage Conditions", Bulletin 6, June 1967, Water Resources Research Center, University of Minnesota, Minneapolis, Minnesota.

Watkins, F.A., Jr., "Effectiveness of Pilot Connector Well in Artificial Recharge of the Floridan Aquifer, Western Orange County, Florida", Water Resources Investigations 77-112, Nov. 1977, U.S. Geological Survey, Tallahassee, Florida.

Watson, K.K. and Whisler, F.D., "Profile Desaturation During Sediment Deposition in a Groundwater Recharge Trench", Journal of Hydrology, Vol. 33, No. 3/4, 1977, pp. 397-401.

Watson, R. and Merritt, J., "Consumptive Use of Ground Water", 1970, Eastern Water Law Center, College of Law, University of Florida, Gainesville, Florida.

Williams, R.E. and Allman, D.W., "Factors Affecting Infiltration and Recharge in a Loess Covered Basin", Journal of Hydrology, Vol. 8, No. 3, July 1969, pp. 265-281.

Winqvist, G. and Mareliun, K., "The Design of Artificial Recharge Schemes", Artificial Groundwater Recharge Conference, University of Reading, England, Sept. 21-24, 1970, Paper No. 3, The Water Research Association, Marlow, England.

Wyatt, A.W., "Secondary Recovery of Capillary Water", Proceedings of the Ogallala Aquifer Symposium II, Texas Tech University, Lubbock, Texas, June, 1984, pp. 500-513.

CHAPTER 3

SITE SELECTION AND EVALUATION

Site selection and evaluation involves the consideration of physical siting criteria, recharge (source) water availability and quality, legal and institutional issues, artificial recharge methods, and potential costs and benefits. The purpose of this chapter is to summarize key information on each of these considerations.

PHYSICAL SITING CRITERIA

The two most-used methods of artificial recharge are spreading basins and recharge wells. Modifications of these methods may be necessary depending on such factors as climate, topography, soil, geology, water quality, purpose of recharge, quantity of water involved, costs, and land use (Kincannon and Tiederman, 1972). Physical siting criteria are basic to both method selection and design.

The success of any artificial recharge project will depend on the ability of the method to effectively recharge the aquifer. With any method, be it by water spreading or by wells, this ability is a function of the physical characteristics of the aquifer itself in addition to its surroundings. The development of artificial recharge as an effective water management tool requires that technical site selection criteria be developed. Wood (1980) stated that the single most important factor causing neglect of artificial recharge as a water management technique is the lack of suitable site selection criteria.

In effect, artificial recharge is essentially an enhancement of natural recharge; therefore, an understanding of the hydrologic cycle preceeds development of site selection criteria. Chidley (1981) listed five facets of the hydrologic cycle that need to be well understood: (1) precipitation; (2) surface infiltration; (3) soil moisture; (4) surface runoff and interflow; and (5) evapotranspiration. Wood (1980) expanded upon this and asked the following six general questions which must be quantified if viable site selection criteria are to be developed.

(1) How much rechargeable water is available, when and at what rate?

(2) What is the quality of the recharge water--dissolved solids content, temperature, organic content, ionic ratios, suspended solids, trace metals, bacterial and virus content?

(3) How much underground storage space is available, and at what depth?

(4) How readily will the aquifer accept the recharge water and how readily can it be recovered?

(5) How will the quality of water change after recharge?

(6) How quickly will the aquifer plug due to chemical, physical, or bacterial processes?

Calculation of the volume of surplus water available from surface water sources is a fairly straightforward process if the source is a permanent stream with a relatively long record of flow. If the water is taken from ephemeral streams, storm sewers, ephemeral lakes or other intermittent sources which are, in fact, likely sources of artificial recharge water, determination of the available volume is more difficult. The use of rainfall-runoff models can aid in arriving at an estimate, but values derived from such models are very sensitive to the amount and type of input data, which may be frequently inadequate.

The procedures for physical, chemical and biological analyses for various constituents in water are well advanced and do not inhibit the development of artificial recharge as a management technique. Predicting bacterial populations or suspended solids in water from an ephemeral stream, playa lake, or storm sewer is complicated; consequently, direct measurement of these parameters in individual sources of water over a wide range of flow conditions is necessary to establish their concentration ranges.

The percent of recharge water that may be recovered can be readily calculated using hydrogeological parameters such as porosity, hydraulic conductivity, and storage coefficient. Thus, the amount of recovery can be predicted if the properties of the aquifer and hydraulic system are known or measured (Wood, 1980).

The receptivity of the aquifer to the recharge water, in addition to its recoverability and whether or not the aquifer will be susceptible to plugging, is dependent on source water quality. Wood (1980) stated that the problem of establishing rational design criteria for artificial recharge rests to a large degree on the prediction of the reduction in hydraulic conductivity as a function of the amount and quality of input water. Water quality changes, too, are dependent on the aquifer characteristics and the native water conditions. A discussion of source water quality is presented in the next section of this chapter.

Following these general considerations, a list of potentially significant site selection criteria can be developed. Cochran (1981) listed the following criteria which must be considered in site selection and determining the appropriate method of artificial recharge:

(1) source of recharge water;

(2) chemical, physical and biological characteristics of recharge water;

(3) availability of an aquifer suitable for recharge;

(4) thickness and permeability of the material overlying and underlying the aquifer;

(5) thickness and permeability of the aquifer (recharge of a relatively impermeable aquifer would not be feasible);

(6) chemical characteristics of the native ground water;

(7) proximity of the potential recharge site to an appropriate well field cone of depression;

(8) water level differences between the aquifer and the recharge site;

(9) topography;

(10) availability of property; and

(11) permission and participation by local landowners.

The suitability of an aquifer for recharge operations is dependent on its hydraulic characteristics. These characteristics include the nature, variation, depth and quality of ground water, as well as flow rates and ground-surface water interactions (Franks, 1981). The receiving formation must have sufficient permeability and thickness to accept recharged water at a designated rate. The previous existence of high yield wells in an area would tend to indicate good potential for artificial recharge (Lehr, 1982). Usage of hand-carried hammer impact seismic timing equipment and similar tools may provide additional insight to aquifer characteristics. This equipment utilizes the physics of seismic wave refraction to explore stratigraphic changes in the subsurface environment (Bianchi and Nightingale, 1975).

An example of another way to determine the ability of an aquifer to receive recharge water involves examining the observed effects of the total hydrologic regimen under controlled conditions, or water levels in the area of interest. Deutsch (1967) discusses some tests conducted in Michigan to demonstrate the feasibility of recharging an artesian glacial-drift aquifer by inducing leakage to it from a surficial sand aquifer. During each of the tests, seven of the city's production wells were pumped at a combined rate of about 2475 gpm. Discharge was maintained at a constant level. The first test involved determining the hydraulic behavior of the well field under "natural" conditions after a small recharge channel was constructed (27,000 feet^2). Rates of recharge were calculated and then compared with rates of leakage determined from a second test in which the water in the recharge channel was maintained at a constant level, and from a third test in which the channel size was increased to 143,000 feet^2. As of the publication of Deutsch's paper, the recharge project had been operating successfully for 6 years after the tests aided in site selection and design.

Johnson and Sniegocki (1967) added that laboratory analysis of water bearing materials can and should be used in conjunction with appropriate field tests in developing a site. A more thorough understanding of the hydrologic system under conditions of induced recharge should result. Futhermore, a study of ground water contours and water quality data, as well as conventional study of geological and soil maps, can aid in identifying potential recharge areas. It is important at the outset for the analyst to develop some concept of how and where water is naturally getting into the aquifer, and how and where the water moves when under the ground (Chidley, 1981).

Topography is an important consideration as it permits or retards runoff, thereby influencing infiltration rates and amounts (Colorado High Plains Technical Coordinating Committee, 1984). Tools such as the ring infiltrometer which measures infiltration rates could become useful for estimating long term infiltration rates of water into soils. However, one study found that it may be debatable whether or not it could be used singularly as a recharge potential reconnaissance tool (Gillespie and Hargardine, 1976).

A promising tool for evaluating recharge potential is mathematical modeling. Manbeck (1975) discussed a Fortran IV model which simulates the movement of water from the soil surface to ground water. Components of the model include precipitation, irrigation amounts, air temperature, saturation soil moisture vs. moisture potential relationship, and thickness and number of strata. The model generates the amount of water from precipitation and irrigation sources moving vertically downward from one soil layer to the next on a monthly basis with given initial conditions. Although the assumption of one-dimensional flow restricts the accuracy of the model, it can be used in field situations with readily available data to assist in objective planning of recharge development (Manbeck, 1975).

In summary then, it is apparent that there are many factors to be considered when determining if a particular site will be receptive to artificial recharge. Field studies, laboratory tests, mathematical modeling, and information gathered from existing soils maps, hydrogeologic studies and precipitation surveys are all integral parts of the final site evaluation. All may not be practical or appropriate for each situation, however, it is important to the future of artificial recharge as a management technique that the validity of site determination from modeling, laboratory tests, and literature-reviewing, all more practical than field studies, be evaluated. In this way, an efficient site determination process can be utilized.

SOURCE WATER CONSIDERATIONS

Considerations related to source water for artificial recharge include the types and quality of source water, treatment methods necessary prior to recharge, and resulting water quality following recharge.

Types of Source Water

The availability of a source of water for artificial recharge is an important factor in site selection. Many geographical areas which have demonstrated needs and many sites which are physically suitable for artificial recharge can never be developed due to a lack of an acceptable water source. The types of water that may be considered for artificial recharge vary with the locality. In areas of high periodic storm flow and dry periods, detention can impede intense storm runoff, giving the soil time to absorb the water for storage during dry periods. Storm runoff may also be directed to artificial recharge in urban areas where development has eliminated natural recharge capabilities.

In industrial and urban areas, reclaimed wastewater can often be a viable source of recharge water. Wastewater disposal is always a problem, and further marginal investment resulting in reclaiming wastes for economic use represents a sound means for achieving both a qualitative and quantitative conservation of resources and eliminates the detrimental effects of discharging partially treated effluent to receiving waters (Frankel, 1967).

Quality of Source Water

For water to qualify for consideration as a source water for artificial recharge, the quality must be evaluated in the context of the native water and the intended use. Due to concern over storm water quality recharging the 3,000(+) recharge basins on Long Island, New York, a study was conducted to evaluate the quality of this recharging water (Ku and Simmons, 1983). Five recharge basins were monitored in five representative land use areas (strip commercial, shopping mall, major highway, low density residential, and medium density residential) over a five year period. The ground water was also monitored. During each sampled storm, runoff to each basin was collected for analysis for arsenic, cadmium, calcium, carbon, chloride, chromium, lead, magnesium, nitrogen, phosphorus, potassium, sodium and sulfate. Results indicted low concentrations of most constituents, generally within the standards for potable water. Most exceptions to this generalization were related to specific land uses and seasonal effects such as runoff from major highways and deicing salts used during the winter.

After movement through the unsaturated zone in the Long Island study, the heavy metals were largely removed, chloride was not removed, and nitrogen was often found in greater concentrations in ground water than in storm water because of the use of cesspools, septic tanks and lawn fertilizers. An analysis for 113 selected priority pollutants revealed that in the five storm water samples and nine ground water samples, three contained one to two constituents in excess of New York state guidelines for organic chemicals in drinking water. The median number of indicator bacteria in the storm water ranged from 10^8 to 10^{10} MPN per 100 ml. Fecal coliforms and fecal streptococci increased by one to two orders of magnitude during the warm season, but total coliforms showed no significant seasonal differences. No bacteria were detected in ground water beneath the basins. Therefore, the results of this study indicated that the recharge basins seem to be a safe method for storm water disposal and ground water replenishment under the conditions found on Long Island (Ku and Simmons, 1983).

Another constituent of concern in runoff water as a source is the concentration of clay particles (sediment). The recharge method employed can suffer drastically as a result of clogging from accumulation of sediment. Kincannon and Tiederman (1972) came to these conclusions regarding clogging:

(1) Clogging of a soil during infiltration by water containing suspended clay particles is basically a surface sealing process.

(2) After a horizontally continuous surface layer is formed, clogging progresses at approximately the same rate for varying concentrations of suspended matter. Although clogging and infiltration loss may progress at the same rate, the accumulated infiltration is, of course, much lower.

(3) Clay particles were observed to penetrate deep into a sand column, but it can be concluded that this deep penetration has no effect on the overall infiltration rate.

(4) Low concentrations of suspended matter will eventually reduce the infiltration rate, but the rate will never completely reach zero.

Total removal of particulate matter would need to be approximated in order to solve the problem of clogging by suspended solids.

In the case of wastewater, Gerba et al., (1980) indicated that land application of domestic wastewater can serve as an economical method for removing a wide range of contaminants so that the water can be reused. Admittedly, to obtain optimum contaminant removal, land treatment systems must be carefully designed. Another worker agreed that most reclaimed water contaminants are substantially reduced during vertical percolation through soil and during horizontal movement in aquifers. Notable exceptions were total dissolved solids, hardness, nitrates, and a few heavy metals (Garrison et al., 1980).

The ability of ground water contamination to reach a state of equilibrium shows that with careful management, there is no reason why effluent recharge should cause any serious ground water pollution (Baxter et al., 1984). However, if water with trace element contents not exceeding the recommended quality levels for public water supply or irrigation use is considered suitable for ground water recharge, the possible magnitude of accumulating micro-contaminants in soils may be estimated by the amounts of water applied (Chang and Page, 1980). The 20-year trace metal impact to soils for a slow rate land application is shown in Table 4 (Chang and Page, 1980). Similar calculations for rapid infiltration rates are shown in Table 5 (Chang and Page, 1980). Table 6 shows the calculated maximum wastewater concentration for twenty years of ground water recharge with trace metal input not to exceed maximum permissible levels (Chang and Page, 1980).

In addition, a quality consideration for any source water type is the air which is drawn into the aquifer with the recharge water. This entrained air can effectively reduce the hydraulic conductivity by reducing the available pore space for conveyance of water (Clyde et al., 1984). In some tests, air entrainment was the most serious cause of clogging (Sniegocki, 1963).

The air bubbles in the aquifer have the same effect as clay particles or sand grains and will effectively retard the passage of water. The forces that resist movement of the bubbles may result from one or more of three causes: (1) simple blocking of the air bubbles by sand grains, (2) a phenomenon called the Jamin effect, and (3) distortion of gas bubbles when they are forced through capillary openings. The principle of the Jamin effect is that a capillary tube containing restrictions and filled with a chain of alternate air and water bubbles is capable of sustaining a finite pressure. The pressure that can be sustained is determined largely by the degree of wetting of the capillary tube, contamination of the inner surface of the tube, the number of air bubbles in the chain, the surface tension of the water, and the distortion of gas bubbles when forced through capillary openings (Sniegocki, 1963).

Treatment Methods

Palmer (1983) states that an essential requirement for injection well recharge is that the injected water be free of suspended solids, to minimize the eventual plugging of the injection interval. Reduction in hydraulic conductivity of the receiving aquifer usually occurs at or near the water-aquifer interface by biological and chemical mechanisms in addition to physical means (Wood, 1980).

Table 4: Calculated 20-Year Trace Metal Impact to Soils for Slow Rate (1.5 m/year) Application Based Upon Water Quality Criteria and Wastewater Composition (Chang and Page, 1980)

Element	Water Quality Criteria			Wastewater Composition		Typical Range in Soils	
	Public Water	Irrigation Water Continuous	Irrigation Water Short Term	Minimum	Maximum	Minimum	Maximum
	Input (kg/ha)						
As	30	30	600	1.5	6.9	0.2	80
B	--	225	600	90	750	4	200
Cd	3	3	15	1.5	66	0.05	1.4
Cr	15	30	300	0.3	30	10	6,000
Cu	300	60	1,500	1.8	15.9	4	200
Pb	15	1,500	3,000	1	105	4	400
Mo	--	3	15	0.3	5.4	0.4	10
Hg	0.6	--	--	0.06	0.30	0.02	0.6
Ni	--	60	600	0.9	180	20	2,000
Se	--	6	6	--	--	0.02	4
Zn	15	600	3,000	1.2	105	20	600

Table 5: Calculated 20-Year Trace Metal Impact to Soils for Rapid Infiltration Systems (80 m/year) Applications Based Upon Water Quality Criteria and Wastewater Composition (Chang and Page, 1980)

	Water Quality Criteria						
Element	Public Water	Irrigation Water Continuous	Irrigation Water Short Term	Wastewater Composition Minimum	Wastewater Composition Maximum	Typical Range in Soils Minimum	Typical Range in Soils Maximum
	Input (kg/ha)						
As	1,600	1,600	32,000	80	368	0.2	80
B	--	12,000	32,000	4,800	40,000	4	200
Cd	160	160	800	80	3,520	0.05	1.4
Cr	800	1,600	16,000	16	1,600	10	6,000
Cu	16,000	3,200	80,000	96	848	4	200
Pb	800	80,000	160,000	48	5,600	4	400
Mo	--	160	800	16	288	0.4	10
Hg	32	--	--	3.2	16	0.02	0.6
Ni	--	3,200	32,000	48	9,600	20	2,000
Se	--	320	320	--	--	0.02	4
Zn	800	32,000	160,000	64	5,600	20	600

Table 6: Calculated Maximum Wastewater Concentration for 20-Years of Ground Water Recharge with Trace Metal Input not to Exceed Maximum Permissible Level for Sludge Cropland Applications (Chang and Page, 1980)

Element	Permissible Metal Input[1]	Calculated Maximum Permissible Wastewater Concentration: Slow Rate	Calculated Maximum Permissible Wastewater Concentration: Rapid Infiltration	Irrigation Water Quality Criteria[2]	Median Concentration of Wastewaters[3]
	- kg/ha -	- - - - - - -	- - - - - - - g/liter	- - - - - - -	- - - - - - -
	5	17	0.31	10	5
Cd	10	33	0.62		
	20	67	1.25		
Cu	125	417	8	200	18
	250	833	16		
	500	1,667	31		
	125	417	8	200	4
Ni	250	833	16		
	500	1,667	31		
	500	1,667	31	5,000	8
Pb	1,000	3,333	62		
	2,000	6,667	125		
	250	833	16	2,000	40
Zn	500	1,667	31		
	1,000	3,333	62		

Table 6: (Continued)

[1]Limits are based upon the cation exchange capacity of the soils. Low value applied to soils with CEC < 5 meq/100 g; intermediate value applies to soils with CEC from 5-15 meq/100 g; and maximum value applies to soils with CEC >15 meq/100 g.

[2]For continuous use.

[3]From Table 4.

Plugging of artifically-recharged aquifers by biological growth is frequently a limiting factor in the success of artificial recharge operations. It probably occurs due to the in-situ growth of bacteria at or near the water-aquifer interface. A large part of this loss in hydraulic conductivity is associated with the change from aerobic to anaerobic conditions near the water-aquifer interface.

Laboratory investigations are needed to predict the onset of anaerobic conditions through evaluation of the input of dissolved oxygen, the sorption isotherm of biodegradable organic material, which is generally assumed to be the factor limiting growth, biological population dynamics, and the hydraulic-flow data. The capacity to predict the onset of bacterial plugging would permit a more rational design of the size and number of recharge structures that would accommodate a given amount of recharge with a specified amount of "down time" for clearing and rejuvenation (Wood, 1980).

Chemical reaction between recharge water and aquifer water is another possible cause of aquifer clogging when recharging through wells. Precipitates of calcium and perhaps magnesium carbonate, and various iron compounds, are thought to be the most likely reaction products. Theoretical analysis suggests that calcium carbonate could be precipitated from mixtures of two waters both initially stable with respect to dissolved calcium carbonate. Dissolved ferrous iron in native ground water could be precipitated as ferric oxide or hydroxide as a result of mixing with oxygen-bearing recharge water. This reaction is time-dependent. Mathematical models that describe the amount of chemical remaining in solution when chemical reaction occurs between recharge and aquifer water can be formulated from the basic dispersion equations if it is assumed that the reaction does not alter the dispersion process.

Reaction between recharge water and aquifer water can be prevented by continuous pretreatment to make recharge water nonreactive. A possible alternative method of preventing reaction in some cases involves the injection of a buffer-zone of nonreactive water between the recharge water and the aquifer water (Warner and Doty, 1967).

Experiments have demonstrated that suspended solids, which are mostly colloidal in size, could be removed before recharge by flocculation with appropriate polymers and chemicals, followed by settling. One treatment includes a combination of a cationic polyelectrolyte and alum. In laboratory tests this combination reduced the suspended solids content from 210 mg/l to less than 20 mg/l with a dose of 0.5 mg/l of cationic polyelectrolyte and 5 mg/l of alum (Hauser and Lotspeich, 1967).

To reduce clogging by air entrainment, the well system should be designed carefully to allow the source water to enter the well through an intake pipe without any leaks to the atmosphere. Also, recharge under constant-head conditions with the water level in the well at the land surface, or installation of a valve at the bottom of the injection pipe, would minimize air entrainment (Sniegocki, 1963).

Unfortunately, another possible source of air entrainment may be more difficult to control. Gases may be liberated from solution in the recharge water if it is warmed by contact with the native water and the aquifer. Heat might also be generated by biological decomposition of suspended organic matter in the injected water (Sniegocki, 1963).

Proper management of spreading and percolation systems can optimize contaminant removals. For example, nitrogen form removal can be optimized by proper management of flooding, percolation, and rest periods. Removals of 50-80 per cent have been demonstrated. Phosphorous removal has been found to be basically dependent upon the distance travelled through the soil. The chief removal mechanism is precipitation as calcium phosphate or magnesium ammonium phosphate. Underground travel distances of 9 m produced a 50 percent reduction, and distances of over 100 m led to a 90 percent reduction (Schmidt, 1978).

Presently, in California under their "Draft Ground Water Recharge Guidelines", water for surface spreading would have to be subjected to the following minimum treatment:

(1) source control of toxic industrial chemical compounds;

(2) adequate oxidation;

(3) carbon adsorption with 30 minutes contact;

(4) percolation through an unsaturated aerobic zone of undisturbed or approved soil for a minimum distance of 10 feet;

(5) dilution by an equal quantity of ground water; and

(6) one years storage.

Pretreatment for injection recharge is suggested to include:

(1) adequate disinfection;

(2) oxidation;

(3) coagulation;

(4) clarification;

(5) filtration;

(6) GAC adsorption;

(7) reverse osmosis; and

(8) decarbonator-type stripping for volatile low molecular weight organics.

Resulting Water Quality

In establishing criteria for the quality of recharge waters it is desirable to consider changes which may take place between the time of recharge and the time of withdrawal and use by the consumer. Some undesirable characteristics of recharge water can be appreciably altered or entirely eliminated after the water enters the ground and before it reaches the user. On the other hand, organic pollutants remaining after wastewater treatment pose a potential health threat if the reclaimed water is to be directed to potable reuse (Roberts et al., 1980). Serious health effects could occur if trace amounts of certain organics are injected for long periods of time.

The following basic phenomena must be taken into account in speculations on the water quality changes likely to occur in ground water recharge (Roberts et al., 1980):

(1) biodegradation by and growth of microorganisms;

(2) chemical oxidation or reduction;

(3) sorption and ion exchange;

(4) filtration;

(5) chemical precipitation or dilution; and

(6) volatilization or photochemical reactions.

By measuring the concentration responses of water quality parameters at observation wells near injection wells, major quality changes and the processes responsible for them can be tentatively identified. By distinguishing between biodegradation and adsorption, it can be judged whether the observed removal of organic solutes is likely to be maintained over long periods of time. Where adsorption is responsible for removal, the aquifer capacity is ultimately exhausted and breakthrough occurs. The field capacity for removal by adsorption can be estimated from the breakthrough behavior of the substance at an observation well. This value can be used to predict the rate of transport of the substance beyond the observation well if the hydrogeologic and adsorption characteristics can be assumed similar to the region near the injection well or can be otherwise estimated (Roberts et al., 1980).

The accumulation of trace organic compounds in the aquifer by adsorption has several implications. The retardation of movement means that strongly adsorbing pollutants can be delayed many years before reaching the point of water extraction for reuse. The long residence times resulting from adsorption provide opportunities for biodegradation to occur to a significant extent, even if it is slow (Roberts et al., 1980).

The movement of pesticides through an aquifer from a recharge well and the concentration of these compounds in water subsequently pumped from the aquifer will be greatly affected by the degree of adsorption of the pesticides by aquifer solids. The adsorption of pesticides on aquifer particles is a complex phenomena involving many variables.

Increasing proportions of clays (particularly montmorillonites) would be expected to result in greater adsorption because of the ionic and large surface areas of these substances. Pore size and pore concentration in an aquifer will affect the adsorption characteristics because of their influence on adsorbate-adsorbent distances and their influence on the rate of flow past any given particle. It would appear from a mass action and solubility standpoint that the greater the volume of water moving past a point in a given unit of time, the greater will be the adsorption. Temperature and pH will inversely affect adsorption. The native characteristics of the adsorbate is a primary factor.

Because of differences in the chemical structures of pesticide molecules and the resulting variations in affinities for adsorbent and solute, it seems probable that within a given family of pesticides the adsorption will be inversely proportional to water solubility (Scalf et al., 1968).

Gerba et al., (1980) discusses the fate of pathogenic microorganisms such as bacteria, virus, protozoa, and parasitic worms. The effectiveness of pathogen removal during land application will be determined by their survival in soil and their retention by soil particles. Climate controls two important factors in determining viral and bacterial survival: temperature and rainfall. The survival of microorganisms is greatly prolonged at low temperature (below 40°C they can survive for months or even years), and survival of many viruses and bacteria is adversely affected during drying of the soil.

Soil properties influence moisture-holding capacity, pH and organic matter--all of which control the survival of bacteria in the soil. Adsorption of viruses to the soil surface plays a major role in determining viral survival time. It seems to offer protection against factors responsible for viral inactivation in the soil.

The nature of the microorganism will also play a role in survival. Resistance to environmental factors varies among different species as well as the particular strain. Most enteric bacterial pathogens will die off very rapidly outside of the host, whereas indicator bacteria such as E. coli will persist longer. Survival times among different types of viruses vary greatly and are difficult to assess without studying each type individually.

The longer survival time of enteric bacteria after inoculation into sterilized soil as compared to non-sterilized soil indicates that antagonism is an important factor. Virus survival is much longer under sterile conditions in aerobic environments but not under anaerobic conditions. This is indicative that aerobic but not anaerobic, soil microorganisms are antagonistic to virus survival. Thus viruses retained near the soil surface would be expected to be more rapidly inactivated than those that penetrate the soil surface (Gerba et al., 1980).

Expected water quality after recharge, therefore, is as much a function of aquifer characteristics as it is affected by source water quality. Many factors should be taken into account when considering a source water and the area in which it will be recharged, not the least of which is travel distance from application to use, and the intended use.

LEGAL/INSTITUTIONAL ISSUES

Legal and institutional issues must also be considered in artificial recharge site selection and design. Federal, state and local perspectives and relationships must be addressed.

General Perspectives

Irrigation water supply augmentation has traditionally been a function of Federal water policies. For decades the Federal government has subsidized surface water impoundment and diversion structures for irrigation, but only recently it has become involved in ground water recharge issues. The U.S. Environmental Protection (EPA) Agency provides input into the ground water recharge issue as summarized by Gaston (1980):

(1) Clean Water Grant Program: There is controversy over the use of these funds for wastewater reclamation projects, including land application and possible recharge.

(2) NPDES Permit Program: All waste treatment operations built with Public Law 92-500 funds must comply with the provisions of the National Pollutant Discharge Elimination System.

(3) Underground Injection Control (UIC) Program: Sections 1421-1424 of the Safe Drinking Water Act require that EPA administer a program, to ensure that underground injections will not "endanger" drinking water sources. No commitment for Federal financial assistance may be entered into for any project which the Administrator determines may contaminate such aquifer through a recharge zone so as to create a significant hazard to public health.

The issue of ground water recharge has led to a Federal project for investigating ways to maintain dependable ground water supplies. On September 28, 1984, President Ronald Reagan signed into law the "High Plains States Ground Water Demonstration Program Act of 1983" (Public Law 98-434). This law directs the Secretary of the Interior to engage in a special study of the potential for ground water recharge.

Public Law 98-434 authorizes the establishment of demonstration projects for ground water recharge to aquifers in the High Plains states of Colorado, Kansas, Nebraska, New Mexico, Oklahoma, South Dakota, Texas and Wyoming, as well as the states of Arizona, California, Idaho, Montana, Nevada, North Dakota, Oregon, Utah and Washington. The U.S. Bureau of Reclamation will direct demonstration project plans for various recharge technologies to be applied to the diverse geologic and hydrologic conditions. Twenty-one demonstration sites located in the states listed will be selected. They must be confined to areas with declining water tables, available surface water supplies for recharge, and a high physical, chemical and economic feasibility for recharge of the ground water reservoir. Public Law 98-434 has three amendments that define the scope of the project; they include: (1) specific line-item appropriations for each demonstration project, giving Congress the opportunity to judge each proposed project on its merits should it choose to do so; (2) the involvement of EPA to the extent that it would assure that none of the projects would degrade ground water resources that are, or may be, used for drinking water; and (3) the prohibition of the use of any funds for the diversion of water from the Great Lakes or the Arkansas River Basin. This last amendment implies that importation of water from great distances for the purpose of recharging aquifers would not be acceptable.

Public Law 98-434 also includes funding provisions. A total of $0.5 million was appropriated for project planning, and an additional appropriation of $20 million was authorized for project construction. Twenty percent of the cost must be absorbed by states, their political subdivisions, or other non-federal entities. With the legislative amendments and economic incentives, Public Law 98-434 should achieve its primary objective of focusing greater attention on a previously little recognized management technique (Aller, 1985).

In the development of a ground water recharge project relevant legal considerations include: (1) the right and associated legal duties of storing water underground; (2) rights to withdraw water stored underground; and (3) management authorities to implement recharge project objectives. Legal rights associated with storing water underground are unclear regarding the right to store water under land not owned or controlled by the recharge entity, and regarding liability for damage caused by such storage.

Three questions are of concern to states developing ground water recharge policies: (1) whether costs should be financed through use charges; (2) whether the recharged ground water will be deliberately managed to achieve supply objectives; and (3) whether recharged ground water should be integrated with water from other sources (Aiken, 1980). In addition to the above legal considerations, supplementing ground water supplies through artificial recharge may also have impacts on land use, population, the local economy and public attitudes (State Water Planning and Review Process, 1984).

Increased availability of ground water for irrigation may lead to increased crop production, additional sales of farm products, and increases in gross farm revenues. Increased ground water storage would probably have little effect on recreation opportunities unless streamflows or wet meadows were positively affected. Changes could include improvement of wildlife habitat and improved streamflows. The demand for both housing and public services may temporarily increase during the construction phase of a project. In addition, energy requirements may increase due to ground water pumping. Thus, long-term increases in public service deliveries may also be required. The local tax base would increase if irrigated acreage increased, or if more productive crops could be grown.

An improved economic base could then lead to increased population stability in an area. However, the amount of population change would depend on the extent of economic activity. The population composition of an area may remain unchanged or could include increased numbers of young adults and children due to reduced out-migration forces.

The price of land would probably increase or stabilize since increased quantities of irrigation water would improve local crop production. Increased or stabilized agricultural production should lead to some expanded employment potential in the local area.

Local residents may experience less concern regarding water shortages. Community cohesiveness may increase if water importation was partially a result of local cooperation and lobbying efforts. However, divisiveness may occur if problems should arise concerning allocation of supplemental water supplies. Area residents may feel an enhanced or improved quality of life if the project were constructed in an area which previously experienced water shortages. Improvements in the local economy would affect the general state of satisfaction of residents. Those who prefer economic and population growth would feel an improved quality of life. However, those who prefer the status quo would probably disapprove.

If public monies are used some people may have their attitudes affected by questions of equity. The attitude of some people may be that such monies should not be used for projects which they perceive as directly benefitting only a small number of people.

Ground Water Doctrines

When states adopted ground water laws, many followed the statutes already adopted for surface water. States employ one of four different ground water doctrines: (1) Absolute Ownership, (2) Reasonable Use, (3) Correlative Rights, and (4) the Doctrine of Prior Appropriation. Because ground water conditions are complex, more than one doctrine may be used (U.S. Department of Interior, 1979).

Absolute Ownership states that the landowner can withdraw as much water as desired from beneath his own land without liability to his neighbors. States using this doctrine include Texas, Louisiana, Arkansas, Missouri, Minnesota, Indiana, Ohio, Pennsylvania, Vermont, Massachusetts, Connecticut, Maine, New Jersey, Rhode Island, South Carolina, Georgia, Alabama, Mississippi, and parts of California.

As more knowledge was acquired about ground water flow many states modified the Absolute Ownership doctrine to limit the landowner to a "Reasonable" Use. This concept recognizes equal rights of all to ground water resources. States employing this system of law include Arizona, Nebraska, Iowa, Illinois, Michigan, Kentucky, Tennessee, Florida, North Carolina, Virginia, Delaware, West Virginia, Maryland, New York, New Hampshire, and Wisconsin.

The State of California has gone one step further with the Reasonable Use doctrine in formulating the Correlative Rights doctrine. The landowner's use must not only be reasonable, but must be correlated with the use of others, particularly during times of water shortage. When supplies become insufficient because of a drought or drawdown effect, each landowner is entitled to water in proportion to the percentage of owned land in relation to all other lands overlying the ground water supply. This doctrine is also followed in Oklahoma and somewhat in Colorado and Nebraska. In Colorado, SB-213 allocates water in a confined aquifer based on one percent of the available instorage supply annually.

Under the doctrine of Prior Appropriation the water that returns to the stream (which had been diverted upstream) becomes part of the natural streamflow to be shared by riparian owners. Riparian rights entitle the owners to the natural flow of the stream, including such portions of the natural flow diverted upstream that are allowed to flow back into the stream after use. The fact that such water has been used on upstream lands does not deprive it of the character of natural flow when it returns to the stream from which it was diverted.

In areas where the doctrine of Prior Appropriation is used, the downstream flows on many streams have been augmented by seepage and return flows from irrigation of upstream lands. This is a common phenomenon in irrigated valleys, and much development has been predicated wholly or partly upon the continued existence of such flows. The increased flow does not consist of new water (unless brought in from another watershed), but is the reappearance of water previously diverted. The water may reenter the stream through surface or subsurface flow. In any event, if the irrigator does not lawfully recapture the return flow water, it becomes a part of the stream and accrues to the benefit of downstream claimants in accordance with their rights to the natural flow.

Return flow waters from a diversion under an appropriated right are a part of the streamflow from the time the waters escape from the premises or works of the appropriator, provided they would ultimately return to the stream system from which originally diverted if not artificially intercepted. Consequently, the waters belong to that stream system and are subject to the rights of appropriators depending upon their priorities. An attempt to recapture the water by the appropriator after it leaves the boundaries of the land to which it was appropriated is not permitted under the doctrine of Prior Appropriation. In other words, such water belongs to the stream, even before it comingles with the water of natural flow.

Virtually all interstate compacts and court decrees between states address return flows. Compacts and decrees generally apportion among the states the beneficial consumptive use of the waters of the stream system (e.g., the Colorado River Compacts). The Supreme Court has defined "beneficial consumptive use" as diversions from the stream less return flow.

The doctrine of Prior Appropriation also provides for abandonment and forfeiture of water rights. Appropriation rights are dependent upon continued beneficial use of the water. One way in which an appropriation water right may be lost is for the owner of the right to discontinue use of the water with intent to abandon the water right. In some states presumption of abandonment has the same practical effect as the doctrine of forfeiture.

Forfeiture occurs when the owner of an appropriation water right fails to use the water for the period prescribed by statute. The most common period of nonuse for purpose of forfeiture is 5 years, although some states have periods as short as 2 years. Unlike abandonment, the doctrine of forfeiture is not concerned with the intent of the owner of the water right. Generally, for forfeiture to occur, non-use must be continuous for the statutory period during times when the water was available for use.

The Riparian doctrine system required very little state administration until recent years because of small depletions of water for transportation, domestic, and industrial uses. But with recent increases of irrigation consumptive uses, many of the States are adopting administrative procedures to enforce the reasonable use aspect of the law through permit systems or other means. Essentially, a priority and property rights system is developing in the eastern States.

In the states using the doctrine of Prior Appropriation, water resources agencies have been established to administer water development, use, and management. A person(s) desiring to use or appropriate water must file an application for a permit with the proper official or state water agency. The agency is empowered to issue a permit with a priority date, or deny it on various grounds according to state statute. Denial is generally related to a determination of the lack of unappropriated water (or water set aside for instream purposes) and an interpretation of the public interest.

Most state statutes provide for a three-fold approach to water administration: (1) appropriation, (2) adjudication, and (3) distribution. The above includes keeping all records associated with the tasks. Many western states have a board or commission that supervises and has decision-making powers while assisting the water administration officer. The commission or board generally also has authority to adjudicate water rights. At least five states in the west, New Mexico, Colorado, Utah, California, and Wyoming, adjudicate water through the state court system.

Much of the water used in the United States comes from sources shared by more than one state. With stream systems crossing state boundaries, and with the increased demand for water, problems in allocation of the interstate water and the administration of water rights can arise. In anticipation of such problems, several states have entered into interstate compacts or have requested court apportionment of the affected waters. Where these agreements result in interstate compacts, the compacts must be consented to by the Congress. Where states could not agree on the allocation of water, courts have been asked to allocate the supply for them by court decree.

Existing water compacts may be grouped into four general categories: water allocation, pollution control, flood control and planning, and comprehensive regulatory and project development. Water allocation compacts primarily are those which have addressed irrigation.

Local Water Institutions

As opportunities for irrigation development by individuals in the western United States became exhausted, and people recognized the potential for developing large tracts of land and quantities of water, attention was turned to institutional arrangements to allow people to organize into groups to carry out such projects. The major water organizations used for group irrigation development are unincorporated mutuals, incorporated mutuals, irrigation or other water districts, and companies or associations. Over 75 percent of the land irrigated in Wyoming, Colorado, Utah, Idaho, Montana, and Washington has water delivered through some type of group water organization. Over 50 percent of the land irrigated in Oregon, California, Arizona, New Mexico, North Dakota, and South Dakota is served by group organizations. Nevada, Nebraska, Kansas, Oklahoma, and Texas have less than 25 percent, and the eastern states less than 5 percent of irrigated land to which water is supplied by organizations.

With more complex facilities involving high construction costs, these organizations expedited development by combining private, state, and Federal capital and labor. State legislation has provided for the creation and operation of many types of districts for irrigation water use and management--mostly single-purpose districts such as irrigation districts, drainage districts, watershed districts, water improvement districts, water control districts, and ground water control districts. In recent years, state statutes have provided multiple-use water districts, such as water conservancy districts and natural resource districts which may comprise an entire watershed and combine the operation of many single-purpose districts for managing water on a multi-purpose basis.

The powers of the organizations vary with the duties or purpose of the group. These powers may include fixing water prices; collecting assessments; planning, constructing, and maintaining waterworks; and allocating district water. But, as with individuals, surface and ground water supplies are still basically allocated on a priority basis and quantities regulated according to the beneficial use concept. Water users in many of the organizations often "pool" their water rights during water-short years. Ground water management districts in Nebraska, Kansas, Colorado, Oklahoma, and Texas provide the means for managing the water supply beneficially other than on a priority basis.

State Examples of Legal/Institutional Issues

Many of the legal requirements to be described herein seem to constrain supplemental water-supply development by requiring that a permit be obtained or that a permit holder be legally liable for damages caused to another. However, these apparent constraints actually facilitate such development in clarifying the developer's water rights. In general, the most serious legal constraints in Nebraska exist where inadequate or no legislative authorization is available to implement certain supplemental water supply alternatives, an example is the lack of authority for agricultural ground water transfers. Legal constraints to be discussed include underground water storage, and ground water transfers (State Water Planning and Review Process, 1984).

In Nebraska, water statutes did not address ground water storage in a manner similar to surface water storage until passage of LB 198 by the State Legislature in 1983. This legal vacuum severely constrained deliberate attempts to store water underground, in that the recharge project operators had no legal security in being able to control withdrawals of water stored underground, or to charge for its withdrawal. A related issue was the status of surface water appropriations when water appropriated for direct irrigation was instead stored underground. A final issue relates to liability for private damages resulting from storing water underground. These significant legal constraints to storing water underground have now been addressed through legislation.

Nebraska statutes explicitly authorize ground water transfers for public water supply and industrial purposes if a Department of Water Resources permit has been obtained. Ground water transfer authorities for agricultural purposes may be inferred from surface water appropriation statutes and natural resources district statutes, although these authorities are unclear. The absence of a clear legislative authorization to transfer ground water for agricultural purposes is a major constraint to supplemental water development for irrigation.

The two aspects of Nebraska surface and ground water law most constraining supplemental water supply development are: (1) the preference for junior natural-flow appropriators over senior storage appropriators during the irrigation season; and (2) the absence of clear agricultural ground water transfer authorities (State Water Planning and Review Process, 1984).

The Nebraska Legislature recognized that, as a result of water project operations, surface water in some areas of the state has been, is, and will be in the future intentionally and incidentally stored in and withdrawn from underground strata. Rights to water intentionally or incidentally stored underground and rights to withdrawal of such water should be formally recognized and quantified. All beneficiaries should proportionately share in the financial obligations necessary for construction, operation, and maintenance of water projects which cause underground water storage. Uses of water for incidental and intentional underground water storage are beneficial uses of water, and comprehensive conjunctive management of surface water and intentional or incidental underground water storage is essential for the continued economic prosperity and well-being of the state.

Several states have developed regulations for land disposal of wastewaters --an indirect method of recharge. By 1976, approximately 26 states had prepared regulations and guidelines for ground water recharge (Schmidt, 1978). The state regulations are intended to control the disposal of wastewater to the land surface in an environmentally acceptable manner, and are not directed specifically to formal ground water recharge operations by either percolation or direct injection.

California courts have affirmed the public's right to store, protect and retrieve imported water in natural underground basins (Gleason, 1976). This was done to alleviate the need for additional dams by providing a ready means to capture, during the rainy season, the excess water which usually escapes unused. California is using underground water system facilities in its multi-billion dollar water system. One legal problem still requiring clarification is the exchange storage situation which results when a ground water right owner foregoes the extraction of a certain amount of ground water by substituting imported surface water supplies. An additional problem is establishing a priority list for underground storage in areas where the storage space is limited. Obviously the overlying communities should get first priority. Perhaps one of the most significant consequences of recent judicial action in California is the current public concern that the state's existing resources, both natural and man-made, should be used to their fullest extent.

California and Washington artificial recharge management policies differ in significant respects, owing in part to different water supply and use characteristics. Both states, however, authorize the control of recharge water withdrawals, although by different means. Both states also permit integrating the use of surface and ground water; in California, to make the most effective use of all water supplies regardless of their sources, and in Washington, to allow management of water stored underground as part of the total reclamation project water supply (Aiken, 1980).

Direct injection of treated wastewater for recharge is regulated on an individual project basis in California and New York, and it requires failsafe provision of drinking water quality. In total, 49 monitoring parameters are included in the California State Health Department (CSHD) regulations. The CSHD has imposed stringent requirements on the direct injection recharge of treated wastewater in Orange County. The high cost of meeting such requirements makes it probable that ground water recharge by direct injection of treated wastewater will find only limited application nationally (Schmidt, 1978).

As with direct injection, the regulation of recharge through spreading basins has been practiced by states on an individual case basis. Wastewater applied to the land evaporates, runs off into surface water, is retained in plant tissue, and percolates into the ground water. Percolation into ground water accounts for over half of the applied wastewater in most land disposal operations.

The CSHD recommendations are based on all relevant aspects of each project, including the following factors: treatment provided; effluent quality and quantity; spreading area operations; soil characteristics; hydrogeology; residence time; and distance to withdrawal. A public hearing is held prior to making the final determination regarding the public health aspects of each ground water recharge project. Final recommendations are submitted to the

Regional Water Quality Control Board (Gaston, 1980). Generally, the requirements are less stringent than those for direct injection because of the added protection provided by the passage of the treated wastewater through the soil above the ground water aquifer.

Morris and Jewell (1976) in their 50-state overview found that a rapidly developing trend is the requirement that all land discharges receive minimum secondary treatment; 21 states now have such a requirement. Some state regulations stipulate minimum buffer zones around the site, maximum application rates, minimum storage capacity, vertical distance to ground water, and monitoring requirements.

In Utah, the Water Research Laboratory made a study to determine the procedure to develop ground water recharge operations in Weber County at some future date (Clyde et al., 1984). Numerous organizations, both public and private, are involved in the management and use of waters of the Weber and Ogden Rivers. There are 63 mutual irrigation companies, 4 water improvement districts, 12 municipalities, and 2 water conservation districts in Weber County; not to mention a number of private water companies which provide culinary water. In addition, there are many irrigation companies and other entities in Morgan and Summit Counties which obtain their water supply from these two rivers.

To establish a long-term ground water recharge system on the Weber River will require some form of organizational arrangement for financing and management. Water rights must be purchased, the abandoned gravel pits to be used for ponding must be acquired, diversion and conveyance facilities must be constructed, the system must be maintained and operated when in place, and the costs of all of these must be properly allocated to and collected from the beneficiaries.

Although large surplus flows in the order of 2000 cfs pass the recharge site for two to three months during most years, to obtain a year around continuous flow of 5 to 10 cfs for recharge would probably require the purchase of existing direct flow or storage rights. These and other costs outlined above should be passed on to the beneficiaries of the recharge operation.

A special improvement district appears to be the most suitable form of institution to finance and manage the Utah recharge system. This is a governmental subdivision with powers to levy taxes on all taxable property within the district, to issue bonds, and collect charges or fees for services rendered. Its powers and authority are limited to the purposes specified in the resolution creating the district. It is created by the board of County Commissioners who can also serve as its board of trustees, or they can appoint or arrange for election of other trustees. It could act as the sole manager and operator of the recharge system, or it could make arrangements with other organizations such as the Weber Basin Water Conservancy District to supply the water and operate the system.

The details of the method of financing the Utah system could be patterned after the ground water replenishment system established and successfully operated for several years in Orange County, California. A replenishment assessment for each acre foot of water extracted from the ground water basin provides the major source of revenue. In the case of the Weber River recharge system, it would be appropriate to determine the boundaries of the district to encompass only that part of the ground water basin benefitted by the recharge operation. Even though the actual construction of the artificial recharge project is still years away, Clyde et al., (1984) stressed the importance of obtaining the land now while it is easily and cheaply available. The land could be utilized for a park, playground, golf course, and/or wetlands habitat.

In Florida, two of the state water management districts have developed permitting systems for artificial recharge that have become part of the Florida Administrative Code. The following regulations are for artificial recharge systems within the Suwannee River Water Management District. Included in the definition are fluids discharged from land application facilities, and disposal, storage, and connector well facilities. Excluded are wastewaters from residential septic tank systems, effluent from certain sewage treatment plants, and facilities for agricultural irrigation. In most instances, work may not commence on an artificial recharge system without a permit. Certain wells, however, are exempt from the permit requirement. There are rules detailing how to file a permit application and under what conditions construction may be exempted. The District may require public hearings before a permit is issued; it is also entitled to give orders relating to drilling techniques designed to prevent waste and pollution. A specific casing and sealing program must be submitted to the District for approval. Periodic well testing is required, as are written records of all construction work performed (Florida Department of Natural Resources, 1978).

The rules of the South Florida Management District provide for the implementation of a permit system designed to regulate and control the use of ground and surface water. Permits are required before construction may begin on any project involving artificial recharge. It is also necessary to acquire a permit for any use, diversion or withdrawal of water from the District which exceeds 100,000 gallons per day. Persons using water in conjunction with oil well drilling projects in Lee, Collier, or Hendry Counties must have a general permit issued for the work. In certain emergency situations, permission to begin use, withdrawal, or divert water may be given before a permit has been issued. The District may declare the existence of a water shortage when insufficient water is available to meet the requirements of the permit system. When this occurs, other regulations may be imposed on those seeking water permits (Karl-Goran, 1978).

Arizona's 1980 Ground Water Management Act essentially prohibits additional ground water development for irrigation, and also requires reduced ground water irrigation withdrawals phased over 45 years. The Act's goal is to balance withdrawals and recharge (natural and artificial) by 2025. The Arizona statute also authorizes purchase of irrigated land and retirement of the associated ground water irrigation right, financed by taxes of up to $2 per acre-foot on ground water withdrawals (Aiken, 1984).

The Arizona legislature has been trying to work out the details of how artificial recharge should be accomplished. The legislature currently is considering a bill that would amend the 1980 Ground Water Management Act to add a Chapter 13 dealing specifically with recharge. The proposed amendment to the ground water law would give Arizona water users a new way to receive a right to pump ground water by storing surplus water into the ground for use during times of shortage. State lawmakers are taking a close look at who might benefit and who might be hurt if a recharge bill were enacted. The Arizona

Department of Water Resources supports the basic concept of ground water recharge in the state, but they want to make sure it benefits the entire state, not just one area.

Additional Perspectives

Many water agencies that operate on both ground water and surface water sources conjunctively are involved in artificial recharge activities. These programs are motivated by periodic attempts to utilize surface water to maximum benefit in excess of operational delivery requirements. Other than saline-water intrusion barriers, recharge can be a haphazard matching of excess surface water sources to available parcels of permeable soils. Independent of technology, any planned use of artificial recharge is initially limited by active local competition for land and water.

If the recharge technology now available is to be properly installed and kept operational, the water industry is going to have to treat the conjunctive operation of its surface- and ground water storage with the same technical level and personnel competency standards as it does wastewater treatment (Bianchi and Cehrs, 1981). Such a program of conjunctive use would:

(1) establish and maintain the priority for fresh water resource management within the local water hierarchy;

(2) plan and site recharge facilities by matching the technology to the soils, geology and ground water storage complex;

(3) monitor quantity, quality, and performance; and

(4) ideally, have its staff compensated according to the acre-feet/year recharged.

A first priority program for conjunctive use would be to establish surface storage as a quality pretreatment step in the artificial recharge process. Because ground water storage is now looked upon as a place to put excess surface water, flood water which is high in sediment is frequently the main recharge water source. Yet many watersheds have multiple-use storage and flood control reservoirs, where winter storage is established as part of surface water management. Storing runoff is the most efficient method of sediment control. A controlled summer release of water yields organic solid loads that are more easily managed than the silts and clays in flood water.

Surface water managers engineer to retain carry-over storage. In theory, if water logistics were truly conjunctive, ground water storage would function as carry-over, recharged in the summer when high water temperature and low suspended loads give high recharge rates. The surface reservoirs would then be left empty for maximum winter catchment and flood control. But, the public acceptance of empty reservoirs and the managers who allow them is very low.

Education on the concepts of conjunctive use is a problem of scale and technical complexity needing expanded public relations treatment. The main problem in public understanding stems from water management's emphasis on cost rather than the resource value of water in dollars. Application of existing recharge technology is more easily justified when all of the hidden costs are included in the value of water placed in ground water storage. If ground water is a renewable resource through artificial recharge, then the priorities within land use, surface water distribution, and wastewater disposal must be reordered. Areas of aquifer recharge must be identified and preserved. Either surface water must be managed to provide high quality water to the recharge facilities, or the basin ground water storage dedicated to disposal. Land disposal of waste water must be consistent with acceptable equilibrium ground water quality standards (Bianchi and Cehrs, 1981).

SELECTION OF ARTIFICIAL RECHARGE METHODS

Once a geographical area of need has been identified, a source of recharge water located, the physical capabilities of the aquifer for recharge established, and the legal and institutional constraints to recharge overcome, the operational design must be chosen to provide for optimum recharge potential. The choice of injection wells versus spreading basins is only a first step--variations in design are extensive and provide for differences in recharge rates, capabilities, and cost.

Initially, the thickness and permeability of the earth materials overlying a dewatered aquifer exert a strong influence on the type of recharge method, in general, that can be applied. When these materials are thick and of low permeability, wells are in order; but where thin or more permeable, spreading basins, large shafts, or irrigation systems may be preferable. Larger amounts of land are required for spreading basins than for injection wells.

An example of variations on the general theme of surface infiltration is the case of design to reduce the effects of sediment accumulation. The Kansas Water Resources Board (1972) evaluated operational designs to reduce the effects of sediment deposition with source waters carrying a high concentration of suspended sediment that consisted mainly of clay size particles; they recommended:

(1) catch sediment in retention reservoirs in uppermost of a series of basins;

(2) maintain sufficient flow velocity to carry suspended sediment through the system;

(3) maintain heavy grass cover to retard erosion and break up sediment layer;

(4) scrape, harrow, or disk bottom of pond after adequate drying period;

(5) remove sediment after drying (could be used to build up levees along channels); and

(6) sluice sediment-laden water into vertical or steep-sided ponds and channels, so the sediment will be deposited on the bottom for later removal, and most of the water will infiltrate through the sides.

This type of configuration may be compared with a mechanical filtration system as a pretreatment; however the filtration system's inefficiency and economic cost generally makes it less acceptable for use in most recharge operations. Proper recharge design and operation in the form of stilling periods of adequate duration to allow for the settling of a large portion of the suspended solids seems to be the most reasonable substitute for mechanical or chemical clarification (Dvoracek and Peterson, 1971).

One promising method of evaluating an area for optimum design configurations is mathematical modeling. Input can include descriptive parameters such as permeability, infiltration rates and natural recharge characteristics. Nonuniform hydrogeologic conditions, however, decrease the predictive capabilities of design calculations, they may have to be followed by field measurements to determine how the actual performance of the system compares with predictions, and what modifications in the operation of the system may be desirable (Bouwer et al., 1972).

Pilot projects in the area of interest are often utilized to evaluate the horizontal and vertical hydraulic conductivity of the aquifer and, in the case of spreading techniques, overlying beds. These observed hydraulic conductivities can be used in the design of an operational system to predict actual underground detention times and water table profiles for various geometries of recharge basins and wells for pumping water. In addition, indications of clogging problems or a decrease in the efficiency of the system can be observed for possible design modifications. Field studies should also include determinations of the quality of the water as it moves to and below the water table (Bouwer et al., 1972).

By accounting for the movement of artificial recharge water, mathematical models can aid in determining the operational policies and benefits of artificial recharge. The models have the advantage that they can transfer the results of demonstration artificial recharge projects to other sets of hydrologic and operational conditions. For example, the models can be used to predict aquifer response to varying recharge rates, and times, or changes in aquifer characteristics (Sunada, Warner and Molden, 1983).

Sunada, Warner and Molden (1983) discuss an analytical model and a numerical model. The analytical model chosen is simple and fast to use compared to the numerical model which is more complex. Each model was calibrated to an actual operation by comparing the calculated aquifer response to the aquifer response measured in the field. Only large deviations in the value assumed for the storage coefficient affected results obtained by the models. Transmissivity was varied until the model calculations gave a best fit to the measured field values. The calibrated models were then used to predict the aquifer response to a different sequence of recharge events. Various operational policies consistent with recharge facilities and water availability were used in the models. The models showed that in this example if water was recharged during the non-irrigation season, excepting those months when water would freeze, one foot of water could be added to storage for a radius of at least 1000 feet from the recharge basins. The models were capable of predicting mound build-up and thereby spreadability of the recharging water.

The development of mathematical models to simulate aquifer behavior is an important factor in the success of artificial recharge practices. Once model parameters and their integration have become well understood through actual field calibration and verification, they will become a viable method for choosing design facilities for optimum recharge in the interest of increasing ground water supplies.

ECONOMIC EVALUATION OF PROPOSED PROJECTS

Any proposed artificial recharge scheme should first be evaluated relative to its potential impact and effectiveness. When evaluating recharge projects utilizing renovated wastewater, for example, the composition of the effluent is question should be determined and trace metal inputs for each particular case should also be carefully examined (Chang and Page, 1980). Because of the benefits obtained from a properly managed and maintained recharge system, it is worthwhile to thoroughly examine the proposed project for design and operational effectiveness. Examples of some benefits are (Kansas Water Resources Board, 1972):

(1) larger and more dependable water supply;

(2) flood and silt control;

(3) insurance against drought;

(4) increased well discharge;

(5) less well interference; and

(6) decreased pumping lifts and costs.

When examining potential projects, an important factor is the intended use in reference to the amount of storage to be obtained and water quality. Irrigation use as opposed to drinking water use will require that different standards be maintained.

Perhaps the most significant factors in deciding whether a project should take place or not is its financial feasibility and economic justification. Financial feasibility refers to the ability of the project beneficiaries to repay the cost of the project. Economic justification, which may be expressed in terms of a cost-benefit ratio, permits comparison of alternative projects to select the most economical (Department of Economics and Social Affairs, 1975).

The Department of Economics and Social Affairs (1975) conducted an in-depth study of the costs and benefits of artificial recharge operations. An examination of the costs of existing projects showed that the total cost of a project is made up of expenditures for land and easements, engineering and construction of facilities, water or rights to water, and operation and maintenance. These costs were shown to vary greatly with:

(1) purpose of project;

(2) method of recharging;

(3) quantity and quality of water available for recharge, and regimen of flow;

(4) surface and subsurface conditions;

(5) location of the artificial recharge project; and

(6) standards and requirements of the agencies involved in recharging operations.

The different types of recharge, were found to have different cost factors. In the case of recharge through injection wells, the largest portion of the total cost will be due to (Department of Economics and Social Affairs, 1975):

(1) the supply facilities;

(2) the injection wells;

(3) the regulating equipment; and

(4) operations and maintenance (redevelopment of wells is particularly expensive).

Most of the total cost of spreading basin projects goes to (Department of Economics and Socia Affairs, 1975):

(1) divert water from streams;

(2) convey water to and from the recharging area;

(3) measure the amount of flow;

(4) contain water within, and control flow through, the recharging area; and

(5) operate and maintain the facilities efficiently and safely.

The cost of the land, of course, varies with location and time of purchase. Included in this cost would be the cost of surveys, maps and acquisitions. Legal and rezoning fees may be involved in some cases. Any natural resources found on purchased lands, however, can be used to defray some of the costs. A common example of this would be sand and gravel.

Engineering costs, such as field surveys, laboratory tests, designs, plans and specifications, and construction inspection and control, are greatly affected by the standards and requirements generally dictated by the type of development in and around the project area. In urban areas, the need for fencing and other protective measures, as well as aesthetic treatment, may add significantly to the total investment.

There are many intangible benefits associated with artificial recharge and the use of ground water reservoir storage. Examples of intangible benefits include (Department of Economics and Social Affairs, 1975):

(1) if an area must import its water supplies, or at least a portion of them, it is necessary that storage reservoirs be a part of the system;

(2) in nearly all areas the ground water reservoirs provide for greater capacity for storage and at a lower cost than would be considered necessary for surface storage;

(3) if it is economical and feasible to use the ground water reservoirs, there is great value in having the water available locally, in case a catastrophe, such as an earthquake, cuts off the imported supply;

(4) the ground water reservoirs eliminate evaporation losses, perhaps not an intangible benefit, as this can be closely measured if desired;

(5) algae growth and growth of similar nuisance plants in surface-water reservoirs may impose an undesirable taste to a water supply; water stored in ground water reservoirs is, of course, not affected by plant growth, and generally retains a high quality during storage; pollution problems are minimized in ground water reservoir storage and less care need be taken to prevent contamination, as compared with surface storage;

(6) rarely is a surface water system free of leakage, and normally this leakage replenishes the ground water; the exploitation of the ground water provides a means of recouping a portion, and perhaps all, of this loss through leakage;

(7) as ground water is withdrawn and an aquifer de-watered, subsidence will often occur, especially in unconsolidated formations; artificial recharge, if practiced on a more-or-less safe-yield basis will minimize subsidence problems; subsidence over a large area, especially in areas with minimum-surface slopes, can be a very expensive problem; gravity drains, such as sewers and storm drains, may have gradients altered so that they no longer flow, or may be broken; differential subsidence can have very adverse effects upon structures; the subsidence problem should be carefully considered before an exploitation plan of ground water mining is adopted; and

(8) artificial recharge provides a treatment and generally improves the quality of the water so recharged as it moves through the rock formations; this improvement, too, may not necessarily be intangible, as a value may be established for the improvement in water quality.

Finally, a cost check list was developed by the Department of Economics and Social Affairs (1975), this list should be considered when evaluating costs of artificial recharge projects:

(1) Planning

- (a) Source of water and pricing policy
- (b) Water rights
- (c) Potential liability (inverse condemnation problems -- flood-out gravel-mining operations, building fresh-water mound too close to ground surface, thus jeopardizing surface improvements)
- (d) Effects on ground water basin management
- (e) Effects on ground water quality

(f) Investigations (generally involving an exploration of the geological and hydrological conditions) to prove feasibility of scheme

(g) Finances

(1) Capital improvements
(2) Water
(3) Operation and maintenance

(2) Operation and Maintenance

(a) Spreading facilities

(1) Grading
(2) Storm-protection
(3) Structure repair and replacement
(4) Equipment maintenance
(5) Fuel for equipment
(6) Equipment rental
(7) Disking or silt-removal
(8) Insect-control
(9) Weed-control
(10) Aesthetic treatment (such as screen planting with adequate watering systems)
(11) Rodent-control
(12) Patrolling
(13) Treatment (flocculents)
(14) Slope maintenance
(15) Vandalism

(b) Injection facilities

(1) Observation of monitoring devices
(2) Measuring water levels
(3) Sampling water
(4) Redevelopment of wells and disposal of wastes
(5) Treatment
-- Chlorine
-- Acid
-- Other
(6) Maintenance of equipment
(7) Structure repair
(8) Fuel
(9) Equipment rental
(10) Patrol
(11) Water analyses
(12) Vandalism

(c) Office

(1) Supervision
(2) Administration
(3) Wages
(4) Office (and field) overhead
-- Rent and utilities

- -- Telephone
- -- Supplies
- -- Maintenance of office equipment

(5) Salaries
(6) Liability (insurance)
(7) Taxes
(8) Interest

SUMMARY

The success of any artificial recharge project will depend on the area's responsiveness to artificial recharge. When determining the potential of an area to be responsive to artifical recharge, it is necessary to consider a large number of factors. These factors are a function of several aspects of the area: physical characteristics, source water availability, legal and institutional considerations, and benefit/cost analyses. Concurrent with the examination of these factors should be an evaluation of the methods of artificial recharge applicable in the area.

Some physical characteristics which are pertinent to artificial recharge potential are the transmission characteristics of the aquifer, the amount of underground storage space available, the depth to the underground storage space, the thickness and permeability of the overlying beds, precipitation rates, and the topography and land use in the area. Field studies, laboratory tests, mathematical modeling, and information gathered from existing soils maps, hydrogeologic studies, and precipitation surveys are integral parts of site evaluation.

Source water availability may be the determining factor in the implementation of artificial recharge projects. Two types of water commonly utilized as a source are storm runoff and reclaimed waste-water. When conducting recharge with these types of sources, quality is an important consideration. With proper pretreatment and management practices to minimize problems of clogging and contamination, both can be a viable source for artificial recharge.

There are several legal/institutional issues that can arise during the planning phase and/or operation of artificial recharge projects. Usually every project, whether demonstration or large-scale, will need to consider the following issues:

(1) Who will finance the project?

(2) Who will benefit from the storage of water?

(3) What organizations/agencies have the authority to construct and operate the project? Should recharge water be integrated with surface water?

(4) What are the rights and duties of entities storing water underground?

(5) Is the required land available and affordable?

(6) What will be the impacts of the project on interstate compacts that deal with water allocation and flood control?

(7) What will be the impacts on land use, population, and the local economy?

Most of the western states enforce the Reasonable Use doctrine through a permit system. State agencies usually have a three-fold approach to water administration--appropriation, adjudication, and distribution; local water institutions have developed water supplies on large tracts of land through the formation of mutuals, irrigation or water districts, and companies or associations. Currently, over 75 percent of the irrigated land in Wyoming, Colorado, Utah, Idaho, Montana, and Washington is delivered through some types of water organization.

Twenty six states have regulations for land disposal of wastewater (an indirect method of artificial recharge), some of which are very strict and may hinder the direct injection of water for artificial recharge. Nebraska and California have taken the lead in dealing with the legal/institutional issues of artificial recharge, followed by Florida and Arizona. In the past state legislation has allowed the formation of irrigation districts, drainage districts and, most recently, water districts. Future trends point to the creation of multi-purpose water conservancy districts to deal not only with the legal/institutional aspects of artificial recharge discussed here--but also to manage surface and ground water conjunctively.

The choice of operational design is an important consideration for optimum recharge potential. The thickness and permeability of the earth materials overlying the aquifer exert a strong influence on the type of recharge system that can be applied. In general, when these materials are thick and of low permeability, wells are in order; but where thin or more permeable, spreading basins, large shafts, or irrigation systems may be preferable. Many variations are available on the theme of surface spreading to allow for optimization of infiltration. Mathematical modeling used in conjunction with field measurements is becoming an effective way to judge design optimization.

Once the site location and design scheme have been developed, every artificial recharge project should be evaluated to determine its impact, effectiveness, and benefits/costs. The impact and effectiveness of any project can be determined through pilot studies, laboratory analyses, or mathematical modeling. The benefit/cost relationship may be more subjective. Costs will be reflected by such things as purpose, methodology, quantity and quality of source water, surface and subsurface conditions, location, and standards and requirements. Benefits include the formation of a larger and more dependable water supply, decreased pumping lifts and costs, prevention of irreversible subsidence, elimination of evaporation losses, high quality of water, and the convenience of locally pumped water over an imported surface water.

SELECTED REFERENCES

Aiken, J.D. "Legal Aspects of Ground Water Recharge", Nebraska University, Department of Agricultural Economics, Staff Paper No. 10, 1980, Lincoln, Nebraska.

Aiken, J.D., "Depleting the Ogallala: High Plains Ground Water Management Options", Proceedings of the Ogallala Aquifer Symposium II, Texas Tech University, Lubbock, Texas, June 1984, pp. 451-464.

Aller, L. "Artificial Ground Water Recharge: Its Time Has Come", Water Well Journal, Jan.-Feb. 1985, Vol. 23, No. 1, pp. 8-9.

Baxter, K.M. et al., "Effluent Recharge: The Effects of Effluent Recharge on Groundwater Quality", TR 199, Feb. 1984, Water Research Center, Medmenham Laboratory, Medmenham, Marlow, Bucks, SL7 2HD.

Bianchi, W.C. and Cehrs, D., "Ground Water Reservoir Management Through Artificial Recharge-Fact or Fantasy", Ground Water, Vol. 22, No. 2, May-June 1981, pp. 266-271.

Bianchi, W.C. and Nightingale, H.I., "Hammer Seismic Timing as a Tool for Artificial Recharge Site Location", Proceedings of the Soil Science Society of America, Vol. 39, No. 4, July-Aug. 1975, pp. 747-751.

Bouwer, H. et al., "Renovating Secondary Sewage by Water Recharge with Infiltration Basins", Water Pollution Control Research Series, 16060DRV, Mar. 1972, U.S. Environmental Protection Agency, Washington, D.C.

Chang, A.C. and Page, A.L., "Fate of Inorganic Micro-Contaminants During Groundwater Recharge", Wastewater Reuse for Groundwater, Proceedings of the Symposium, Sept. 6-7, 1980, at the Kellogg West Center for Continuing Education, California Polytechnic University, Ed., Asano, T. et al., Pomona, California, pp. 118-135.

Chidley, T.R.E., "Assessment of Groundwater Recharge", in Lloyd, J.W. (Ed.), Case Studies in Groundwater Resources Evaluation, Clarendon Press, Oxford, 1981, pp. 133-149.

Clyde, C.G. et al., "Management of Groundwater Recharge Areas in the Mouth of Weber Canyon", UWRL/H-84/01, May 1984, Utah Water Research Laboratory, Utah State University, Logan, Utah.

Cochran, R., "Artificial Recharge and Its Potential for Oklahoma's Ogallala Aquifer", Oklahoma Water Resources Board, May 1981, Oklahoma City, Oklahoma.

Colorado High Plains Technical Coordinating Committee, "Artificial Aquifer Recharge in the Colorado Portion of the Ogallala Aquifer", Draft, Sept. 1984, Denver, Colorado.

Department of Economics and Social Affairs, United Nations, "Groundwater Storage and Artificial Recharge", Natural Resources/Water Series No. 1, ST/ESA/13, 1975, New York, New York.

Deutsch, M., "Artificial Recharge by Induced Interaquifer Leakage", Artificial Recharge and Management of Aquifers, Proceedings of the Symposium at Haifa, Publication No. 72, International Association of Scientific Hydrology, United Nations Educational Scientific Cultural Organization, March 1967.

Dvoracek, M.J. and Peterson, S.H., "Development of Systems for Ground Water Recharge into the Ogallala Formation", Agricultural Engineering Department, June 1971, Texas Tech University, Lubbock, Texas.

Florida Department of Natural Resources, "Permitting of Artificial Recharge Systems", Florida Administrative Code Chapter 16 H5, Secs.5.01-5.31, 1978, Suwannee River Water Management District, White Springs, Florida.

Frankel, R.J., "Economics of Artificial Recharge for Municipal Water Supply", Artificial Recharge and Management of Aquifers, Proceedings of the Symposium at Haifa, Publication No. 72, International Association of Scientific Hydrology, United Nations Educational Scientific Cultural Organization, Mar. 1967, pp. 289-302.

Franks, B.J., "Land Application of Domestic Wastewater in Florida-Statewide Assessment of Impact on Ground Water Quality", Water Resources Investigations 81-3, 1981, U.S. Geological Survey, Tallahassee, Florida.

Garrison, W.E. et al., "A Study on the Health Effects of Groundwater Recharge in Southern California", Wastewater Reuse for Groundwater Recharge, Proceedings of the Symposium, Sept., 6-7, 1979, at the Kellogg West Center for Continuing Education, California Polytechnic University, Ed., Asano, T. et al., Pomona, California, 1980, pp. 33-39.

Gerba, C.P. et al., "Pathogen Removal from Wastewater During Groundwater Recharge", Wastewater Reuse for the Groundwater Recharge", Proceedings of the Symposium, Sept. 6-7, 1979, at the Kellogg West Center for Continuing Education, California State Polytechnic University, Ed., Asano, T. et al., Pomona, California, 1980, pp. 137-144.

Gillespie, J.B. and Hargardine, G.D., "Results of Infiltration Tests-New Scott City, Western Kansas", Water-Resources Investigations 76-12, Apr. 1976, U.S. Geological Survey, Lawrence, Kansas.

Gleason, U.E., "Water Projects Go Underground", Ecology Law Quarterly, Vol. 5, No. 4, 1976, pp. 625-668.

Hauser, V.L. and Lotspeich, F.B., "Artificial Ground Water Recharge Through Wells", Journal of Soil and Water Conservation, Vol. 22, No. 1, Jan.-Feb. 1967, pp. 67-72.

Johnson, A.I. and Sniegocki, R.T., "Comparison of Laboratory and Field Analyses of Aquifer and Well Characteristics at an Artificial Recharge Well Site", Artificial Recharge and Management of Aquifers, Proceedings of the Symposium at Haifa, Publication No. 72, International Association of Scientific Hydrology, United Nations Educational Scientific Cultural Organization, Mar. 1967.

Kansas Water Resources Board, "Natural and Artificial Ground-Water Recharge-Wet Walnut Creek", Bulletin No. 17, 1972, Topeka, Kansas.

Karl-Goran, M., "Permitting Uses of Water", Florida Administrative Code, CH 16K-2, Secs. 2.01-2.16, 1978, South Florida Water Management District, Palm Beach, Florida.

Kincannon, D.F. and Tiederman, W.G., "Water Reclamation for Ground Water Recharge", Final Report A-034, 1972, Water Resources Research Institute, Oklahoma State University, Stillwater, Oklahoma.

Ku, H.F.H. and Simmons, D.L., "Use of Recharge Basins for Storm Water Management on Long Island, New York", Proceedings of the National Water Well Association Eastern Regional Conference on Ground Water Management, 1983, Orlando, Florida, pp. 17-43.

Lehr, J.H., "Artificial Ground Water Recharge: A Solution to Many of the U.S. Water Supply Problems", Ground Water, Vol. 20, No. 3, May-June 1982, pp. 262-266.

Manbeck, D.M., "Recharge Simulation Model", June 1975, Water Resources Research Institute, University of Nebraska, Lincoln, Nebraska.

Morris, C.E. and Jewell, W.J. "Land Application of Wastes: A 50-State Overview", Public Works, Oct. 1976, pp. 89-92.

Palmer, P.L., "Artificial Recharge of Fresh Water into Florida's Saline Aquifers", Proceedings of the National Water Well Association Eastern Regional Conference on Ground Water Management, Orlando, Florida, 1983, pp. 3-16.

Roberts, P.V., "Water Reuse for Groundwater Recharge: An Overview", American Water Works Association Journal, Vol. 72, No. 7, July 1980, pp. 375-379.

Roberts, P.V. et al., "Field Study of Organic Water Quality Changes During Groundwater Recharge in the Palo Alto and Baylands", Wastewater Reuse for Groundwater Recharge, Proceedings of the Symposium, Sept. 6-7, 1979, at Kellogg West Center for Continuing Education, California State Polytechnic University, Ed., Asano, T. et al., Pomona, California, 1980, pp. 283-316

Roberts, P.V., et al., "Organic Contaminant Behavior During Groundwater Recharge", Journal Water Pollution Control Federation, Vol. 52, No. 1, Jan. 1980, pp. 161-172.

Scalf, M.R. et al., "Fate of DDT and Nitrate in Ground Water", Robert S. Kerr Water Research Center, Apr. 1968, Ada, Oklahoma.

Schmidt, C.J., Clements, E.J. and Skelton, S.P., "A Survey of Practices and Regulations for Reuse of Water by Groundwater Recharge", Journal of the American Water Works Association, Vol. 70, No. 3, Mar. 1978, pp. 140-147.

Sniegocki, R.T., "Problems in Artificial Recharge Through Wells in the Grand Prairie Region, Arkansas", Geological Survey Water-Supply Paper 1615-F, 1963, Washington, D.C.

State Water Planning and Review Process, "Supplemental Water Supplies Policy Issues Study", Jan. 1984, Nebraska Natural Resources Commission, Lincoln, Nebraska.

Sunada, D.K., Warner, J.W. and Molden, D.J., "Artificial Groundwater Recharge-San Luis Valley, Colorado", Completion Report No. 123, Mar. 1983, Colorado Water Resources Research Institute, Fort Collins, Colorado.

U.S. Department of Interior, U.S. Department of Agriculture, U.S. Environmental Protection Agency, "Irrigation Water Use and Management", An Interagency Task Force Report, June 1979, Washington, D.C.

Warner, D.L. and Doty, L.F., "Chemical Reaction Between Recharge Water and Aquifer Water", Artificial Recharge and Management of Aquifers, Proceedings of the Symposium at Haifa, Publication No. 72, International Association of Scientific Hydrology, United Nations Educational Scientific Cultural Organization, Mar. 1967, pp. 278-288.

Wood, W.W., "Development of Technical Site Selection Criteria for Artificial Recharge", Wastewater Reuse for Groundwater Recharge, Proceedings of the Symposium, Sept. 6-7, 1979, at the Kellogg West Center for Continuing Education, California State Polytechnic University, Ed., Asano, T. et al., Pomona, California, 1980, pp. 73-79.

CHAPTER 4

CASE STUDIES OF ARTIFICIAL RECHARGE

Information on artificial recharge projects in the United States may be described as being randomly scattered in the literature. The purpose of this chapter is to summarize information on actual recharge projects by describing the status of artificial recharge nationwide, and presenting more detailed information on selected projects. Brief information on some international projects will also be presented.

NATIONAL ARTIFICIAL RECHARGE ACTIVITY

Artificial recharge has been recently given much attention in the United States. In fact, many projects have substantially increased water supplies and prevented environmental damage from declining water levels in many parts of the country. Where aquifers have high storage capacity, as in parts of western United States, it is feasible in many localities to utilize artificial recharge by pumping large quantities of ground water during the summer low-flow period of heavy water demand, thereby creating a deep cone of depression. The cone of depression is subsequently filled during humid, heavy runoff periods by using a method of artificial recharge appropriate to the specific site. Among the areas in western United States where artificial recharge projects are now successfully practiced are localities in Los Angeles, Orange County and Fresno, California; the High Plains States; Arizona; and eastern New Mexico (Motts et al., 1983).

Depletion of the Ogallala aquifer has introduced many problems which threaten the economic futures of Colorado, Kansas, Nebraska, New Mexico, Oklahoma, and Texas. A bill recently passed by Congress allocates funds to aid in artificial recharge studies and demonstration projects in the Plains States (Hydrologic Engineering Center, 1984; U.S. Congress, 1983). In addition, projects have recently been initiated in Arkansas, California, Colorado, Florida, Louisiana, New Mexico, New York, North Dakota, Oregon, Texas, and Virginia (CH_2M-Hill, 1984).

Artificial recharge activities include feasibility studies, experiments, demonstration projects, and actual systems in operation. At the present time, feasibility studies, experiments, and demonstration projects occupy most recharge efforts.

All artificial recharge projects involve either well injection, surface spreading, or inducing recharge from rivers, streams, or canals. Injection wells are most often preferred where carbonate formations, characterized by high permeability and secondary solution openings, are present (CH_2M-Hill, 1984). They are used where the water table is relatively deep and is separated from the surface by an impermeable layer. Surface spreading is the most widely used recharge system. It is applied where a relatively permeable land surface material and shallow water table is present.

Recharge waters used in artificial recharge activities include untreated surface and ground water, and treated waste water effluent. The quality of the recharge waters depend greatly on the objective of the recharge project. Artificial recharge in the United States has been used to build salinity barriers to stop migration of high chloride waters into freshwater supplies, increase freshwater aquifers, develop underground storage of drinking water, or, in some areas, to control subsidence.

Table 7 presents a brief description and reference to sites of artificial recharge activity throughout the United States. Figure 2 shows the location of the sites mentioned in Table 7. While artificial recharge technology is well developed, the need to develop physically and economically feasible recharge systems has hindered construction of full-scale recharge projects. Relatively few operational projects are in existence which provide significant recharge to aquifers. Several of these are located in California. To provide detailed information on the types of artificial recharge projects, 13 sites have been chosen and are briefly described in the following section.

Table 7: National Artificial Recharge Activity (Projects, Demonstration/Pilot Projects, Experiments, Studies)

<u>Alaska</u>

1. Anchorage - Experiments with spreading basins allowing infiltration of diverted creek water (Anderson, 1977; Guyman, 1972).

<u>Arizona</u>

2. Lower Oak Creek Basin - Study of recharge potential of stock tanks capturing storm water runoff.*

3. Flushing Meadows, Phoenix - Pilot project to study feasibility of renovating secondary effluent with spreading basins (Pettyjohn, 1981).

4. Phoenix - Urban study, by the U.S. Army Corps of Engineers, to explore potential recharge.*

5. Salt River - Study of potential for recharge.*

6. Gila Bend - Debris pool of Painted Rock Reservoir released for artificial recharge through basins.*

7. Superior - Study of management considerations for artificial recharge along Queen Creek.*

8. Tucson - Experimental study with pit recharge.*

9. Tucson - Urban study, by the U.S. Army Corps of Engineers, to explore potential for recharge.*

Arkansas

10. Arkansas County - Experimental recharge with injection wells to rejuvenate aquifer, discontinued 1969 (International Association of Scientific Hydrology, 1970).

11. Newport - Disposal of groundwater, pumped for cooling system, through injection wells (International Association of Scientific Hydrology, 1970).

12. Fayetteville - Study of potential methods of artificial recharge in the Grand Prairie (Griffis, 1976).

California

13. Butte Valley*

14. Santa Clara Valley*

15. Livermore*

16. Gilroy - Hollister Valley*

17. Salinas Valley*

18. Santa Maria Valley*

19. San Juaquin Valley*

20. Santa Clara River Valley*

21. San Fernando Valley*

22. Los Angeles Coastal Plain*

23. San Gabriel Valley*

24. Orange County Coastal Plain*

25. Upper Santa Ana Valley (Koehler, 1983)

26. San Jacinto Basin*

27. San Gorgonio Pass Area - Underground storage of imported water (Bloyd, 1969)

28. Santa Cruz and Monterey Counties, Pajaro Valley - Artificial recharge and sea water intrusion prevention (Muir, 1974).

29. Upper Coachella Valley - Artificial recharge effects (Swain, 1978).

Colorado

30. Brush - Recharge of irrigation water, during non-irrigation seasons, through pits.*

31. Denver - Study of potential recharge methods for the South Platte River Basin.*

32. Fort Morgan - Recharge of irrigation water, during non-irrigation seasons, through pits.*

33. Fort Garland - Recharge of irrigation water, during non-irrigation seasons, through pits.*

34. Morgan County - Proposed artificial recharge project diverting water from the South Platte River (Burns, 1980).

35. Prewitt - Recharge of irrigation water, during non-irrigation seasons, through pits.*

36. Arikaree River Near Cope (Brookman and Sunada, 1968).

37. San Luis Valley - Artificial groundwater recharge (Sunada et al., 1983).

38. Sterling - Recharge of wastewater effluent during winter months.*

39. El Paso County - Tests with recharge pits in Upper Squirrel Creek Basin (Emmons, 1977).

40. Kit Carson County - Evaluation of water availability (Jenkins and Hofstra, 1970).

Delaware

41. New Castle County - Possibility of a large-scale artificial groundwater recharge project to insure adequate water supply (University City Science Institute, 1971).

Florida

42. Okaloosa County and adjacent area - Consideration of artificial recharge (Trapp et al., 1977).

43. Hillsborough County - Feasibility of artificial recharge (Sinclair, 1974).

44. East Orange County - Experiments and studies of connector well to allow water from shallow aquifer to recharge lower Floridan aquifer (Bush, 1979).

45. Orlando - Recharge through drainage wells (Kimrey, 1978).

46. St. Petersburg - Feasibility tests with subsurface injection of storm water runoff through wells (Hickey and Barr, 1979).

47. Gainesville - Study of potential of combining flood control with artificial recharge in Peninsular Florida.*

48. Cocoa - Feasibility tests with recharge through injection wells to prevent saltwater intrusion (Tibbals and Frazee, 1976).*

49. Tampa Bay - Experiments to determine optimum method of recharge (Sinclair, 1977).

50. Lake Manatee - Recharge recovery (CH_2M-Hill, 1984)

51. St. Lucie County - Artificial recharge of fresh water into brackish water aquifer (Wedderburn and Knapp, 1983).

Hawaii

52. Kahului (Maui) - Recharge of wastewaters through injection wells (Hargis and Peterson, 1974).

53. Hanapepe (Kauai) - Recharge of streamwater through wells (Hargis and Peterson, 1970).

54. Island of Hawaii - Recharge of streamwater through wells (Hargis and Peterson, 1970).

55. Pearl Harbor - Recharge of streamwater (Hirashima, 1971).

Idaho

56. Snake River - Proposed project to divert waters from the Snake River for artificial recharge (Norvitch et al., 1969).*

57. Big Lost River - Disposal of low-level aqueous radioactive wastes through injection wells (International Association of Scientific Hydrology, 1970).

Illinois

58. Peoria - Recharge with pits to prevent decline of groundwater levels (Pettyjohn, 1981; Harmeson et al., 1968).

59. Chicago Suburbs - Cost effective water supply alternative for the Chicago suburban area - Artificial recharge (Nielsen and Aller, 1983).

Kansas

60. Groundwater Management District No. 1 - Pilot recharge site using impoundments.*

61. Groundwater Management District No. 2 - Pilot recharge site using pit infiltration basin with sedimentation trap.*

62. Groundwater Management District No. 3 - Several pilot recharge sites using various methods.*

63. Groundwater Management District No. 4 - Numerous pilot recharge sites using various methods.*

64. Groundwater Management District No. 5 - Pilot recharge site using channel modification.*

<u>Louisiana</u>

65. Baton Rouge - Recommendation to institute a program to construct reservoirs for direct water supply and artificial recharge.*

66. Jefferson Parish - Underground storage of treated water, a field study (Smith and Hanor, 1975).

<u>Massachusetts</u>

67. Statewide - Feasibility of increasing water supplies and preventing environmental damage by artificial recharge in Massachusetts (Motts et al., 1983).

<u>Michigan</u>

68. Kalamazoo - Recharge by induced infiltration to prevent decline of water table (Pettyjohn, 1981).

69. East Lansing-Meridian Township - Surface water-esker recharge study (Sutherland and Bruce, 1977).

70. Clinton, Eaton, and Ingham Counties - Consideration of artificial recharge in water management (Vanlier, Wood, and Burnett, 1973).

<u>Minnesota</u>

71. St. Paul - Tests to determine feasibility of recharge with injection wells.*

<u>Montana</u>

72. Old West Regional Commission - Artificial recharge project (Smith, 1977).

<u>Nebraska</u>

73. Aurora - Investigation of potential for recharge through injection wells (Lichtler, Stannard, and Kouma, 1979).

74. Little Blue Natural Resources District - Project with recharge of floodwaters under construction.*

75. Lincoln - Economic evaluation of the feasibility of artificial groundwater recharge in Nebraska (Supalla, 1981). Artificial recharge through injection wells (Marlette, 1968).

76. Tryon - Recharge with stock tanks.*

77. Upper Big Blue Natural Resources District - Proposed project to divert Platte River water for recharge.*

Nevada

78. Cold Spring Valley - Study to determine feasibility of recharge by injection methods.*

New Jersey

79. Malbaro - Storage of municipal water in a recharge well.*

80. Wildwood - Injection of freshwater during off-season to meet peak demands during the summer months.*

81. Princeton - Artificial recharge through augmented bank storage (Dewiest, 1967).

New Mexico

82. Santa Clara Indian Reservation - Artificial aquifer used to retain waters infiltrating from surface (Pettyjohn, 1981).

83. Southern High Plains of New Mexico - Study of potential artificial recharge systems using playa lake water (Brown et al., 1978).

New York

84. Bay Park - Recharge by injecting reclaimed waters through wells (Ku et al., 1975; Ragone, 1977).

85. Long Island, Nassau County - Stormwater basins used for recharge (Aronson, 1978).

North Carolina

86. Coastal Plain - Artificial recharge in aquifer management (Sherwani, 1980).

North Dakota

87. Glenburn - Subsurface dam retains groundwater in sand and gravel aquifer (Pettyjohn, 1981).

88. Minot - Injection of river water through shafts (Pettyjohn, 1981).

89. Valley City - Water diverted from river to recharge pit (Pettyjohn, 1981).

90. Old West Regional Commission - Artificial recharge project (Smith, 1977).

Ohio

91. Canton - Recharge with collector well connecting upper and lower aquifers (Pettyjohn, 1981).

92. Dayton - Lagoons and ditches flooded to allow infiltration of diverted river water (Pettyjohn, 1981).

93. Mill Creek Valley - Potential of injection well recharge studied (Fidler, 1970).

Oklahoma

94. Oklahoma Panhandle - Examination of potential for recharging the Ogallala Aquifer.*

95. Southwest Oklahoma - Recharge to dilute flouride groundwater proposed (Pettyjohn et al., 1982).*

Oregon

96. Dalles - Minor recharge project, closed down recently.*

97. Salem - Tests to determine feasibility of injection well recharge (Foxworthy, 1970).

South Dakota

98. Sioux Falls - Modification of channel diverting waters from a flood control dam (International Survey of Scientific Hydrology, 1970).

99. Old West Regional Commission - Artificial recharge project (Smith, 1977).

Texas

100. Dell Valley - Project under construction to inject impounded floodwaters through wells.*

101. El Paso - Potential for injection well recharge with treated sewage effluent appraised (Garza et al., 1980).

102. High Plains of Texas - Numerous sites where recharge with playa lake water has been attempted and appears to be economically infeasible (International Association of Scientific Hydrology, 1970).*

103. Houston - Studies of the feasibility of preventing land subsidence with artificial recharge (Garza, 1977).

104. San Antonio - Proposed project to recharge Edwards Aquifer (Green, 1967).*

105. Llano Estacado - Recharge to the Ogallala aquifer from playa lake basins (Wood and Osterkamp, 1984).

Utah

106. Salt Lake Valley - Recommendation to purchase land and begin studies of potential for artificial recharge.*

107. Utah Valley - Potential for artificial recharge to protect and purify water for municipal and industrial uses exists.*

108. Wasatch Aquifer - Potential for artificial recharge along western front of Wasatch Mountains.*

Virginia

109. Norfolk - Tests with injection of fresh water into a brackish-water aquifer (Brown and Silvey, 1977).

110. Roanoke - Disposal of stormwater runoff through drainage wells.*

Washington

111. Walla-Walla - Recharge project no longer in operation.*

112. Columbia Basin Project - Economic analysis of artificial recharge in a conjunctive irrigation plan (Karlinger and Hansen, 1983).

113. Eastern Washington - Feasibility of artifically recharging basalt aquifers (Garret and Londquist, 1972).

Wisconsin

114. Washura County - Demonstration project using an infiltration pond.*

Wyoming

115. Old West Regional Commission Project (Smith, 1977).

---

*Hydrologic Engineering Center, 1984.

SELECTED ARTIFICIAL RECHARGE ACTIVITIES

The intention of this section is to present summary information on selected case studies, demonstrations, or projects. The factors considered in selecting the recharge activities included method of artificial recharge, size and location of activity, and information availability. Table 8 provides a list of the selected sites.

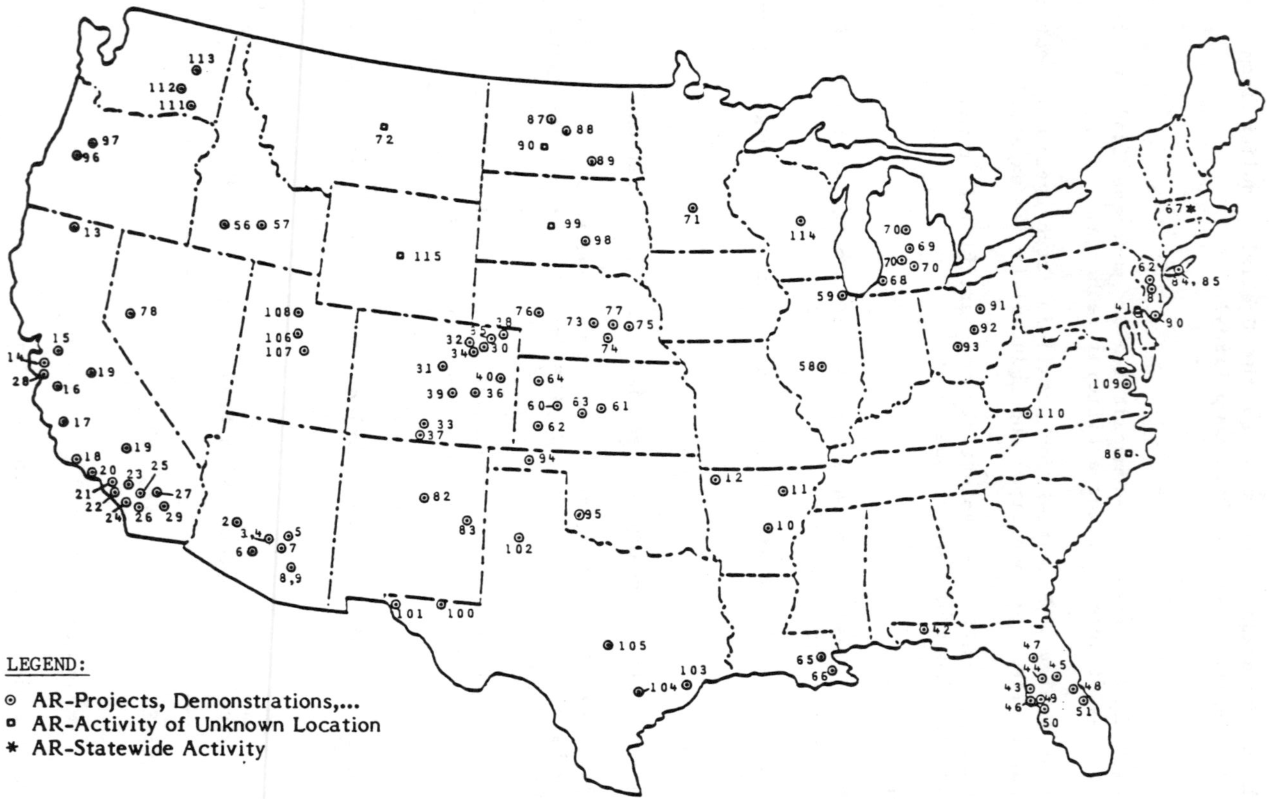

LEGEND:

- ⊙ AR-Projects, Demonstrations,...
- ▫ AR-Activity of Unknown Location
- * AR-Statewide Activity

Figure 2: National Artificial Recharge Activity (Projects, Demonstrations/Pilot Projects, Experiments, Studies). Refer to Table 7.

Table 8: Selected Artificial Recharge Activities

1. Oak Creek Basin, Arizona
2. Grand Prairie, Arkansas
3. Los Angeles County, California
4. Orange County, California
5. San Bernardino Valley, California
6. North High Plains, Colorado
7. Manatee County, Florida
8. North Hillsborough County, Florida
9. Wet Walnut Creek, Kansas
10. Blue Basin, Nebraska
11. Sioux Falls, South Dakota
12. Dell Valley, Texas
13. Columbia Basin, Washington

Oak Creek Basin, Arizona

The Hydrologic Engineering Center (1984) stated that flood control and artificial recharge with small man-made impoundments could become an important water resources management method in northern Arizona and other arid areas in the southwest. Stock tanks, or ponds, occupying small drainage areas collect stormwater runoff during the wet season, thereby providing drinking water for cattle. These ponds are constructed and utilized by private land owners, and vary in size and shape. Storage capacities range from less than one to more than eleven acre-feet (1,200-14,000 m^3). Almost all are lined with bentonite or clay to reduce seepage. Most recharge from the stock ponds is therefore considered unintentional and insignificant.

Investigations of the effect of stock ponds on the hydrology of lower Oak Creek Basin between Flagstaff and Phoenix imply that more extensive and scientific use of ponds for flood control and artificial recharge could prove beneficial (Hydrologic Engineering Center, 1984). Most of the annual precipitation (15 inches) in northern Arizona occurs during July and August. The stormwater rapidly runs out of drainage areas, allowing little infiltration into the permeable alluvial soils. The Salt River receives all the runoff, creating flood pulses which ultimately reach Phoenix. Therefore, by retaining stormwater runoff, stock ponds can attenuate flood pulses and reduce downstream impacts (flood control). A large number of stock ponds constructed in alluvial soils without linings could promote significant quantities of infiltration into the underlying aquifers (artificial recharge).

Artificial recharge would greatly benefit agricultural activities in northern Arizona, which, in the absence of perennial streams, relies entirely on ground water for irrigation. Water rights need to be considered in such a project. Extensive impoundment of runoff could violate legal water rights of downstream users (Hydrologic Engineering Center, 1984).

Grand Prairie, Arkansas

The Grand Prairie Region of Arkansas is an area where the major source of water is underground, and where the supply of ground water is depleted faster than it is naturally replenished (Griffis, 1976). The Grand Prairie is located in east-central Arkansas between the White River and Bayou Meto. Land usage in the area is primarily agricultural, with rice being one of the major crops. Rice crops require a considerable amount (22 inches) of water each year, and most of the water for this purpose is pumped from wells. As a result, the supply of water in the water-bearing strata is being depleted faster than it is being replaced. For this reason, some means of artificially replenishing ground water is needed.

Griffis (1976) noted that previous investigation showed only two technologically feasible solutions to the ground water depletion problem: (1) drastic conservation efforts; or (2) artificial recharge through wells. Griffis (1976) reported on the establishment of a gravity-flow recharge well to allow surface runoff water to flow through it and into the aquifer.

An 8-inch Plexiglas tube partially filled with a layer of gravel was used to simulate the recharge well (Griffis, 1976). The rest of the tube was filled with one of four grades of washed sand. The coarse sand consisted of particles that were 1.19 to 2.38 mm in diameter; the medium sand had particles in the range of 0.59 to 1.19 mm; the fine sand had particles from 0.25 to 0.59 mm in diameter; and the very fine sand had particles from 0.149 to 0.25 mm.

The bench-scale tests showed that a sand-filled, gravity-flow recharge well was not suitable for recharging untreated runoff water in the Grand Prairie. A series of studies were then made to search for a low-cost water treatment system. The water treatment techniques studied included the use of alum, cationic and anionic polyelectrolytes, lime, and magnesium hydroxide. In addition, some studies included combinations of two or more of the chemicals, and in one series of tests the sludge formed in one batch of treated water was used as seed material for the next batch. No treatment method was able to clarify the water for a cost less than the target cost of $5 per acre foot (Griffis, 1976).

Other sources for recharge water were examined and the most likely candidates were found to be the drainage water from flooded rice fields, and rainwater collected on the fields during the off-season for rice production (Griffis, 1976).

Los Angeles County, California

A network of 19 dams and over 2,000 miles of drainage channels, developed through cooperation between the Los Angeles County Flood Control District and the U.S. Army Corps of Engineers, provides flood protection for the residents of Los Angeles County. Furthermore, almost all the storm runoff impounded by these facilities is routed into underground reservoirs, where it is stored for municipal, agricultural, and industrial use. Only 5 percent of the precipitation falling on the watershed upstream from major conservation facilities, and only 15 percent of all rainfall in Los Angeles County, reaches the Pacific Ocean each year (Hydrologic Engineering Center, 1984).

The Hydrologic Engineering Center (1984) noted that a large portion of the impounded floodwaters is conserved in Los Angeles County's 3,126 acres of spreading grounds, 2,000 of which are owned by the Flood Control District. An annual average of 250,000 acre-feet of water, valued at $20 million, recharges the aquifers underlying Los Angeles County through the District's spreading grounds alone. Since artificial recharge activity first began in 1919, more than 11 million acre-feet of water have been percolated via spreading grounds, in unlined channels, and behind dams and reservoirs.

The largest recharge facility in Los Angeles County is the Rio Hondo Coastal Basin spreading grounds, a series of basins paralleling the Rio Hondo River (Hydrologic Engineering Center, 1984). These basins, along with the San Gabriel River system, retain stormwaters released from Whittier Narrows Dam, a U.S. Army Corps of Engineers flood control project spanning across the Rio Hondo and the San Gabriel Rivers. Flocculants are added to the waters just as they exit the dam to reduce the concentration of sediment in waters diverted to the Rio Hondo and San Gabriel spreading grounds. While outflows from the Whittier Narrows Dam area are intermittent, recharge of imported and reclaimed water has necessitated year round use of the recharge facilities. Insect problems associated with continuous use of spreading grounds have been limited by control measures.

The U. S. Corps of Engineers has changed the outflow schedules for the spillway gates of Whittier Narrows Dam to permit increased use of floodwater for recharge. In the past, the Corps operated the Dam's San Gabriel outlets to release waters at 5,250 cfs, a rate far exceeding the 300 cfs intake capacity of the San Gabriel spreading grounds. Any waters building up behind the dam were diverted via a flood flow channel to the Rio Hondo conservation pool. There they would be released through the Rio Hondo outlet gates at rates below the 700 cfs intake capacity of the Rio Hondo spreading grounds. This operating procedure allowed 15,000 acre-feet of water to bypass the spreading grounds and flow into the Pacific Ocean every year. With the recent enlargement of the Rio Hondo conservation pool to 2,500 acre-feet, releases into the San Gabriel and Rio Hondo Rivers can be held at 385 and 700 cfs, respectively. Only when the conservation pool is full are waters released at rates exceeding the intake capacities of the San Gabriel and Rio Hondo spreading grounds. By changing outflow schedules and the size of the Rio Hondo conservation pool, the Corps has enabled the Flood Control District to conserve 6,000 acre-feet of releases from Whittier Narrows Dam annually (Hydrologic Engineering Center, 1984).

Studies to retain larger quantities of water for artificial recharge are being made throughout Los Angeles County. The potential exists for recharging 100,000 acre-feet per year of Los Angeles River water. However, none of the proposed recharge schemes, which include injecting water through wells adjacent to the river and redirecting water to the Rio Hondo spreading grounds, appear cost-effective at this time (Hydrologic Engineering Center, 1984). In addition, remedial work on the Flood Control District's dams and reservoirs, which, when completed, will allow increased storage of stormwaters, has already begun. The greatest extension of current conservation efforts will most likely occur through changes in operations of the five U.S. Army Corps of Engineers' dams in Los Angeles County (Hydrologic Engineering Center, 1984).

Orange County, California

Stormwater and floodwater runoff are used to recharge the aquifer underlying the Santa Ana River basin in Orange County in southern California. Floodflows, starting in the flood plain east of Orange County, are collected by the U.S. Corps of Engineers' Prado Control Reservoir, an earthern structure with 196,240 acre-feet of storage capacity (Hydrologic Engineering Center, 1984). Waters leave the dam to flow into the Santa Ana River, where they are routed into multiple recharge facilities covering 1,100 acres of Orange County.

The Santa Ana River itself has been transformed into a spreading facility by the Orange County Water District (Hydrologic Engineering Center, 1984). A levee built by the District divides the river in half longitudinally. The western section contains a sequence of spreading basins, each of which fills before allowing water to pass to the next basin downstream. The basins, 200 to 400 feet wide and up to 10 feet deep, can retain large amounts of water and percolate up to 4 vertical feet of water daily. Water not held in the spreading basins weaves through a maze of dikes in the eastern section of the Santa Ana River. These temporary sand structures, often destroyed during heavy storm flows, slow down the river, thus allowing infiltration of water which otherwise would pass into the Pacific Ocean.

Several off-channel basins provide additional ground water recharge in Orange County (Hydrologic Engineering Center, 1984). Anaheim Lake, Warner Basin, Kraemer Basin, Placentia Basin, and Burris Pit, a former sand and gravel site, all retain Santa Ana River runoff for infiltration purposes. Waters imported from the Colorado River and northern California, by the State Water Project, are also percolated through these basins. While recharge is their chief function, a few of these basins also serve as recreational facilities; for example, Anaheim Lake is stocked with trout and open for fishing during much of the year.

Prevention of sea water intrusion into ground water in Orange County has been extremely successful due to recharge by infiltration and wastewater injection systems. In fact, Orange County was able to survive decreased supplies from the State Water Project during California's 1975-77 drought (Hydrologic Engineering Center, 1984). Recharge through on and off-channel basins adds an average of 250,000 acre-feet of water to the aquifer annually, helping to meet expanding water needs in the Santa Ana River Basin.

San Bernardino Valley, California

The Kern River Intertie consists of a sedimentation basin and a gated 3,500 cfs gravity flow connection between the Kern River and the California Aqueduct east of Los Angeles. This flood control project was completed by the U.S. Army Corps of Engineers in 1977 (Hydrologic Engineering Center, 1984).

Normally, the flow diverted from the Kern River would be impounded in surface reservoirs throughout southern California. These storage facilities were full in 1978 due to local runoff conditions. Therefore, waters were routed for two months to spreading lands in the Mojave River Valley and the San Bernardino Basin (Hydrologic Engineering Center, 1984). No efforts were made to measure the quantity of water entering the ground water basins, although project officials estimated that about 22,500 acre-feet were conserved. The primary purpose of the project was to explore and evaluate the nature of negotiations and agreements between the California Department of Water Resources and local water agencies necessary to increase conservation of State Water Project waters by artificial recharge. No long term recharge using State Water Project waters, however, will be implemented in the near future in the San Bernardino Basin or Mojave River Valley (Hydrologic Engineering Center, 1984).

Northern High Plains, Colorado

The major ground water system in the High Plains of Colorado is the Ogallala aquifer. Average annual natural recharge to the Ogallala has been estimated to be less than 1 inch. This low rate of natural recharge, combined with intensive use of ground water for irrigation, has resulted in mining of the ground water in the Ogallala aquifer. Three methods of artificial recharge have been investigated to prevent further depletion of the aquifer (Warner et al., 1984).

The northern High Plains of Colorado, located along the northeastern border, have an area of approximately 9,300 square miles. The topography of the area is characterized by an eastward sloping, gently rolling plain. Large parts of the area have low undulating sand dunes which contrast with the relatively flat surrounding area. Most of the High Plains of Colorado is covered by a loose silty and sandy soil. However, there are several portions with tight clay soils. The major streams in the area within Colorado are the North and South Forks of the Republican River and the Arikaree River. All three rivers are ephemeral and influent except near their lower reaches near their points of flow into either Kansas or Nebraska. Precipitation in the High Plains of Colorado averages about 16 to 18 inches per year. Most of this precipitation occurs during the spring and summer months, generally in the form of scattered thundershowers (Warner et al., 1984).

One of the artificial recharge methods studied involved building a series of earth dams and dikes in stream channels, thereby spreading the flood flow over a larger area and increasing the potential for infiltration (Warner et al., 1984). This method was very effective in increasing recharge, but had a serious disadvantage in that these dams were subject to damage from very large floods. Warner et al. (1984) also evaluated the use of recharge pits. They reported that the classic use of pits is limited to areas where no low-permeability clay or silt layers occur between the pit and the ground water table. This is generally not the case in the High Plains of Colorado. It was demonstrated that recharge pits are still feasible when a layer of low permeability exists beneath the pit if the low permeability layer is overlain by a much more permeable layer of sand or gravel which can act as an underground spreading basin. This is the case in many areas of the High Plains in Colorado. It was demonstrated that the water moved principally through the sides of the pit into the coarse sand and gravel layers where it was spread over a large area. Sediment accumulations in the pit and also behind the dams were a major problem.

The third method of artificial recharge investigated involved terracing the land surface (Warner et al., 1984). This is one of the more elementary methods of capturing runoff water to induce additional recharge. The amount of additional recharge caused by terracing was determined by monitoring changes in soil moisture with depth and time. The effect of terracing was also studied using modeling techniques. It was concluded that terracing can significantly increase recharge to the aquifer.

The demonstration recharge projects described show that artificial recharge in the High Plains of Colorado is technically feasible (Warner et al., 1984). The factor limiting artificial recharge on a large scale is probably the availability of water. Infrequent flood flows that result from local high-intensity rainstorms are the only source of excess water.

Manatee County, Florida

Manatee County is south of Tampa, Florida and includes the City of Bradenton. Manatee County relies upon a surface water supply fed by Lake Manatee. This surface water is treated at the Manatee County Water Treatment Plant, a 54 mgd facility (CH_2M-Hill, 1984). At this site, water is diverted from the treatment plant after filtration and just prior to final pH stabilization and chlorination. The diverted water is stored in a permeable limestone aquifer between 400 and 700 feet below the site. Water is introduced through a 700 foot deep, 16 inch, gravity-flow well, which is cased to a 400 foot depth and an open hole for the remainder of its depth. The storage site is several miles east of Bradenton.

Nine test cycles of well usage have been conducted, ranging from about three days to a year in duration. As the underground stored water is recovered, water quality changes were noted based on the mixture of injected water and native ground water. The quality of the recovered water approaches background concentrations as more than 100 percent of the injected water is recovered. With each successive cycle of injection and recovery, the water in the storage zone improves in quality with the flushing action so that a natural underground reservoir is developed.

This test project demonstrated that by using the aquifer as a storage facility, the peak water supply demands for Manatee County could be met. As a result, the County can continue to maximize the effective use of their surface water facility by using an aquifer that normally contains poor quality ground water as a storage zone. A major advantage of this system is that the County can produce the improved ground water at 20 percent below the current surface water cost, and half the cost of any other considered alternative for water supply expansion (CH_2M-Hill, 1984).

Northwest Hillsborough County, Florida

Four experiments were designed and conducted to investigate possible recharge alternatives in a 30 square mile area in northwest Hillsborough County in Florida (Sinclair, 1977). The County includes the City of Tampa. The experiments included four different artificial recharge methods: (1) sinkhole recharge, (2) water spreading, (3) connector wells, and (4) subsurface-tile drainage.

The terrain in the study area is characterized by low relief and numerous sinkhole swamps and lakes. The low gradients of the swales and streams result in sluggish discharge. The sinkhole swamps and lakes fill when precipitation is large, then spill one into another as the water moves generally southwestward toward the Gulf or Tampa Bay. Runoff accounts for nearly one-third of the total precipitation, and evapotranspiration accounts for about two-thirds. Natural recharge to the underlying Floridan aquifer is about 1 percent of precipitation. Streamflow records for two small streams in the study area indicate an average runoff of 96,000 $ft^3/d/mi^2$. The two watersheds are typical of the terrain throughout the study area. The low topographic relief and high rates of evaporation combine with the irregular and seasonal nature of the precipitation to make surface retention of water infeasible.

The methods of recharge which were studied are described as follows (Sinclair, 1977):

(1) Sinkhole recharge--Those sinkholes having good hydraulic connection with the Floridan aquifer are an important means of natural recharge. Recharge could be increased by about 3.65 million cubic feet per year by maintaining high water levels in the sinkholes.

(2) Water-spreading--Testing indicated that recharge to the surficial aquifer could be increased by water spreading. Recharge to the Floridan aquifer, however, would not be increased appreciably unless a good hydraulic connection exists between the surficial and Floridan aquifer at the water-spreading site.

(3) Connector wells--Connector-well experiments, designed to drain the surficial aquifer to the Floridan by gravity flow, have a potential to recharge about 535,000 cubic feet per year per well under the prevailing geologic and hydrologic conditions in the study area.

(4) Subsurface-tile drainage--It is estimated that recharge by subsurface-tile drainage to a connector well open to the Floridan aquifer would recharge about 1.3 million cubic feet per year per 1,000 feet of tile.

Sinclair (1977) concluded that the recharge experiments indicated that all four of the methods investigated can be effective. However, the sinkhole recharge experiment resulted in moving the greatest volume of water into the Floridan aquifer. The drain-tile experiment indicated the greatest potential for draining the surficial aquifer. Combinations of the four methods could be used in areas where there is potential for downward movement of water and a sufficient thickness of unsaturated aquifer for water storage.

Wet Walnut Creek, Kansas

Wet Walnut Valley in west central Kansas is a narrow alluvial valley in which an interrupted meandering stream dominates the hydrologic system (Gillespie and Slagle, 1972). Wet Walnut Creek is generally a losing stream in most of the reach through the study area. The average annual streamflow leaving the area is 50,530 acre-feet, most of which occurs as storm runoff (Gillespie and Slagle, 1972).

The alluvial deposits in Wet Walnut Valley are primarily of fluvial origin and can be divided into an upper alluvium of lenticular silt, clay, and sand; and a lower alluvium of coarse sand and gravel that comprises the major aquifer of the area (Gillespie and Slagle, 1972). Fine-grained colluvial deposits are present along the edges of the Valley. Ground water in the alluvium is unconfined, or semiconfined in areas where the potentiometric surface is in the fine-grained materials of the upper alluvium. The general movement of ground water is eastward down the Valley. Owing to heavy pumping for irrigation and water moving from the stream to the aquifer, the gradient normal to the Valley is generally away from the Creek. The estimated total volume of ground water that can be stored in the alluvium of Wet Walnut Valley in Rush County was 240,000 acre-feet in 1960. The average annual pumpage for irrigation is about 15,000 acre-feet. The ground-water system is dependent on periodic recharge from flow in the ephemeral tributaries and high flow in the main stream. Over the long term the hydrologic system is in balance; however, significant increases or decreases in the amount of water in storage occur for any single year or short periods.

Gillespie and Slagle (1972) reported that artificial recharge experiments at the edge of the Valley showed an infiltration rate of about 0.5 foot per day. The infiltrated water is temporarily perched above a discontinous zone of tight claying silt at a depth of 40 feet, from where it is assumed to percolate to the aquifer. Recharge experiments in three ponds assumed within a tributary channel indicated an infiltration rate of 3 feet per day on the sixth day after water from wells was pumped into one of the ponds. Infiltration rates in the other two ponds ranged from about 0.3 to 0.1 foot per day after six days due to the fact that the channel bottoms had been previously trampled by livestock, and to the lack of vegetation. A small temporary dam across the inner channel of Wet Walnut Creek created a pond whose surface rose about 5 feet above the pool level of the stream for 25 days. The water level in a well located about 250 feet from the channel rose about 2.25 feet during the same period.

Gillespie and Slagle (1972) also reported that 53 floodwater retarding structures were proposed for construction on tributaries of Wet Walnut Creek. These structures will greatly increase the recharge potential because they will hold storm runoff in the watershed for a much longer period of time. Recharge could be increased without these structures, but the quantity of water recharged would be considerably smaller. A series of dams in the main channel and tributary channels of Wet Walnut Creek would enhance the natural recharge system. The major detriment to enhancement of recharge is sediment; however, this problem can be overcome by proper construction, management, and maintenance of an artificial recharge system. It was concluded that the maximum amount of water available for artificial recharge is the streamflow, which averages about 50,000 acre-feet per year. The average amount probably will be less than the potential for recharge after the floodwater retarding structures and artificial recharge system have been placed in service (Gillespie and Slagle, 1972).

Blue Basin, Nebraska

Recharge with floodwaters is being considered as a solution to severe ground water overdraft problems in south central Nebraska. Studies conducted at two reservoir sites in the Upper Blue River Basin suggest that seepage from flood-control reservoirs, encouraged by chisel plowing or cultivating in the reservoirs during dry periods, could significantly aid ground water recharge. In addition to recharging the underlying portions of the Ogallala aquifer, waters stored in flood-control reservoirs could relieve pressure on ground water supplies by helping to provide irrigation water necessary to support Nebraska's agricultural economy (Hydrologic Engineering Center, 1984; and Lichtler, Stannard, and Kouma, 1980).

Two projects incorporating both flood control and artificial recharge are currently being planned. The Big Sandy Creek Project in the Little Blue Natural Resources District will capture and allow infiltration of floodwaters in the Little Blue Basin. One reservoir has been completed and will be used to make further studies of recharge potential. Four or five more reservoirs may be built. The proposed Big Blue Irrigation Project in the Upper Big Blue Natural Resources District consists of six reservoirs which can be operated to impound flood flows during May through July storms. The primary source of water, however, will be Platte River waters diverted from Plum Creek. In addition to aiding in recharge, water supply, and flood control, the two projects will also increase water-based recreation and benefit wildlife.

Sioux Falls, South Dakota

Flood control facilities in Sioux Falls, South Dakota have been used to artificially recharge the underlying aquifer. The U.S. Army Corps of Engineers facilities, completed in 1961, include a dam and a diversion channel which provide flood protection for Sioux Falls by diverting floodflows passing through the Big Sioux River. Flooding occurs when heavy rains fall in the Sioux River Valley north of Sioux Falls (Hydrologic Engineering Center, 1984).

The U.S. Army Corps of Engineers and the City of Sioux Falls Water Department have implemented several measures to induce infiltration of waters entering the diversion channel from the Big Sioux River and from Silver Creek (Hydrologic Engineering Center, 1984). The Corps also operates the floodgates of the dam to insure that water for recharge is provided whenever possible, often closing all ten gates during low-flow periods. Sedimentation problems are addressed by the City Water Department, which occasionally dredges out the section of the channel between the dam and the weir. Without dredging operations, silts and clays deposited by diverted waters would prevent recharge. The city last dredged out the channel in 1976. Infiltration of diverted water has prevented wells from going dry during years of low rainfall. By maintaining ground water levels, artificial recharge should allow ground water to continue to fulfill the water needs of Sioux Falls.

Dell Valley, Texas

An artificial recharge project, planned for the Dell Valley, is considered a secondary purpose of a U.S. Soil Conservation Service project. The project area is located along the Texas-New Mexico border, two miles east of El Paso (Hydrologic Engineering Center, 1984). Collected floodwaters will be injected through wells into the Victoria Peak aquifer. The artificial recharge project is estimated to add about 6,000 acre-feet of water to the aquifer annually. The original plans consisted of impounding floodwaters behind five earthern dams and then routing the water through overland pipelines and channels to the Valley. The cost of implementing such a project would have rendered the entire project uneconomical. However, the present plan suggests that the impounded floodwaters will exit the dams through spillways, pass through filter fields to screen out sediment, and then flow by gravity into a group of wells located 300 to 3,000 feet downstream from the dams (Hydrologic Engineering Center, 1984).

The project is in the construction stage. Only one dam has been built, and injection wells are still being drilled (Hydrologic Engineering Center, 1984). Wells were sited using a photogeological process. Fracture trace intersections were identified with the aid of aerial photographs. Field work subsequently determined optimal well locations. Each well will include a 26 inch diameter surface hole 40 feet into the limestone formation, followed by a 20 inch diameter production hole extending between 1,200 and 1,400 feet below the surface. The project is expected to cost $16 million upon completion. However, the ground water availability in the Valley will increase significantly and flood damages will be reduced by 85 percent. A decreasing decline in the water table level is estimated at 15 percent per year.

Salts leaching from irrigated soils and percolating from the dry salt lakes have been deteriorating the underlying aquifer. This artificial recharge project will enhance the quality of the existing ground water and reduce the degradation rate as well (Hydrologic Engineering Center, 1984).

Columbia Basin, Washington

Two major types of irrigation systems in a part of the Columbia Basin project in south central Washington were analyzed for their cost-effectiveness. The first is a surface water irrigation system in which farm units are supplied with water through canals and lateral drainage from a main-trunk canal. The second system is a plan in which surface water is recharged via wells and delivered to the farm units using the transmissive properties of the aquifer and subsequent pumping (Karlinger and Hansen, 1983). The area selected to study the feasibility of artificial recharge is in the southern part of the Columbia Basin project, where development of irrigable land is currently in the planning stages.

The present plan is to import water from Roosevelt Lake to irrigation blocks via an extension of the East Low Canal, and to apportion the water to the farm units using a lateral distribution system. An artificial recharge system alternative would, using the same water source and the extension of the East Low Canal area, divert water from the lateral system and deliver it to recharge facilities. The transmissive and storage properties of the aquifer would then be utilized to convey the water to the individual agricultural sections to be pumped for irrigation at the appropriate time. The deep basalt aquifer would be used to obtain ground water for irrigation in this area. From a project perspective, the economic tradeoffs between alternatives would be the savings from a reduced surface-conveyance system versus the costs of well construction, recharge facilities, and, most importantly, additional energy needs for pumping.

Based on the previous considerations and the system design, Karlinger and Hansen (1983) suggested that an artificial-recharge-irrigation operation is not an economical alternative to a surface-irrigation system in the study area. Study of systems utilizing shallower aquifers and different types of recharge may be worthwhile to pursue. A shallower aquifer would require less expensive wells and pumps and reduced electric power, and a water-spreading type of recharge might be possible.

INTERNATIONAL ARTIFICIAL RECHARGE ACTIVITIES

Artificial recharge projects are currently active in over 20 countries besides the United States. Such countries include Algeria, Australia, Bulgaria, Czechoslovakia, Mexico, Monaco, Egypt, France, Germany, Great Britain, Holland, Hungary, Iran, Israel, Jamaica, Japan, Rumania, Senegal, Spain, South Africa, Sweden, Switzerland, and the Union of Soviet Socialist Republics (CH_2M-Hill, 1984; and International Association of Scientific Hydrology, 1970). In Australia, Germany, Holland, and Switzerland, recharge basins and canals are preferred for infiltration. In Great Britain, Israel, Jamaica, and Japan, wells are most often preferred, reflecting the presence of carbonate formations that have secondary solution openings and great permeability. In France, both basins and wells are used (CH_2M-Hill, 1984). Appendix A includes abstracts on international artificial recharge activities. Two international artificial recharge case histories were selected for illustration purposes. These case histories include projects in Great Britain and Israel.

Great Britain

According to current forecasts the demand for water resources in the midlands of England is expected to increase, but at a lower rate than previously anticipated. Despite the economic advantages, the increased use of some aquifers in this area is limited by environmental and technical constraints, including the inadequacy of natural recharge from rainfall (CH_2M-Hill, 1984). With regard to the principal aquifer of the midlands, the Triassic Sandstones, the design and operational uncertainties of an artificial recharge plant need to be resolved before this method may be counted upon to enhance natural recharge. The preliminary design standards for the principal components of a 9 Ml/d (2.38 MGD) prototype recharge plant at Stourbridge in the west midlands have been developed (Jones, 1982). A test well, measuring 36 inches in diameter and 561-feet deep, is cased to 279 feet and constructed open-hole to total depth. The well is fitted with four small-diameter injection pipes, a 2.1 mgd submersible pump, and a sampling tube for data collection. Recharge and recovery rates have been 2.4 and 2.1 mgd, respectively.

An extensive study was conducted, prior to construction of the well, regarding potential operational problems such as air entrainment and aquifer plugging. To reduce the likelihood of air entrainment caused by cascading flow and vacuum effects, the small diameter injection pipes were fitted with a bucket device attached to the downhole end of the pipe to reduce discharge velocities and direct the discharge stream upward away from the borehole, thus allowing the entrained air to separate and move up the annulus. To prevent borehole plugging, the following limits are placed on recharge water quality (CH_2M-Hill, 1984):

Turbidity	$<$0.5 NTU
Particulate Matter	$<$5 microns
Chlorination	2 mg/l residual free chlorine
pH	7.0 - 7.5
Residual Coagulants (Insoluble Aluminum or Iron)	$<$0.05 mg/l

The well is periodically backflushed to remove any clogging that should occur. In spite of these limits, well operation will be monitored to evaluate the effectiveness of these limits and changes will be made to optimize the system. Automated facilities are planned to dispense Calgon (a dispersive agent), gaseous chlorine, and possibly a polyelectrolyte. All system controls and monitoring equipment will also be automated.

Israel

Well-recharge is the predominant method of artificial recharge in Israel, recharging a total of 264 million gallons in 1968 via 135 wells. Two types of wells are generally employed, single-use injection wells and dual-use injection/pumping wells. In single-purpose wells injection rates decreased over a period of time from 25 to 80 percent due to clogging. In dual-purpose wells injection rate decreases were less than 25 percent. Research into well clogging showed that the accumulation of silt or organic matter on the face of the borehole and in the close formation seems to be the predominant clogging factor (CH_2M-Hill, 1984). When large quantities of lake-water (even if potable), containing minute quantities of silt (less than 2 ppm) and organic matter (150 to 800 ASU), are introduced into a well, these particles are filtered out by the gravel-pack or the porous aquiferous material and form an impermeable mat, mainly organic, at the borehole face (Harpaz, 1971). Harpaz (1971) reported that a two-phase redevelopment, that is, a short rigorous pumping followed by continuous withdrawal of water for supply in the summer months has resulted in almost full restoration of the recharge capacity in limestone wells, and has retained the operational capacity of sandstone wells for years. This procedure has become a routine practice in Israel's dual-purpose wells.

As regards water quality considerations, Harpaz (1971) emphasized the need to exert control over its injected water spreading and over the movement of accompanying chemical, physical and biological constitutents, and discussed the mixing zone which develops at the injected/native water interface. In limestone aquifers spreading is predominantly elongated, and the width of the dispersion zone may reach hundreds of meters over a spreading distance of a few kilometers. In dual-purpose wells, Harpaz (1971) suggested that the initial repumped water will likely be contaminated by decaying organic matter and coliform bacteria, and will be highly turbid, and suggested that this initial water be pumped to waste.

The estimated cost of well-recharging in Israel ranges from \$0.11 to \$0.18 per 1,000 gallons (1971 costs), including the price of the recharge water (Harpaz, 1971).

PERSPECTIVES ON CASE STUDIES

Two critical and pertinent questions regarding artificial recharge application are: (1) is artificial recharge practical and feasible for increasing water supplies and/or preventing salt water intrusion, and (2) is artificial recharge an appropriate method/technique? That is, even if the hydrology and geology of the area under consideration are favorable for artificial recharge, is this method preferred to other water supply alternatives because of greater economic, environmental, and social values? According to the case studies presented herein, artificial recharge as a methodology is technically achievable. However, when it comes to feasibility, several factors must be considered. Such factors include the recharge water quality and quantity, water availability, and effectiveness of water management.

Appendix A consists of an annotated bibliography on artificial recharge and includes 64 abstracts on case studies. Seventy-five percent of these abstracts suggest that artificial recharge is either feasible or potentially feasible for increasing water supplies and/or controlling salt water intrusion. Only 3 percent of these abstracts indicated that artificial recharge is not feasible or successful. The remaining 22 percent of the abstracts do not provide enough information for a conclusion. Furthermore, among the thirteen case histories (projects, demonstrations, studies, experiments) reported in this chapter, eleven indicated that artificial recharge is either feasible or potentially feasible. Only two case studies concluded that artificial recharge is not feasible under the conditions of the area.

While artificial recharge is an important method for obtaining an optimum water supply, it or any other method should not be implemented without proceeding with an exploration and development program that insures a water supply of lowest economic cost and least negative effects on a short and long term basis. The feasibility and cost-effectiveness of artificial recharge projects can be demonstrated in terms of dollars per volume of recovered water. Bianchi and Lang (1974) reported that artificial recharge can be accomplished into the Fresno depression at a total cost of \$25.00-\$25.50/acre-feet (at the mains) by water spreading in the urbanized area. This is about one-half of filtration and treatment costs. The actual cost of water spreading amounts to only \$2.50-\$3.00 acre-feet, with the rest going to water purchase and pumping costs to put the water into the city's mains. The estimated costs of artificial recharge in eastern Washington range from about \$6 per acre-feet for recharging 2,000 acre-feet per year, to about \$4 per acre-feet (1970 values) for recharging 6,000 acre-feet per year if the source and injection sites were both on the valley floor (the most favorable situation) (Garrett and Londquist, 1972). In the East Lansing-Meridian, Michigan, artificial recharge study, Sutherland and Bruce (1977) noted that potential savings in treatment and pumping costs, estimated at \$30,000 per year for a present demand of 5 mgd, are attractive incentives for a surface water-esker recharge program. CH_2M-Hill (1984) reported the costs for providing supplemental water for Manatee County in Florida is approximately 33 cents per 1000 gallons; this cost figure is more than 20 percent below the county's current cost for water, and less than half the cost of other regional water supply alternatives.

In summary, it can be concluded that artificial recharge is a viable method that is cost effective and environmentally advantageous at many localities in the United States.

SELECTED REFERENCES

Anderson, G.S., "Artificial Recharge Experiments on the Ship Creek Alluvial Fan, Anchorage, Alaska", USGS/WRI-77-38, June 1977, U.S. Geological Survey, Anchorage, Alaska.

Aronson, D.A., "The Long Island Experience with Water Retention and Recharge Management", Proceedings of a Workshop on Managing the Demand for

Recreation and Tourism held in Ocean City, Maryland, Sept. 27-29, 1978, 1978, Co-sponsored by the Cooperative Extension Services of Delaware, Maryland, New Jersey and Virginia, pp. 48-51.

Bianchi, W.C. and Lang, G.J., "City of Fresno's Leaky Acres Groundwater Recharge Project Construction and Performance", Journal American Water Works Association, Mar. 1974, pp. 176-179.

Bloyd, R.M., Jr., "Underground Storage of Imported Water in the San Gorgonio Pass Area, California", Open File Report, July 1969, U.S. Geological Survey, Menlo Park, California.

Brookman, J. and Sunada, D.K., "Artificial Ground Water Recharge on the Arikaree River Near Cope, Colorado-Completion Report", CER68-69JB-DKS16, Nov. 1968, Civil Engineering Department, Engineering Research Center, Colorado State University, Fort Collins, Colorado.

Brown, R.F., Signor, D.C. and Wood, W.W., "Artificial Ground-Water Recharge as a Water-Management Technique on the Southern High Plains of Texas and New Mexico", Report 220, Aug. 1978, Texas Department of Water Resources, Austin, Texas.

Brown, D.L. and Silvey, W.D., "Artificial Recharge to a Freshwater-Sensitive Brackish-Water Sand Aquifer, Norfolk, Virginia", Professional Paper 939, 1977, U.S. Geological Survey, Reston, Virginia.

Burns, A.W., "Hydrologic Analysis of the Proposed Badger-Beaver Creeks Artificial Recharge Project, Morgan County, Colorado", Water Resources Investigations 80-46, July 1980, U.S. Geological Survey, Lakewood, Colorado.

Bush, P.W., "Connector Well Experiment to Recharge the Floridan Aquifer, East Orange County, Florida", Water Resources Investigations 78-73, 1979, U.S. Geological Survey, Tallahassee, Florida.

CH_2M-Hill, "Recharge Recovery at Lake Manatee: Phase II", Southwest Florida Management District, Mar. 1984, Gainesville, Florida.

DeWiest, R.J.M., "Artificial Recharge Through Augmented Bank Storage", Artificial Recharge and Management of Aquifers, Proceedings of the Symposium at Haifa, Publication No. 72, International Association of Scientific Hydrology, United Nations Educational Scientific and Cultural Organization, Mar. 1967.

Emmons, P.J., "Artificial-Recharge Tests in Upper Black Squirrel Creek Basin, Jimmy Camp Valley, and Fountain Valley, El Paso County, Colorado", Water Resources Investigations 77-11, July 1977, U.S. Geological Survey, Lakewood, Colorado.

Fidler, R.E., "Potential Development and Recharge of Groundwater in Mill Creek Valley, Butler and Hamilton Counties, Ohio, Based on Analog Model Analysis", Water Supply Paper 1893, 1970, U.S. Geological Survey, Washington, D.C.

Foxworthy, B.L., "Hydrologic Conditions and Artificial Recharge Through a Well in the Salem Heights Area of Salem, Oregon", Water Supply Paper 1594-F, 1970, U.S. Geological Survey, Tacoma, Washington.

Garrett, A.A. and Londquist, C.J., "Feasibility of Artificially Recharging Basalt Aquifers in Eastern Washington", Open File Report, 1972, U.S. Geological Survey, Tacoma, Washington.

Garza, S., Weeks, E.P. and White, D.E., "Appraisal of Potential for Injection-Well Recharge of the Hueco Bolson with Treated Sewage Effluent--Preliminary Study of the Northeast El Paso Area, Texas", Open File Report 80-1106, Sept. 1980, U.S. Geological Survey, Austin, Texas.

Gillespie, J.B. and Slagle, S.E., "Natural and Artificial Ground-Water Recharge, Wet Walnut Creek, Central Kansas", Kansas Water Resources Board Bulletin No. 17, 1972, U.S. Geological Survey, Lawrence, Kansas.

Green, M.G., "Artificial Recharge to the Edwards Limestone Aquifer in South Texas", Proceedings of Dubrovnik Symposium (Oct. 1965), Hydrology of Fractured Rocks, Volume 2, Publication 74, 1967, International Association of the Science of Hydrology, pp. 465-481.

Griffis, C.L., "Artificial Recharge in the Grand Prairie, Arkansas", Bulletin 810, Aug. 1976, Agricultural Experiment Station, University of Arkansas, Fayetteville, Arkansas.

Guyman, G.L., "Application of Artificial Recharge Technology for Managing the Water Resources--Anchorage, Alaska", IWR-24, June 1972, Institute of Water Resources, University of Alaska, Anchorage, Alaska.

Hargis, D.R. and Peterson, F.L., "Artificial Recharge Practices in Hawaii", OWRR-A-028-HI(1) Oct. 1970, Water Resources Research Center, University of Hawaii, Honolulu, Hawaii.

Hargis, D.R. and Peterson, F.L., "Effects of Well Injection on a Basaltic Ghyben-Herzberg Aquifer", Ground Water, Vol. 12, No. 1, Jan.-Feb. 1974, pp. 4-9.

Harmeson, R.H., Thomas, R.L. and Evans, R.L., "Coarse Media Filtration for Artificial Recharge", Journal of the American Water Works Association, Vol. 60, No. 12, Dec. 1968, pp. 1396-1403.

Harpaz, Y., "Artificial Ground-Water Recharge by Means of Wells in Israel", Journal of the Hydraulics Division, American Society of Civil Engineers, Vol. 97, No. HY12, 1971, pp. 1947-1964.

Hickey, J.J. and Barr, G.L., "Hydrogeologic Data for the Bear Creek Subsurface-Injection Test Site, St. Petersburg, Florida", Open File Report 78-853, 1979, U.S. Geological Survey, Tallahassee, Florida.

Hirashima, G.T., "Availability of Streamflow for Recharge of the Basal Aquifer in the Pearl Harbor Area, Hawaii", Water Supply Paper 1999-B, 1971, U.S. Geological Survey, Washington, D.C.

Hydrologic Engineering Center, "Survey of Conjunctive Use and Artificial Recharge Activity in the United States", Jan. 1984, U.S. Army Corps of Engineers, Davis, California.

International Association of Scientific Hydrology, "International Survey on Existing Water Recharge Facilities", Publication No. 87, 1970, Gentbrugge, Belgium.

Jenkins, C.T. and Hofstra, W.E., "Availability of Water for Artificial Recharge, Plains Groundwater Management District, Colorado", Groundwater Series Circular No. 13, 1970, Colorado Water Conservation Board, Denver, Colorado.

Jones, H.H., "Investigations for Artificial Recharge of the Triassic Sandstones Aquifer Near Stourbridge, U.K.", Presented at a meeting of the United Kingdom Institute of Water Engineers and Scientists, Mar. 1982.

Karlinger, M.R. and Hansen, A.J., Jr., "An Engineering Economic Analysis of Artificial Recharge in a Conjunctive Irrigation Plan in the Columbia Basin Project, Washington", Water Resource Investigation 581-67, 1983, U.S. Geological Survey, Tacoma, Washington.

Kimrey, J.O., "Preliminary Appraisal of the Geohydrologic Aspects of Drainage Wells, Orlando Area, Central Florida", Water Resources Investigations 78-37, May 1978, U.S. Geological Survey, Tallahassee, Florida.

Koehler, J.H., "Artificial Recharge in the Northern Part of Chino Ground-Water Basin, Upper Santa Ana Valley, California", Water Resources Investigations Report 82-4122, June 1983, U.S. Geological Survey, Menlo Park, California.

Ku, H.F.H., Vecchioli, J. and Ragone, S.E., "Changes in Concentration of Certain Constituents of Treated Waste Water During Movement Through the Magothy Aquifer, Bay Park, New York", _Journal of Research of the U.S. Geological Survey_, Vol. 3, No. 1, Jan.-Feb. 1975, pp. 89-92.

Lichtler, W.F., Stannard, D.I. and Kouma, E., "Artificial-Recharge Investigation Near Aurora, Nebraska, Two-Year Progress Report", Open File Report 79-1492, 1979, U.S. Geological Survey, Lincoln, Nebraska.

Lichtler, W.F., Stannard, D.I. and Kouma, E., "Investigation of Artificial Recharge of Aquifers in Nebraska", Water Resources Investigations 80-93, Sept. 1980, U.S. Geological Survey, Lincoln, Nebraska.

Marlette, R.R., "Artificial Recharge Through Injection Wells in a Sandstone Aquifer", _Groundwater, Proceedings of General Assembly of Bern (Sept.-Oct. 1967)_, Publication No. 77, 1968, International Association of Scientific Hydrology, pp. 300-308.

Motts, W.S. et al., "Feasibility of Increasing Water Supplies and Preventing Environmental Damage by Artificial Recharge in Massachusetts", Publication No. 132, Jan. 1983, Water Resources Research Center, University of Massachusetts, Amherst, Massachusetts.

Muir, K.S., "Seawater Intrusion, Ground-Water Pumpage, Ground-Water Yield, and Artificial Recharge of the Pajaro Valley Area, Santa Cruz and Monterey

Counties, California", Water Resources Investigations 9-74, Oct. 1974, U.S. Geological Survey, Menlo Park, California.

Nielsen, D.M. and Aller, L., "A Cost-Effective Water Supply Alternative for the Chicago Suburban Area-Artificial Ground Water Recharge", Proceedings of the National Water Well Association Eastern Regional Conference on Ground Water Management, Orlando, Florida, 1983, pp. 85-121.

Norvitch, R.F., Thomas, C.A. and Madison, R.J., "Artificial Recharge to the Snake Plain Aquifer in Idaho; An Evaluation of Potential and Effect", Water Information Bulletin No. 12, Aug. 1969, U.S. Geological Survey, Washington, D.C.

Pettyjohn, W.A. et al., "Preliminary Estimate of Effective Ground-Water Recharge Rates in Central Oklahoma", Final Report, Nov. 1982, Department of Geology, Oklahoma State University, Stillwater, Oklahoma.

Pettyjohn, W.A., "Introduction to Artificial Ground-Water Recharge", EPA-600/2-81-236, Oct. 1981, Robert S. Kerr Environmental Research Laboratory, U.S. Environmental Protection Agency, Ada, Oklahoma.

Ragone, S.E., "Geochemical Effects of Recharging the Magothy Aquifer, Bay Park, New York, with Tertiary-Treated Sewage", Professional Paper 751-D, 1977, U.S. Geological Survey, Mineola, New York.

Sherwani, J.K., "Public Policy for the Management of Groundwater in the Coastal Plain of North Carolina", Report No. 158, July 1980, Water Resources Research Institute, North Carolina State University, Raleigh, North Carolina.

Sinclair, W.C., "Hydrogeologic Characteristics of the Surficial Aquifer in Northwest Hillsborough County, Florida", Information Circular No. 86, 1974, Florida Bureau of Geology, Tallahassee, Florida.

Sinclair, W.C., "Experimental Study of Artificial Recharge Alternatives in Northwest Hillsborough County, Florida", Water Resources Investigations 77-13, May 1977, U.S. Geological Survey, Tallahassee, Florida.

Smith, A., "Nebraska Studies Ways to Recharge Aquifer Storage", Johnson Drillers Journal, Vol. 49, No. 5, Sept.-Oct. 1977, pp. 1-3.

Smith, C.G., Jr. and Hanor, J.S., "Underground Storage of Treated Water: A Field Test", Ground Water, Vol. 13, No. 5, 1975, pp. 410-417.

Sunada, D.K., Warner, J.W. and Molden, D.J., "Artificial Groundwater Recharge-San Luis Valley, Colorado", Completion Report No. 123, Mar. 1983, Colorado Water Resources Research Institute, Fort Collins, Colorado.

Supalla, R.J. and Comer, D.A., "An Economic Evaluation of the Feasibility of Artificial Ground Water Recharge in Nebraska", OWRT-A-065-NEB(1), Jan. 1981, Office of Water Research and Technology, U.S. Department of the Interior, Washington, D.C.

Sutherland, J.C. and Bruce, R.M., "Surface Water-Esker Recharge Study East Lansing-Meridian Township Michigan", Water Resources Bulletin, Vol. 13, No. 6, Dec. 1977, pp. 1215-1225.

Swain, L.A., "Predicted Water-Level and Water-Quality Effects of Artificial Recharge in the Upper Coachella Valley, California, Using a Finite-Element Digital Model", Water Resources Investigations 7-29, Apr. 1978, U.S. Geological Survey, Menlo Park, California.

Tibbals, C.H. and Frazee, J.M., Jr., "Ground-Water Hydrology of the Cocoa Well-Field Area, Orange County, Florida", Open File Report FL 75-676, Jan. 1976, U.S. Geological Survey, Tallahassee, Florida.

Trapp, H., Jr., Pascale, C.A. and Foster, J.B., "Water Resources of Okaloosa County and Adjacent Areas, Florida", Water Resources Investigation 77-9, Aug. 1977, U.S. Geological Survey, Tallahassee, Florida.

University City Science Institute, "A Study to Evaluate Potential Solutions to the Problem of Insuring an Adequate Supply of Water for New Castle County, Delaware", Submitted to the Committee of 100, Mar. 1971, Philadelphia, Pennsylvania.

Vanlier, K.E., Wood, W.W. and Burnett, J.O., "Water-Supply Development and Management Alternatives for Clinton, Eaton, and Ingham Counties, Michigan", Water Supply Paper 1969, 1973, U.S. Geological Survey, Washington, D.C.

Warner, J.W., Sunada, D.K. and McWhorter, D.B., "Potential for Artificial Recharge of the Northern High Plains of Colorado", Proceedings of the Ogallala Aquifer Symposium II, Texas Tech University, Lubbock, Texas, June 1984, pp. 350-366.

Wedderburn, L.A. and Knapp, M.S., "Artificial Recharge of Fresh Water into a Brackish Water Aquifer, St. Lucie County, Florida", Proceedings of the National Water Well Association Eastern Regional Conference on Ground Water Management, Orlando, Florida, 1983, pp. 44-71.

Wood, W.W. and Osterkamp, W.R., "Recharge to the Ogallala Aquifer from Playa Lake Basins on the Llano Estacado", Proceedings of the Ogallala Aquifer Symposium II, Texas Tech University, Lubbock, Texas, June 1984, pp. 337-349.

CHAPTER 5

POTENTIAL AREAS FOR ARTIFICIAL RECHARGE

This chapter is focused on the aggregation of information on potential areas for artificial recharge within the United States. Three levels of consideration are included: (1) national; (2) state; and (3) site specific. The national level information is based on a methodology developed for use herein. The methodology is focused on ground water usage for agricultural purposes; however, the general findings can be applied relative to any area of extensive ground water usage. Information from a state-wide study in Massachusetts is used as an illustration of state-level information. Information for determining the potential for artificial recharge at specific sites is largely summarized from Chapter 3.

NATIONAL INFORMATION ON POTENTIAL AREAS FOR ARTIFICIAL RECHARGE

The basic geographical area used for determining potential areas for artificial recharge is the water resource aggregated subarea (ASA) delineated in 1975 by the U.S. Water Resources Council. The continental United States is divided into 99 ASA's as shown in Figure 3. A list of counties by state and ASA is in Appendix B. It should be noted that some ASA's occur in two or more states.

Description of Procedure

The need for artificial recharge for irrigation purposes for 1985; 2000, and 2030 was established based on the following three items of information:

(1) identified ground water availability problems in the ASA as determined from the 1983 National Water Summary prepared by the U.S. Geological Survey;

(2) acres of cropland by ASA irrigated by ground water (this information is included in Table 9);

(3) total water usage for irrigation by state for 1985, 2000, and 2030 (this information is included in Table 10).

Table 9 is from state water use and socio-economic data related to the Second National Water Assessment. The data were developed by the Iowa State University Center for Agricultural Research and Development (CARD) using a linear programming model to solve for the least cost of production. As the starting point, CARD used acres in the 1969 census of agriculture as modified by the Second National Water Assessment work groups. Table 10 projections were done by the U.S. Department of Agricultural using a shift-share method involving extension of historical trends. The latest census information was used to make these National Interregional Agricultural Projections (NIRAP).

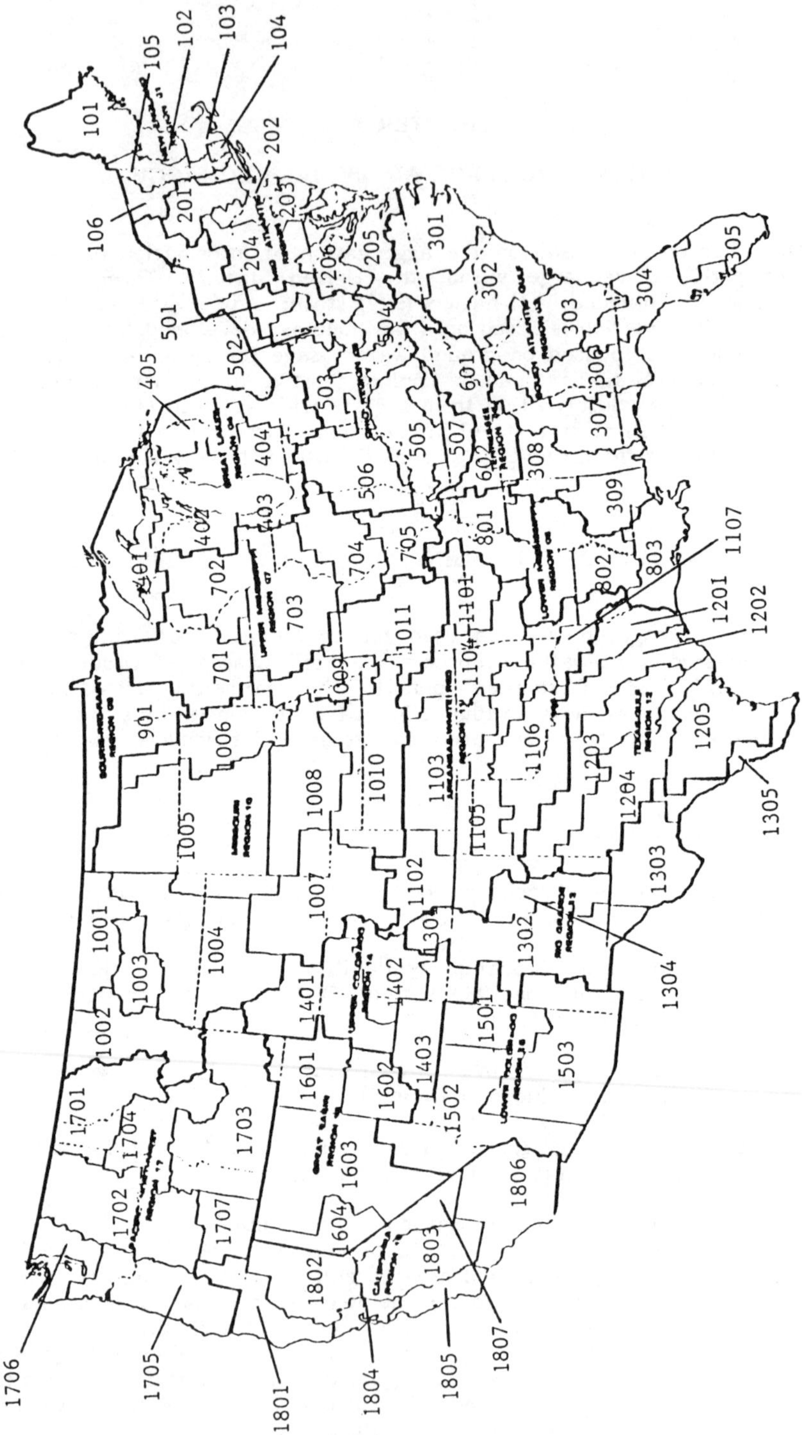

Figure 3: ASA's in the Continental United States

Table 9: Total Surface Area and Irrigated Acres for Each ASA

ASA	Total Surface Area	Total Irrigated Lands
	(1000 Acres)	
101	19968.4	3.0
102	4147.9	11.1
103	4253.1	21.9
104	3088.1	0.7
105	7936.3	9.1
106	4783.5	1.4
201	9712.5	4.1
202	3114.7	40.3
203	9908.7	127.4
204	17236.5	18.2
205	16240.6	100.3
206	9634.1	18.9
301	25668.9	306.2
302	27126.9	96.2
303	22800.8	303.9
304	18788.3	1060.7
305	12535.9	1502.2
306	18325.1	619.5
307	20889.8	75.9
308	15349.4	11.7
309	12840.0	0.0
401	17152.4	5.0
402	11227.7	119.2
403	5304.6	59.1
404	16845.5	389.9
405	8665.2	36.2
406	10468.7	25.1
407	5415.8	5.7
408	10952.4	13.5
501	9759.9	0.0
502	17932.5	4.8
503	13062.3	18.4
504	8312.9	0.0
505	20414.4	19.8
506	21538.2	65.1
507	11377.1	5.0
601	15764.5	10.8
602	11491.7	8.8

Table 9: (Continued)

701	28683.4	354.2
702	19735.9	185.2
703	35630.6	111.1
704	19601.6	138.0
705	12056.2	15.5
801	17959.6	2855.6
802	31195.2	1422.5
803	17252.9	1052.7
901	34987.9	94.7
1001	16406.8	205.8
1002	24043.2	998.3
1003	11017.7	193.3
1004	45828.6	1286.8
1005	61064.5	399.4
1006	23914.8	389.6
1007	38238.9	2079.2
1008	29950.1	3892.2
1009	14269.8	200.1
1010	35291.4	4003.4
1011	26261.8	120.2
1101	12810.9	281.6
1102	15841.7	519.2
1103	29680.1	2836.8
1104	25504.3	138.6
1105	29951.8	2051.0
1106	25824.5	1367.2
1107	16483.8	80.6
1201	11273.5	155.4
1202	17037.9	629.2
1203	29795.3	3535.8
1204	29988.5	1628.7
1205	25363.0	654.4
1301	5254.4	618.7
1302	34637.0	669.2
1303	21175.4	280.5
1304	13039.9	253.1
1305	10551.1	755.0
1401	29675.6	856.7
1402	16523.9	502.6
1403	19671.9	313.2
1501	17042.8	113.6
1502	40523.4	162.1
1503	44506.9	1154.9

Table 9: (Continued)

1601	17369.8	902.4
1602	13069.3	436.1
1603	46896.2	805.4
1604	11878.3	264.7
1701	23183.8	708.1
1702	38458.1	2250.2
1703	44014.2	4100.5
1704	20249.7	307.6
1705	24921.9	524.6
1706	10565.6	74.4
1707	11896.1	492.1
1801	15272.5	481.7
1802	20692.3	2297.5
1803	21022.4	5608.0
1804	4807.6	301.8
1805	7201.3	471.5
1806	27473.6	806.3
1807	9028.7	62.2

Table 10: Water Usage for Irrigation

State and Region	1978	Projections: 1982	Total 1985	Irrigation 2000	2030
Maine	7097	7097	7097	7097	7097
New Hampshire	1802	1802	1802	1802	1802
Vermont	1595	1595	1595	1595	1595
Massachusetts	16933	16933	16933	16933	16933
Rhode Island	2985	2985	2985	2985	2985
Connecticut	6953	6953	6953	6953	6953
New York	56444	56444	56444	56444	56444
New Jersey	77750	79871	81336	87511	96514
Pennsylvania	16500	16500	16500	16500	16500
Delaware	33726	42105	48389	79811	142654
Maryland	28724	35563	40692	66338	117631
*Northeast	250508	267847	280726	343968	467107
Michigan	226930	293187	342879	591343	1088269
Wisconsin	234556	276070	307206	462885	774243
Minnesota	271909	346404	402276	481634	1240351
*Lake States	733394	915661	1052361	1735862	3102862
Ohio	26317	27986	29172	34477	43118
Indiana	75352	113994	142975	287883	577698
Illinois	130174	165356	191743	323676	587542
Iowa	100856	114439	124627	175564	277438
Missouri	344332	427806	490412	803440	1429498
	677031	849580	978928	1625040	2915294
North Dakota	141435	173499	197546	521785	728541
South Dakota	341106	406999	456419	864065	1358266
Nebraska	5696214	6256437	6676605	8645441	12517115
Kansas	2684742	2949956	3148867	4150547	6139656
*Northern Plains	8863497	9786891	10479437	14181837	20743577
Virginia	43897	46791	48853	58112	73308
West Virginia	1236	1236	1236	1236	1236
North Carolina	93340	93340	93340	93340	93340
Kentucky	14537	19872	23873	43878	83889
Tennessee	14117	17624	20255	33407	59711
*Appalachian	167127	178863	187556	229973	311484
South Carolina	32639	61020	82306	188737	401599
Georgia	463523	605057	711208	1241960	2303464
Florida	1990335	1990335	1990335	1990335	1990335
Alabama	58923	73654	84702	139942	250423
*South East	2545420	2730066	2868551	3560974	4945821
Mississippi	309505	408546	482827	854233	1597044
Arkansas	1682954	1983481	2208877	3335855	5589810
Louisiana	681440	702357	716927	779347	873209
*Delta States	2673898	3094384	3408631	4969435	8060064

Table 10: (Continued)

Oklahoma	602326	602326	602326	605521	605521
Texas	6069158	6069158	6069158	6072608	6072608
*Southern Plains	6671483	6671483	6671483	6678128	6678128
Montana	2085601	2087897	2089619	2248514	2278514
Idaho	3508203	3688463	3823658	4481842	5773190
Wyoming	1684945	1710757	1730116	1730116	1730116
Colorado	3458143	3738143	3948143	3948143	3948143
New Mexico	904191	904191	904191	1026786	1026786
Arizona	1112118	1112118	1112118	1143618	1143618
Utah	1524302	1524302	1524302	1541852	1541852
Nevada	899540	907428	913344	913344	913344
*Mountain	15177044	15673300	16045491	17034216	18355564
Washington	1681377	1798830	1886921	2473098	3463802
Oregon	1920332	1973072	2012627	2195867	2562346
California	8603898	9316534	9851011	9851011	9851011
*Pacific	12205607	13088437	13750559	14519976	15877159
*Total: 48 States	49965010	53256514	55723726	64879409	81457061
Alaska	919	955	981	1091	1258
Hawaii	159317	159317	159317	159317	159317
*Revised Pacific	12365843	13248710	13910858	14680384	16037734
*Total: U.S.	50125246	53416787	55884024	65039818	81617636

Detailed information on the methodology developed for determining the needs and potential for artificial recharge is contained in Appendix C.

The left portion of Appendix C was used to delineate the need for artificial recharge based on the above-three information sources. Information in Appendix C is organized by ASA and associated state(s). If a quantity problem (ground water availability problem) was indicated in the 1983 National Water Summary, an X was placed in the first column in Appendix C. The quantity of irrigated lands in the ASA, in 1000 acres, as listed in Table 9 was entered in the second column in Appendix C. These two factors were then considered relative to whether or not there is a need for artificial recharge in 1985. The following rationale was used:

(1) If a quantity problem was indicated in any state occupied by the ASA, and if any irrigated lands occurred in the ASA, a P was assigned in the third column (first Sum column) in Appendix C. The P signifies a problem and thus a potential need for artificial recharge.

(2) If a quantity problem was not indicated in any state occupied by the ASA, irrespective of the acres of irrigated land, an X was assigned in the third column in Appendix C. The X signifies no identified problem and thus no potential need for artificial recharge for enhancing ground water resources for irrigation purposes.

The needs in the years 2000 and 2030 were based on calculating the percentage change from 1985 in the water usage for irrigation based on the information in Table 10. The Δ_1 column in Appendix C refers to the percentage change in water usage in 2000 as compared to 1985 usage. The fifth column (second Sum column) in Appendix C reflects whether an increase in ground water usage, and thus an increase in the need for artificial recharge, is expected for the ASA by 2000. If the calculated percentage change in ground water usage was less than 5 percent, an N was assigned to the ASA, with the N denoting no significant increase. If the calculated change was equal to or greater than 5 percent, an I was assigned to denote a significant increase in ground water usage, and thus an increased need for artificial recharge.

The right portion of Appendix C was used to delineate the technical feasibility of artificial recharge by ASA. This "technically feasible potential" for artificial recharge was based on three items of information:

(1) the Heath region(s) occupied by the ASA (the 12 Heath regions were categorized in accordance with hydrogeological features conducive to artificial recharge; this classification is described in the next sub-section);

(2) published literature describing previous, current, or planned artificial recharge projects by ASA (see Table 7); and

(3) numerous telephone contacts describing artificial recharge studies by ASA.

The first column under potential in Appendix C identifies the Heath region(s) by ASA and state, with these regions divided into those with Low(L), Moderate(M), and High(H) potential for artificial recharge. The rationale for these divisions is described in the next sub-section. If more than one region was identified, the highest potential was used in the right hand column. If field studies or verbal information was found, an X was placed in the respective columns. The Sum column for technical potential for artificial recharge uses the following codes:

L = low technical potential for artificial recharge based on Heath region characteristics

M = moderate technical potential based on Heath region

H = high technical potential based on Heath region

S = published information exists on field studies

V = verbal information existing on field studies or technical potential for artificial recharge.

Based on these codes, the following combinations could occur in the right-most column of Appendix C:

L, LS, LSV
M, MS, MSV
H, HS, HSV

Heath Region Classification

When determining the potential of an area to be responsive to artificial recharge, it is necessary to consider a great number of decision factors. These were discussed in Chapter 3 and include physical (hydrogeological) characteristics, source water availability, legal and institutional considerations, and socioeconomics. Most of the factors are highly site-specific; to obtain an overall national perspective on the potential for artificial recharge, it is necessary to narrow the decision factors considered.

Source water availability, legal and institutional considerations, and socioeconomics are all dependent on various situations and circumstances, and are subject to change. Although each of these could well be the determining factor for implementing a proposed project, it is beyond the scope of this chapter to examine all of the limitations imposed by such variable conditions on a nationwide scale. Water can be imported, needs can offset economic barriers, and laws can be changed. The physical characteristics of an area, though also subject to some degree of change, are more stable and a more reliable national indicator of artificial recharge potential.

In order to judge the potential responsiveness of an area to artificial recharge, a classification of High, Moderate, or Low was assigned to given divisions of the continental United States. The basis for the divisions was hydrogeological--Heath (1984) devised a methodology for delineating fifteen ground water regions for the United States, Puerto Rico and the Virgin Islands. The discussion hereafter will focus on the development of Heath's ground water region classification. Figures 4 and 5 display the regional classifications (Heath, 1984).

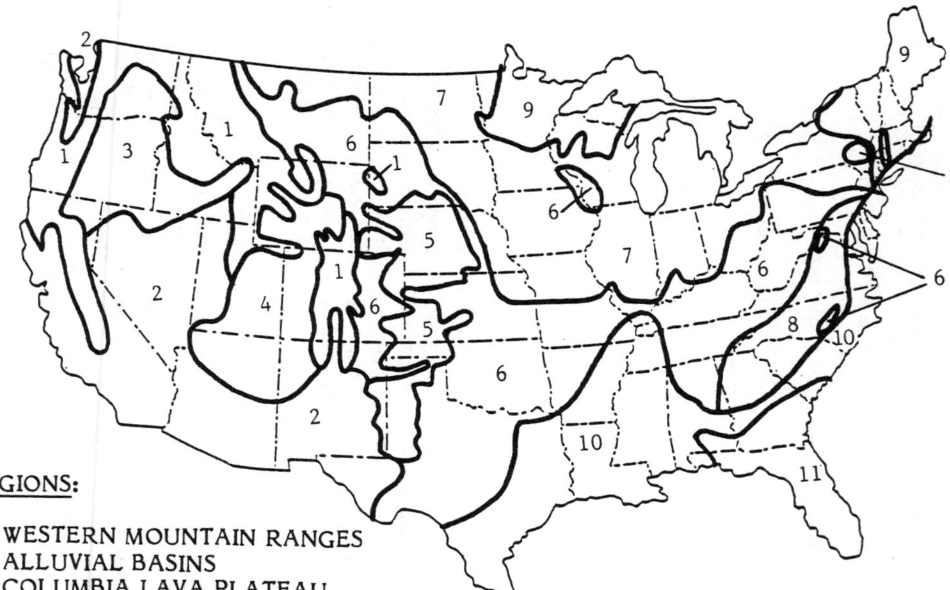

Figure 4: Heath's Ground Water Regions (See Also Figure 5) (Heath, 1984)

Figure 5: Alluvial Valleys Ground Water Region (Region 12) (Heath, 1984)

Heath (1984) identified four types of water-bearing rocks: sands and gravels; carbonates (limestone and dolomite); basalts; and sandstone. Sands and gravels are often the most common source of ground water, while limestone and dolomite aquifers tend to provide the largest discharges. The high solubility of carbonates allows for formation of large cavities in the rock where a large amount of water may be stored. Basalts are the most abundant of the volcanic rocks and form extensive lava flows in many regions. Small basaltic intrusives such as dikes are also commonly found. The composition of basalt is such that it is often brecciated, and contains many cracks or joints. They can be very productive in terms of ground water. Sandstones are most productive when they are only partially cemented. Fractures are also significant water conduits in these rocks.

The occurrence and availability of ground water is dependent on five features of the ground water system (Heath, 1984): (1) components of the hydrological system and their arrangement; (2) nature of the water-bearing openings; (3) mineral composition of the water-bearing rock; (4) water storage and transmission characteristics; and (5) nature and location of recharge and discharge areas. These features, the nature and extent of the dominant aquifers, and their relations to other units of the ground water system, were the criteria Heath used in delineating the regions.

In Table 11 each aspect of the five features, along with their range of conditions, is presented in addition to a brief comment related to each feature's significance to ground water (Heath, 1984). The first two of the features were primary criteria in the region classification. The remaining three were secondary criteria that were used to subdivide otherwise large and cumbersome areas into more homogeneous regions.

Each of the 12 regions, excluding Regions 13, 14, and 15, are presented in Table 12 according to where they fit among the range of conditions described in Table 11 (Heath, 1984). For each aspect of each feature, a dot (·) is placed under the condition which best describes the region. The last 3 regions are left out because only Regions 1 through 12 are found in the continental United States.

Table 13 provides a quantitative description of the regions, showing the geologic situation of the ranges of hydraulic characteristics for each region. For the purpose of determining artificial recharge potential, the features involving system components and their storage and transmissive characteristics were utilized as the principal indicators. As for the others, the nature of water-bearing openings is a determinant of porosity and transmissivity, mineral composition affects solubility which again affects porosity and transmissivity, and the nature and location of recharge and discharge areas, although important to keep in mind, will be more significant when choosing the specific site rather than in a national overview.

In Table 14, for simplicity, each condition for each aquifer type was assigned a letter:

	Aquifer Type	Condition
1.	Unconfined	(A) dominant aquifer (B) minor or not very productive (C) hydrologically insignificant

Table 11: Features of Ground Water Systems Useful in the Delineation of Ground Water Regions (Heath, 1984)

Feature	Aspect	Range in Conditions	Significance of Feature
Component of the system	Unconfined aquifer	Thin, discontinuous, hydrologically insignificant. Minor aquifer, serves primarily as a storage reservoir and recharge conduit for underlying aquifer. The dominant aquifer.	Affect response of the system to pumpage and other stresses. Affect recharge and discharge conditions. Determine susceptibility to pollution.
	Confining beds	Not present, or hydrologically insignificant. Thin, markedly discontinuous, or very leaky. Thick, extensive, and impermeable. Complexly interbedded with aquifers or productive zones.	
	Confined aquifers	Not present, or hydrologically insignificant. Thin or not highly productive. Multiple thin aquifers interbedded with nonproductive zones. The dominant aquifer--thick and productive.	
	Presence and arrangement of components	A single, unconfined aquifer. Two interconnected aquifers of essentially equal hydrologic importance. A three-unit system consisting of an unconfined aquifer, a confining bed, and confined aquifer.	

Table 11: (Continued)

Feature	Aspect	Range in Conditions	Significance of Feature
		A complexly interbedded sequence of aquifers and confining beds.	
Water-bearing openings of dominant aquifer	Primary openings	Pores in unconsolidated deposits. Pores in semiconsolidated rocks. Pores, tubes, and cooling fractures in volcanic (extrusive-igneous) rocks.	Control water-storage and transmission characteristics. Affect dispersion and dilution of wastes.
	Secondary openings	Fractures and faults in crystalline and consolidated sedimentary rocks. Solution-enlarged openings in limestones and other soluble rocks.	
Composition of rock matrix of of dominant aquifer	Insoluble	Essentially insoluble. Both relatively insoluble and soluble constituents.	Affects water-storage and transmission characteristics. Has major influence on water quality.
	Soluble	Relatively soluble.	
Storage and transmission characteristics of dominant aquifer	Porosity	Large, as in well-sorted, unconsolidated deposits. Moderate, as in poorly-sorted unconsolidated deposits and semiconsolidated rocks. Small, as in fractured crystalline and consolidated sedimentary rocks.	Control response to pumpage and other stresses. Determine yield of wells. Affect long-term yield of system. Affect rate at which pollutants move.

Table 11: (Continued)

Feature	Aspect	Range in Conditions	Significance of Feature
	Transmissivity	Large, as in cavernous limestones, some lava flows, and clean gravels. Moderate, as in well-sorted, coarse-grained sands, and semiconsolidated limestones. Small, as in poorly-sorted, fine-grained deposits and most fractured rocks. Very small, as in confining beds.	
Recharge and discharge conditions of dominant aquifer	Recharge	In upland areas between streams, particularly in humid regions. Through channels of losing streams. Largely or entirely by leakage across confining beds from adjacent aquifers.	Affect response to stress and long-term yields. Determine susceptibility to pollution. Affect water quality.
	Discharge	Through springs or by seepage to stream channels, estuaries, or the ocean. By evaporation on flood plains and in basin "sinks." By seepage across confining beds into adjacent aquifers.	

Table 12: Summary of the Physical and Hydrologic Characteristics of the Ground Water Regions of the United States (Heath, 1984)

Region No.	Name	Components of the system														Characteristics of the dominant aquifers																				
																Water-bearing openings					Composition			Storage and transmission properties							Recharge and discharge conditions					
		Unconfined aquifer			Confining beds			Confined aquifers				Presence and arrangement				Primary			Secondary		Degree of solubility			Porosity			Transmissivity				Recharge		Discharge			
		Hydrologically insignificant	Minor aquifer or not very productive	Dominant aquifer	Hydrologically insignificant	Thin, discontinuous, or v. leaky	Interlayered with aquifers	Hydrologically insignificant	Not highly productive	Multiple productive aquifers	The dominant productive aquifer	Single unconfined aquifer	Two interconnected aquifers	Unconfined aquifer, confining bed, confined aquifer	Complex interbedded sequence	Pores in unconsolidated deposits	Pores in semiconsolidated rocks	Tubes and cooling cracks in lava	Fractures and faults	Solution-enlarged openings	Insoluble	Mixed soluble and insoluble	Soluble	Large (>0.2)	Moderate (0.01–0.2)	Small (<0.01)	Large (>2,500 $m^2 day^{-1}$)	Moderate (250–2,500 $m^2 day^{-1}$)	Small (25–250 $m^2 day^{-1}$)	Very small (<25 $m^2 day^{-1}$)	Uplands between streams	Losing streams	Leakage through confining beds	Springs and surface seepage	Evaporation and basin sinks	Into other aquifers
1	Western Mountain Ranges		●		●				●				●			●			●		●				●				●		●	●		●		
2	Alluvial Basins			●			●			●					●	●					●			●			●					●			●	●
3	Columbia Lava Plateau		●				●			●					●	●		●	●		●					●	●					●		●		
4	Colorado Plateau and Wyoming Basin	●					●			●					●		●		●		●					●			●		●		●	●		
5	High Plains			●	●			●				●				●					●			●			●					●		●		
6	Nonglaciated Central Region		●				●			●					●		●		●	●		●				●		●			●		●	●		●
7	Glaciated Central Region		●				●			●					●	●	●		●	●		●			●			●					●			●
8	Piedmont and Blue Ridge		●		●				●				●			●			●		●					●				●	●			●		
9	Northeast and Superior Uplands		●			●			●				●			●			●		●					●			●		●			●		
10	Atlantic and Gulf Coastal Plain		●				●			●					●	●				●		●			●			●			●		●	●		●
11	Southeast Coastal Plain		●				●				●			●		●				●			●	●			●				●		●	●		
12	Alluvial Valleys			●		●		●				●				●					●			●			●				●			●		

Table 13: Common Ranges on the Hydraulic Characteristics of Ground Water Regions of the United States (All Values Rounded to one Significant Figure) (Heath, 1984)

			Common Ranges in Hydraulic Characteristics of the Dominant Aquifers							
			Transmissivity		Hydraulic Conductivity		Recharge Rate		Well Yield	
Region No.	Region	Geologic Situation	$m^2 day^{-1}$	$ft^2 day^{-1}$	$m\ day^{-1}$	$ft\ day^{-1}$	$mm\ yr^{-1}$	$in.\ yr^{-1}$	$m^3 min^{-1}$	$gal\ min^{-1}$
1	Western Mountain Ranges	Mountains with thin soils over fractured rocks, alternating with narrow alluvial and, in part, glaciated valleys	- 100	5-5,000,000	0.0003- 15	0.001- 50	3 - 50	0.1 - 2	0.04- 0.4	10- 100
2	Alluvial Basins	Thick alluvial (locally glacial) deposits in basins and valleys bordered by mountains	20 - 20,000	2,000- 200,000	30 - 600	100 - 2,000	0.03- 30	0.001- 1	0.4 -20	100- 5,000
3	Columbia Lava Plateau	Thick sequence of lava flows interbedded with unconsolidated deposits and overlain by thin soils	2,000 -500,000	20,000-5,000,000	200 -3,000	500 -10,000	5 -300	0.2 -10	0.4 -80	100-20,000
4	Colorado Plateau and Wyoming Basin	Thin soils over fractured sedimentary rocks	0.5- 100	5- 1,000	0.003 - 2	0.01 - 5	0.3 - 50	0.01 - 2	0.04- 2	10- 1,000
5	High Plains	Thick alluvial deposits over fractured sedimentary rocks	1,000 - 10,000	10,000- 100,000	30 - 300	100 - 1,000	5 - 80	0.2 - 3	0.4 -10	100- 3,000
6	Nonglaciated Central Region	Thin regolith over fractured sedimentary rocks	300 - 10,000	3,000- 100,000	3 - 300	10 - 1,000	5 -500	0.2 -20	0.4 -20	100- 5,000
7	Glaciated Central Region	Thick glacial deposits over fractured sedimentary rocks	100 - 2,000	1,000- 20,000	2 - 300	5 - 1,000	5 -300	0.2 -10	0.2 - 2	50- 500
8	Piedmont and Blue Ridge	Thick regolith over fractured crystalline and metamorphosed sedimentary rocks	9 - 200	100- 2,000	0.001 - 1	0.003- 3	30 -300	1 -10	0.2 - 2	50- 500
9	Northeast and Superior Uplands	Thick glacial deposits over fractured crystalline rocks	50 - 500	500- 5,000	2 - 30	5 - 100	30 -300	1 -10	0.1 - 1	20- 200
10	Atlantic and Gulf Coastal Plain	Complexly interbedded sands, silts, and clays	500 - 10,000	5,000- 100,000	3 - 100	10 - 400	50 -500	2 -20	0.4 -20	100- 5,000
11	Southeast Coastal Plain	Thick layers of sand and clay over semiconsolidated carbonate rocks	1,000 -100,000	10,000-1,000,000	30 -3,000	100 -10,000	30 -500	1 -20	4 -80	1,000-20,000
12	Alluvial Valleys	Thick sand and gravel deposits beneath flood plains and terraces of streams	200 - 50,000	2,000- 500,000	30 -2,000	100 - 5,000	50 -500	2 -20	0.4 -20	100- 5,000

Table 14: Receptivity to Artificial Recharge by Heath Region

Heath Region	System Description*			Storage and Transmission Characteristics of Dominant Aquifer*		Summation	Receptivity to Artificial* Recharge
	Unconfined Aquifers	Confined Aquifers	Aquifer Combination Value	Porosity	Transmissivity	Σ	
1	B	C	4	M	S	4MS	LOW
2	A	B	1	L	L	1LL	HIGH
3	B	B	3	S	L	3SL	MODERATE
4	C	B	3	S	S	3SS	LOW
5	A	D	2	L	L	2LL	HIGH
6	B	B	3	S	M	3SM	MODERATE
7	B	B	3	M	M	3MM	MODERATE
8	B	C	4	S	V	4SV	LOW
9	B	C	4	S	V	4SV	LOW
10	B	B	3	M	M	3MM	MODERATE
11	B	A	3	L	L	3LL	HIGH
12	A	D	2	L	L	2LL	HIGH

*Key:

1. System Description

	Aquifer Type	Condition
	1. Unconfined	(A) Dominant Aquifer
		(B) Minor or not very productive
		(C) Hydrologically insignificant

Table 14: (Continued)

	2. Confined	(A) Dominant aquifer (thick) (B) Multiple thin and productive (C) Not highly productive (D) Hydrologically insignificant

	Aquifer Type	Combinations of Conditions				
	1. Unconfined	(A) with	(A) with	(B)/(C) with	(B) with	(C) with
	2. Confined	(A)/(B)	(C)/(D)	(A)/(B)	(B)/(C)	(C)/(D)
	Combination Value:	1	2	3	4	5
		most favorable			least favorable	

2. Storage and Transmission Characteristics

Porosity	(L) Large (>0.2) (M) Moderate (0.01-0.2) (S) Small (<0.01)
Transmissivity	(L) Large (>2500 M^2/day) (M) Moderate (250-2,500 M^2/day) (S) Small (25-250 M^2/day) (V) Very Small (<25 M^2/day)

3. Receptivity Determination

Variables	High	Moderate	Low
System	1,2,3 with	1,2,3,4,5 with	3,4,5 with
Storage and transmission	LL,ML	LM,LS,LV, MM,SL,SM	MS,MV SS,SV

2. Confined
   - (A) dominant productive aquifer
   - (B) multiple, thin, and productive
   - (C) not highly productive
   - (D) hydrologically insignificant

Each region has a condition defined for each aquifer type. The combination of conditions for the unconfined and confined aquifers in a region give an indication of aquifer receptivity. For example, an area where the unconfined aquifer is dominant (A) and the confined aquifer is multiple, thin and productive (B) indicates that this area has been a high yield region. If the water in storage needs replenishing, this area will likely readily accept artificially-induced recharge. On the other hand, where neither the unconfined or confined aquifers in a region are hydrologically significant ((B) and (D), respectively), it is to be assumed that the region would not easily support an artificial recharge operation.

Other combinations are possible, and these have been evaluated for most favorable to least favorabale conditions for artificial recharge. Unconfined aquifers have been assumed to be better than confined aquifers for artificial recharge because of the ease of infiltration methods over injection.

Aquifer Type	Combination of Conditions				
1. Unconfined	(A) with	(A) with	(B) or (C) with	(B) with	(C) with
2. Confined	(A) or (B)	(C) or (D)	(A) or (B)	(B) or (C)	(C) or (D)
Combination Value	1 (most favorable)	2	3	4	5 (least favorable)

The porosity and transmissivity descriptions for Table 14 were taken directly from Table 12 with an assignment from large (L) to very small (V) (see key, Table 14). The letter from each was then placed together along with the combination value from the system description. For example, "1LL" indicates highly productive aquifers with large porosity and transmissivity values in that particular Heath region.

The receptivity of an area was then judged depending on these three indicators. It was arbitrarily decided where the divisions of HIGH, MODERATE and LOW were to be assigned, depending on the combination of system description and storage and transmission characteristics. For example; "3LL" would describe an area of high receptivity whereas "3MM" would be of moderate receptivity (see key in Table 14).

Summary of Findings

Table 15 summarizes the code information in Appendix C. The right side of Table 15, entitled status, combines the codes from both need and potential. For example, for ASA 101, the following explanations relate to the three status columns:

Table 15: Summary of Codes by ASA

ASA	Need 1985	Need 2000	Need 2030	Potential	Status 1985	Status 2000	Status 2030
101	P	N	N	L	PL	PNL	PNL
102	P	N	N	L	PL	PNL	PNL
103	P	N	N	MSV	PMSV	PNMSV	PNMSV
104	P	N	N	M	PM	PNM	PNM
105	X	N	N	L	XL	XNL	XNL
106	X	N	N	M	XM	XNM	XNM
201	P	N	N	M	PM	PNM	PNM
202	P	I	I	MSV	PMSV	PIMSV	PIMSV
203	P	I	I	MSV	PMSV	PIMSV	PIMSV
204	X	N	N	M	XM	XNM	XNM
205	P	I	I	MS	PMS	PIMS	PIMS
206	P	I	I	M	PM	PIM	PIM
301	P	N	N	MS	PMS	PNMS	PNMS
302	P	I	I	H	PH	PIH	PIH
303	P	I	I	H	PH	PIH	PIH
304	P	I	I	HSV	PHSV	PIHSV	PIHSV
305	P	N	N	HS	PHS	PNHS	PNHS
306	P	I	I	H	PH	PIH	PIH
307	P	I	I	HS	PHS	PIHS	PIHS
308	P	I	I	M	PM	PIM	PIM
309	P	I	I	M	PM	PIM	PIM
401	P	I	I	L	PL	PIL	PIL
402	P	I	I	M	PM	PIM	PIM
403	P	I	I	MSV	PMSV	PIMSV	PIMSV
404	P	I	I	MS	PMS	PIMS	PIMS
405	X	N	N	MS	XMS	XNMS	XNMS
406	P	I	I	M	PM	PIM	PIM
407	P	N	N	M	PM	PNM	PNM
408	X	N	N	M	XM	XNM	XNM
501	P	N	N	M	PM	PNM	PNM
502	P	I	I	M	PM	PIM	PIM
503	P	I	I	MS	PMS	PIMS	PIMS
504	P	I	I	MS	PMS	PIMS	PIMS
505	P	I	I	MV	PMV	PIMV	PIMV
506	P	I	I	M	PM	PIM	PIM
507	P	I	I	M	PM	PIM	PIM
601	P	I	I	M	PM	PIM	PIM
602	P	I	I	M	PM	PIM	PIM
701	P	I	I	MS	PMS	PIMS	PIMS
702	P	I	I	MS	PMS	PIMS	PIMS
703	P	I	I	MV	PMV	PIMV	PIMV
704	P	I	I	MSV	PMSV	PIMSV	PIMSV
705	X	I	I	M	XM	XIM	XIM
801	P	I	I	MS	PMS	PIMS	PIMS
802	P	I	I	M	PM	PIM	PIM
803	P	I	I	MS	PMS	PIMS	PIMS
901	P	I	I	MSV	PMSV	PIMSV	PIMSV

Table 15: (Continued)

1001	P	I	I	M	PM	PIM	PIM
1002	X	I	I	M	XM	XIM	XIM
1003	X	I	I	M	XM	XIM	XIM
1004	P	I	I	MS	PMS	PIMS	PIMS
1005	P	I	I	HS	PHS	PIHS	PIHS
1006	P	I	I	MS	PMS	PIMS	PIMS
1007	P	N	N	HSV	PHSV	PNHSV	PNHSV
1008	P	N	N	HSV	PHSV	PNHSV	PNHSV
1009	P	I	I	MV	PMV	PIMV	PIMV
1010	P	I	I	HSV	PHSV	PIHSV	PIHSV
1011	P	I	I	MV	PMV	PIMV	PIMV
1101	X	I	I	MS	XMS	XIMS	XIMS
1102	P	N	N	HSV	PHSV	PNHSV	PNHSV
1103	P	I	I	HSV	PHSV	PIHSV	PIHSV
1104	P	I	I	MS	PMS	PIMS	PIMS
1105	P	I	I	HSV	PHSV	PIHSV	PIHSV
1106	P	N	N	HSV	PHSV	PNHSV	PNHSV
1107	P	I	I	M	PM	PIM	PIM
1201	P	I	I	M	PM	PIM	PIM
1202	P	N	N	MS	PMS	PNMS	PNMS
1203	P	I	I	HSV	PHSV	PIHSV	PIHSV
1204	P	I	I	HSV	PHSV	PIHSV	PIHSV
1205	P	N	N	MS	PMS	PNMS	PNMS
1301	P	N	N	HS	PHS	PNHS	PNHS
1302	P	I	I	HS	PHS	PIHS	PIHS
1303	P	N	N	HS	PHS	PNHS	PNHS
1304	X	I	I	HS	XHS	XIHS	XIHS
1305	P	N	N	M	PM	PNM	PNM
1401	X	N	N	HS	XHS	XNHS	XNHS
1402	X	N	N	L	XL	XNL	XNL
1403	P	I	I	L	PL	PIL	PIL
1501	P	I	I	H	PH	PIH	PIH
1502	P	N	N	H	PH	PNH	PNH
1503	P	I	I	HS	PHS	PIHS	PIHS
1601	P	I	I	HSV	PHSV	PIHSV	PIHSV
1602	P	N	N	H	PH	PNH	PNH
1603	P	N	N	H	PH	PNH	PNH
1604	P	N	N	H	PH	PNH	PNH
1701	X	I	I	M	XM	XIM	XIM
1702	P	I	I	MS	PMS	PIMS	PIMS
1703	P	I	I	HS	PHS	PIHS	PIHS
1704	X	I	I	M	XM	XIM	XIM
1705	P	I	I	HSV	PHSV	PIHSV	PIHSV
1706	P	I	I	H	PH	PIH	PIH
1707	X	I	I	M	XM	XIM	XIM
1801	X	I	I	HS	XHS	XIHS	XIHS
1802	P	N	N	H	PH	PNH	PNH
1803	P	N	N	HSV	PHSV	PNHSV	PNHSV
1804	P	N	N	HS	PHS	PNHS	PNHS
1805	P	N	N	HS	PHS	PNHS	PNHS
1806	P	N	N	HSV	PHSV	PNHSV	PNHSV
1807	P	N	N	H	PH	PNH	PNH

1985: PL = present problem indicating need for artificial recharge; low technical potential for recharge based on Heath region.

2000: PNL = present problem but not worsening to year 2000; low technical potential for recharge based on Heath region.

2030: PNL = present problem but not worsening to year 2030; low technical potential for recharge based on Heath region.

Table 15 provides information on the current and future status of artificial recharge by aggregated subareas (ASA). The information can also be shown by several figures. Figure 6 illustrates the ground water availability problems in the continental United States. The future increases (≥ 5% in irrigated water usage for years 2000 and 2030 are presented in Figure 7. The artificial recharge studies, demonstrations, pilot projects, and projects (Figure 20), obtained from the literature and/or from the telephone contacts with various agencies, (Appendix C), are illustrated in Figure 8. Note that Figure 8 only shows that there is an indication of artificial recharge activity in the ASA.

The data in Appendix C and Table 15 is basically divided into two major categories. The first category reflects the need for artificial recharge, including current ground water availability problems and future demands for irrigated water usage. The Heath region classification as the second category consists of the physical characteristics of the ASA (hydraulic conductivity, porosity, type of aquifer, etc.). The first question one must ask is which of the two categories is more important, and how much more? The answer to this question is a key factor that affects the final conclusion regarding artificial recharge potentiality in a given ASA, and especially those that fall close to the border line between two different levels of artificial recharge potential. Various factors and considerations, discussed earlier in Chapter 3, must be accounted for to reach an acceptable decision with respect to a given artificial recharge project. However, to delineate artificial recharge potential at the national level, a six-group classification scheme by ASA was developed. The scheme ranges from A to F, with A being the most favorable and F the least favorable. The category importance issue is dealt with by analyzing and interpreting the data based on three separate considerations: (1) the need is equally important to the physical characteristics (Heath region classification); (2) the need is more important than the physical characteristics by a ratio of 1 to 1/2; and (3) the need is less important than the physical characteristics by a ratio of 1/2 to 1. Using the paired comparison technique (Dean and Nishry, 1965), each of the various factors (P,X,I,N,H,M,L) discussed earlier is given a factor of importance (weight) as shown in Table 16. Table 17 illustrates all the possible combinations (status) between the factors for 1985, 2000, and 2030. Then, for each combination, the weights of the factors involved are added. The combinations (status) are then classified (A to F) according to their values as shown in Tables 18 and 19. Note that judgment is used to delineate between two different classes; for example, A and B.

The data regarding artificial recharge studies, demonstrations, pilot projects, projects and telephone contacts were not utilized in the category importance calculation because it is felt that some bias might be caused. For example, a study might show that artificial recharge is not feasible or

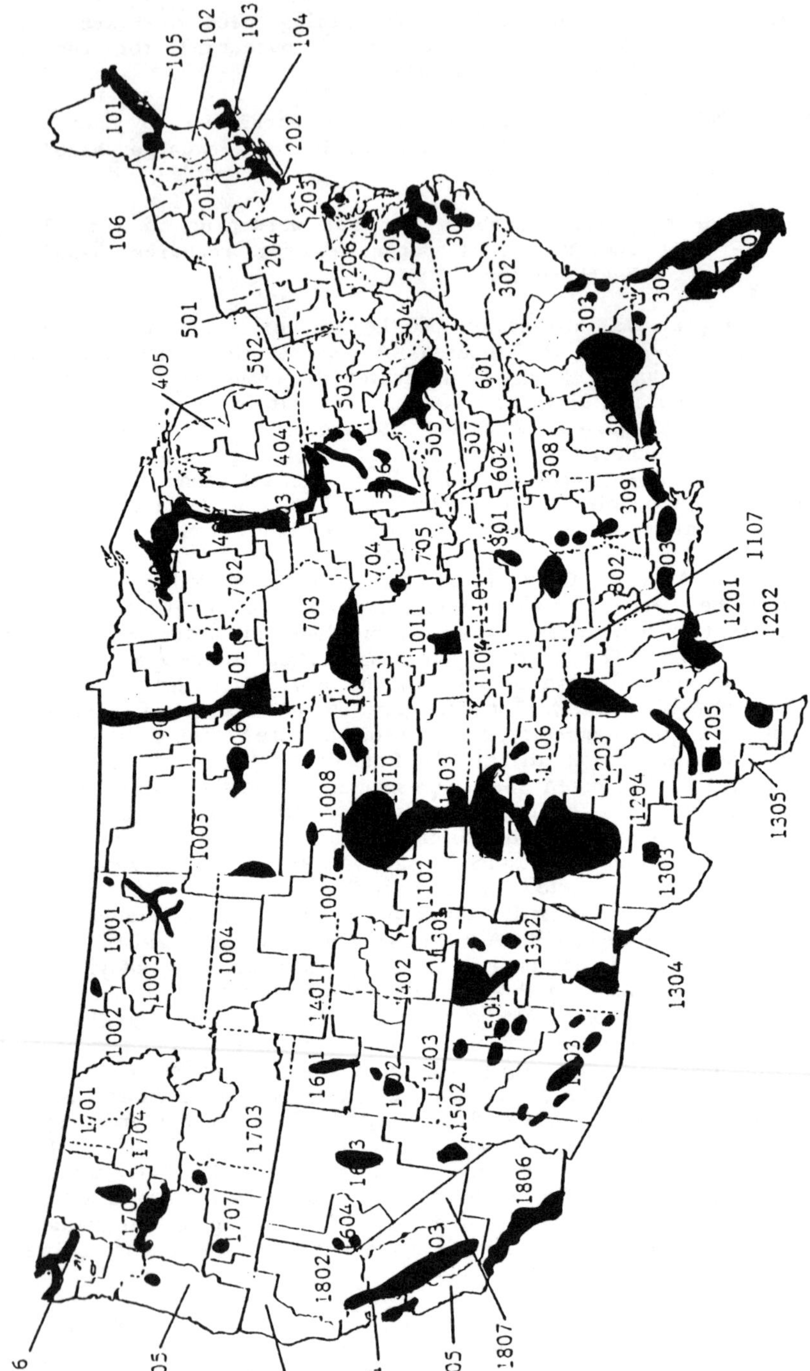

Figure 6: Ground Water Availability Problems in the Continental United States (Information is Obtained from the National Water Summary, 1983)

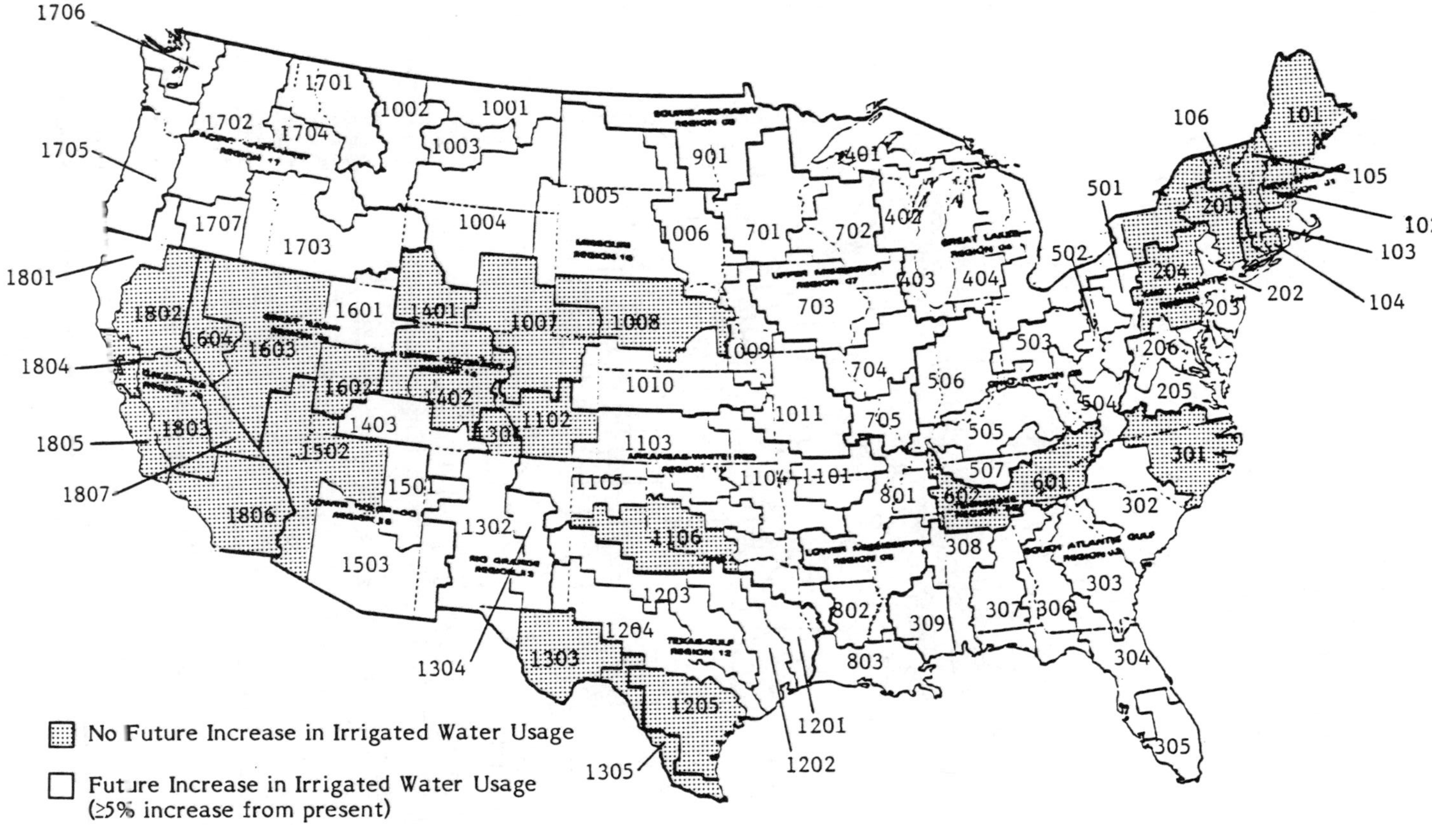

Figure 7: Expected Changes in Irrigated Water Usage

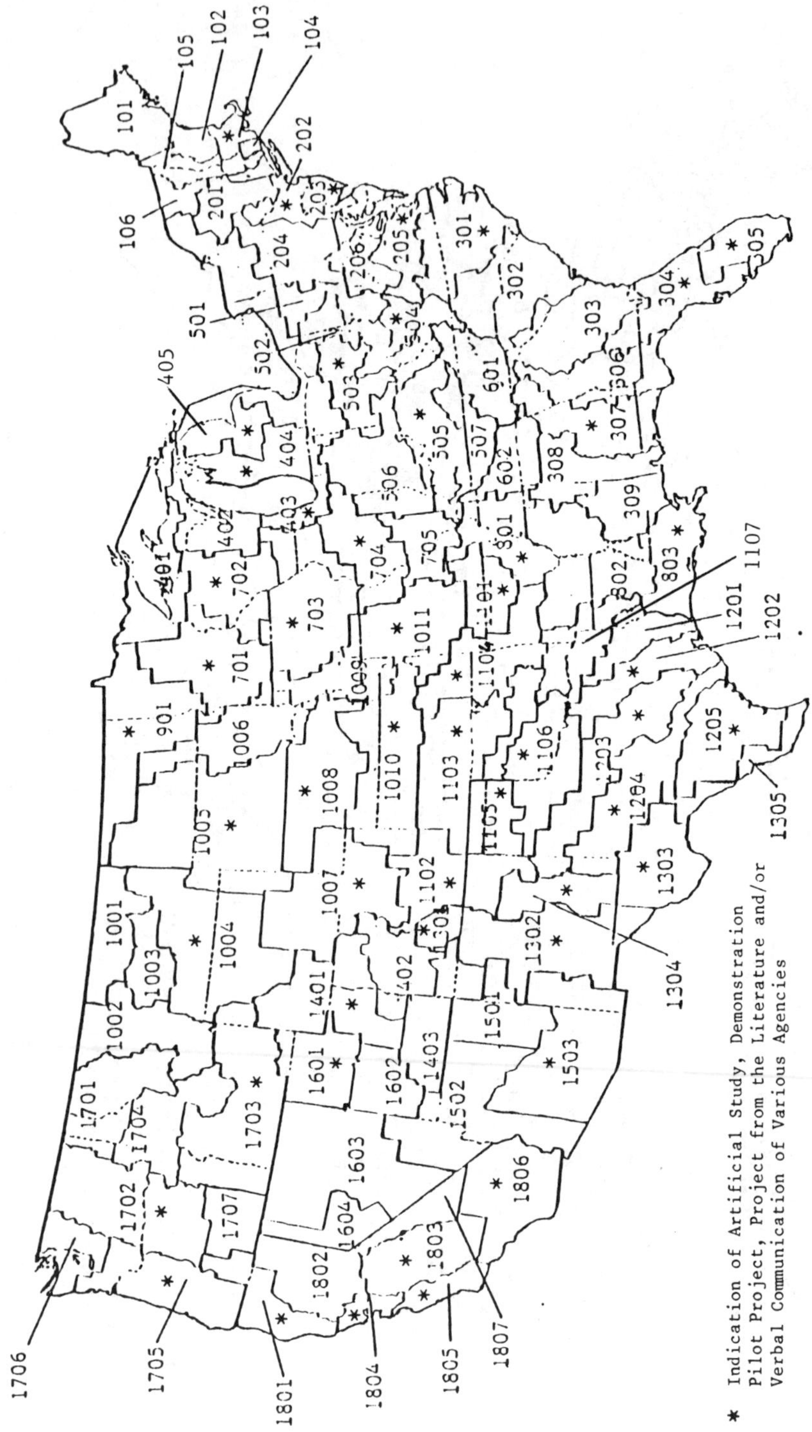

Figure 8: Indications of Artificial Recharge Activities

Table 16: Importance Weights for Methodology Factors

Need Characteristics												Weight	
	Factors Assignment of Weights										Sum	Physical Characteristics are Equally Important	Physical Characteristics are More Important (Ratio 1/.5)
P	1	1	1	1							4	0.40	0.20
X	0				0	½	1				1.5	0.15	0.08
I		0			1			1	1		3	0.30	0.15
N			0			½		0		1	1.5	0.15	0.07
0				0			0		0	0	0	0.00	0.00

Physical Characteristics (Heath Regions)

	Factors Assignment of Weights					Sum	Need is Equally Important	Need is More Important (Ratio 1/.5)
H	1	1	1			3	0.50	0.25
M	0			1	1	2	0.33	0.17

L	0		0		1	1	0.17	0.08
O		0		0	0	0	0.00	0.00

O is a dummy and is the least important of all factors.
P = Ground water availability problems.
X = No ground water availability problems.
I = Future increase in irrigated water usage.
N = No future increase in irrigated water usage.
H = High potential for artificial recharge.
M = Moderate potential for artificial recharge.
L = Low potential for artificial recharge.

Example

P is more important than X, therefore P scores 1
H is more important than L, therefore H scores 1
X is equally important as N, therefore X and N both score ½

Table 17: Possible Combinations of Factors P,X,I,N,H,M,L

1985	PH, XH, PM, XM, PL, XL,
2000	PIH, XIH, PNH, XNH, PIM, XIM, PNM, XNM, PIL, XIL, PNL, XNL
2030	PIH, XIH, PNH, XNH, PIM, XIM, PNM, XNM, PIL, XIL, PNL, XNL

unsuccessful in a specific area; however, artificial recharge could still be successful in a nearby area within the same ASA. Therefore, this information could be regarded as additional indications for the sites considered in the studies (or demonstrations, pilot projects, etc.).

Application of the classifications shown in Tables 18 and 19 to the summary information in Table 15 was then made. The results are summarized in Table 20. Tables 21 through 26 summarize the classifications by ASA's and irrigated lands. The total irrigated land in the continental United States is about 65.5 million acres. Table 27 summarizes the number of ASA's in each classification group.

If classifications A and B are considered as highly favorable for artificial recharge, when the need is equal in importance to the physical characteristics, 79 ASA's are highly favorable for 1985, and 70 ASA's for the years 2000 and 2030. The total acreages encompassed are 60.0 million in 1985, and 57.7 million in 2000 and 2030. When the need is greater in importance than the physical charcteristics, 79 ASA's are highly favorable for 1985, and 72 ASA's for the years 2000 and 2030. The total acreages are again 60.0 million for 1985, while 2000 and 2030 encompass 57.9 million acres. When the need is considered to be less in importance than the physical characteristics, 38 ASA's are highly favorable for 1985, and 37 ASA's for the years 2000 and 2030. The total acreages for these conditions are 47.2 million for 1985, and 43.1 million for 2000 and 2030.

In summary, and depending upon the assumptions, it appears that between 43 and 70 million acres of irrigated land are located in areas that have highly favorable conditions for artificial recharge. The locations of the ASA's for the various assumptions are shown in Figures 9 through 14.

STATE APPROACH FOR DETERMINING POTENTIAL AREAS FOR ARTIFICIAL RECHARGE

Artificial recharge potential can be determined for more finite areas than the continental United States. One illustration is a statewide study in Massachusetts (Motts et al., 1983). The objective of the study was two-fold; first, to evaluate the feasibility of artificial recharge on a statewide basis; and second, to study in depth site-specific artificial recharge methods. The study was designed to fulfill these objectives by three levels of study-intensive research called "level approximation methods". The First Level Approximation consisted of a statewide survey evaluation (an overview). A more detailed investigation was done at the Second Level Approximation, including a greater in-depth study of the towns in the Ipswich basin. Finally, the Third Level

Table 18: Status Classification for 1985

Need and Physical Characteristics are Equally Important

Importance Weight Sum	Status Classification*
PH = 0.40 + 0.50 = 0.90	PH A
XH = 0.15 + 0.50 = 0.65	PM B
PM = 0.40 + 0.33 = 0.73	XH C
XM = 0.15 + 0.33 = 0.48	PL D
PL = 0.40 + 0.17 = 0.57	XM E
XL = 0.15 + 0.17 = 0.32	XL F

Need is More Important than the Physical Characteristics (1/0.5)

Importance Weight Sum	Status Classification
PH 0.40 + 0.25 = 0.65	PH A
XH 0.15 + 0.25 = 0.40	PM B
PM 0.40 + 0.17 = 0.57	PL C
XM 0.15 + 0.17 = 0.32	XH D
PL 0.40 + 0.08 = 0.48	XM E
XL 0.15 + 0.08 = 0.23	XL F

Need is Less Important than the Physical Characteristics (0.5/1)

Importance Weight Sum	Status Classification
PH 0.20 + 0.50 = 0.70	PH A
XH 0.08 + 0.50 = 0.58	XH B
PM 0.20 + 0.33 = 0.53	PM C
XM 0.08 + 0.33 = 0.41	XM D
PL 0.20 + 0.17 = 0.37	PL E
XL 0.08 + 0.17 = 0.25	XL F

*A = most favorable for artificial recharge
F = least favorable for artificial recharge

Table 19: Status Classification for 2000 and 2030

Need and Physical Characteristics are Equally Important

Importance Weight Sum	Status Classification*	
PIH = 0.40 + 0.30 + 0.50 = 1.20	PIH	A
XIH = 0.15 + 0.30 + 0.50 = 0.95	PNH	B
PNH = 0.40 + 0.15 + 0.50 = 1.05	PIM	B
XNH = 0.15 + 0.15 + 0.50 = 0.80	XIH	C
PIM = 0.40 + 0.30 + 0.33 = 1.03	PNM	C
XIM = 0.15 + 0.30 + 0.33 = 0.78	PIL	C
PNM = 0.40 + 0.15 + 0.33 = 0.88	XNH	D
XNM = 0.15 + 0.15 + 0.33 = 0.63	XIM	D
PIL = 0.40 + 0.30 + 0.17 = 0.87	PNL	E
XIL = 0.15 + 0.30 + 0.17 = 0.62	XNM	E
PNL = 0.40 + 0.15 + 0.17 = 0.72	XIL	E
XNL = 0.15 + 0.15 + 0.17 = 0.47	XNL	F

Need is More Important than the Physical Characteristics (1/0.5)

Importance Weight Sum	Status Classification	
PIH = 0.40 + 0.30 + 0.25 = 0.95	PIH	A
XIH = 0.15 + 0.30 + 0.25 = 0.70	PIM	B
PNH = 0.40 + 0.15 + 0.25 = 0.80	PNH	B
XNH = 0.15 + 0.15 + 0.25 = 0.55	PIL	B
PIM = 0.40 + 0.30 + 0.17 = 0.87	PNM	C
XIM = 0.15 + 0.30 + 0.17 = 0.62	XIH	C
PNM = 0.40 + 0.15 + 0.17 = 0.72	PNL	D
XNM = 0.15 + 0.15 + 0.17 = 0.47	XIM	D
PIL = 0.40 + 0.30 + 0.08 = 0.78	XNH	E
XIL = 0.15 + 0.30 + 0.08 = 0.53	XIL	E
PNL = 0.40 + 0.15 + 0.08 = 0.63	XNM	E
XNL = 0.15 + 0.15 + 0.08 = 0.38	XNL	F

Need is Less Important than the Physical Characteristics (0.5/1)

Importance Weight Sum	Status Classification	
PIH = 0.20 + 0.15 + 0.50 = 0.85	PIH	A
XIH = 0.08 + 0.15 + 0.50 = 0.73	PNH	B
PNH = 0.20 + 0.07 + 0.50 = 0.77	XIH	B
XNH = 0.08 + 0.07 + 0.50 = 0.65	PIM	C
PIM = 0.20 + 0.15 + 0.33 = 0.68	XNH	C
XIM = 0.08 + 0.15 + 0.33 = 0.56	PNM	D
PNM = 0.20 + 0.07 + 0.33 = 0.60	XIM	D
XNM = 0.08 + 0.07 + 0.33 = 0.48	PIL	D
PIL = 0.20 + 0.15 + 0.17 = 0.52	XNM	E

Table 19: (Continued)

XIL = 0.08 + 0.15 + 0.17 = 0.40	PNL	E
PNL = 0.20 + 0.07 + 0.17 = 0.44	XIL	E
XNL = 0.08 + 0.07 + 0.17 = 0.32	XNL	F

*A = most favorable for artificial recharge
F = least favorable for artificial recharge

Table 20: Summary of Classification Scheme for Artificial Recharge Potential

	1985				2000 and 2030			
		Need/Physical Importance				Need/Physical Importance		
ASA	Status	=	>	<	Status	=	>	<
101	PL	D	C	E	PNL	E	D	E
102	PL	D	C	E	PNL	E	D	E
103	PMSV	B	B	C	PNMSV	C	C	D
104	PM	B	B	C	PNM	C	C	D
105	XL	F	F	F	XNL	F	F	F
106	XM	E	E	D	XNM	E	E	E
201	PM	B	B	C	PNM	C	C	D
202	PMSV	B	B	C	PIMSV	B	B	C
203	PMSV	B	B	C	PIMSV	B	B	C
204	XM	E	E	D	XNM	E	E	E
205	PMS	B	B	C	PIMS	B	B	C
206	PM	B	B	C	PIM	B	B	C
301	PMS	B	B	C	PNMS	C	C	D
302	PH	A	A	A	PIH	A	A	A
303	PH	A	A	A	PIH	A	A	A
304	PHSV	A	A	A	PIHSV	A	A	A
305	PHS	A	A	A	PNHS	B	B	B
306	PH	A	A	A	PIH	A	A	A
307	PHS	A	A	A	PIHS	A	A	A
308	PM	B	B	C	PIM	B	B	C
309	PM	B	B	C	PIM	B	B	C
401	PL	D	C	E	PIL	C	B	D
402	PM	B	B	C	PIM	B	B	C
403	PMSV	B	B	C	PIMSV	B	B	C
404	PMS	B	B	C	PIMS	B	B	C
405	XMS	E	E	D	XNMS	E	E	E
406	PM	B	B	C	PIM	B	B	C
407	PM	B	B	C	PNM	C	C	D
408	XM	E	E	D	XNM	E	E	E
501	PM	B	B	C	PNM	C	C	D
502	PM	B	B	C	PIM	B	B	C
503	PMS	B	B	C	PIMS	B	B	C
504	PMS	B	B	C	PIMS	B	B	C
505	PMV	B	B	C	PIMV	B	B	C
506	PM	B	B	C	PIM	B	B	C
507	PM	B	B	C	PIM	B	B	C
601	PM	B	B	C	PIM	B	B	C
602	PM	B	B	C	PIM	B	B	C
701	PMS	B	B	C	PIMS	B	B	C

Table 20: (Continued)

	1985				2000 and 2030			
		Need/Physical Importance				Need/Physical Importance		
ASA	Status	=	>	<	Status	=	>	<
702	PMS	B	B	C	PIMS	B	B	C
703	PMV	B	B	C	PIMV	B	B	C
704	PMSV	B	B	C	PIMSV	B	B	C
705	XM	E	E	D	XIM	D	D	D
801	PMS	B	B	C	PIMS	B	B	C
802	PM	B	B	C	PIM	B	B	C
803	PMS	B	B	C	PIMS	B	B	C
901	PMSV	B	B	C	PIMSV	B	B	C
1001	PM	B	B	C	PIM	B	B	C
1002	XM	E	E	D	XIM	D	D	D
1003	XM	E	E	D	XIM	D	D	D
1004	PMS	B	B	C	PIMS	B	B	C
1005	PHS	A	A	A	PIHS	A	A	A
1006	PMS	B	B	C	PIMS	B	B	C
1007	PHSV	A	A	A	PNHSV	B	B	B
1008	PHSV	A	A	A	PNHSV	B	B	B
1009	PMV	B	B	C	PIMV	B	B	C
1010	PHSV	A	A	A	PIHSV	A	A	A
1011	PMV	B	B	C	PIMV	B	B	C
1101	XMS	E	E	D	XIMS	D	D	D
1102	PHSV	A	A	A	PNHSV	B	B	B
1103	PHSV	A	A	A	PIHSV	A	A	A
1104	PMS	B	B	C	PIMS	B	B	C
1105	PHSV	A	A	A	PIHSV	A	A	A
1106	PHSV	A	A	A	PNHSV	B	B	B
1107	PM	B	B	C	PIM	B	B	C
1201	PM	B	B	C	PIM	B	B	C
1202	PMS	B	B	C	PNMS	C	C	D
1203	PHSV	A	A	A	PIHSV	A	A	A
1204	PHSV	A	A	A	PIHSV	A	A	A
1205	PMS	B	B	C	PNMS	C	C	D
1301	PHS	A	A	A	PNHS	B	B	B
1302	PHS	A	A	A	PIHS	A	A	A
1303	PHS	A	A	A	PNHS	B	B	B
1304	XHS	C	D	B	XIHS	C	C	B
1305	PM	B	B	C	PNM	C	C	D
1401	XHS	C	D	B	XNHS	D	E	C
1402	XL	F	F	F	XNL	F	F	F

Table 20: (Continued)

	1985				2000 and 2030			
		Need/Physical Importance				Need/Physical Importance		
ASA	Status	=	>	<	Status	=	>	<
1403	PL	D	C	E	PIL	C	B	D
1501	PH	A	A	A	PIH	A	A	A
1502	PH	A	A	A	PNH	B	B	B
1503	PHS	A	A	A	PIHS	A	A	A
1601	PHSV	A	A	A	PIHSV	A	A	A
1602	PH	A	A	A	PNH	B	B	B
1603	PH	A	A	A	PNH	B	B	B
1604	PH	A	A	A	PNH	B	B	B
1701	XM	E	E	D	XIM	D	D	D
1702	PMS	B	B	C	PIMS	B	B	C
1703	PHS	A	A	A	PIHS	A	A	A
1704	XM	E	E	D	XIM	D	D	D
1705	PHSV	A	A	A	PIHSV	A	A	A
1706	PH	A	A	A	PIH	A	A	A
1707	XM	E	E	D	XIM	D	D	D
1801	XHS	C	D	B	XIHS	C	C	B
1802	PH	A	A	A	PNH	B	B	B
1803	PHSV	A	A	A	PNHSV	B	B	B
1804	PHS	A	A	A	PNHS	B	B	B
1805	PHS	A	A	A	PNHS	B	B	B
1806	PHSV	A	A	A	PNHSV	B	B	B
1807	PH	A	A	A	PNH	B	B	B

Table 21: Summary of 1985 Artificial Recharge Potential (Need Equally Important to Physical Characteristics)

A	Irrigated Land*	B	Irrigated Land*	C	Irrigated Land*	D	Irrigated Land*	E	Irrigated Land*	F	Irrigated Land*
302	96.2	103	21.9	1304	253.1	101	3.0	106	1.4	105	9.1
303	303.9	104	0.7	1401	856.7	102	11.1	204	18.2	1402	502.6
304	1,060.7	201	4.1	1801	481.7	401	5.0	405	36.2		
305	1,502.2	202	40.3			1403	313.2	408	13.5		
306	619.5	203	12.0					705	15.5		
307	75.9	205	100.3					1002	938.3		
1005	399.4	206	18.9					1003	193.3		
1007	2,079.2	301	306.2					1101	281.6		
1008	3,892.2	308	11.7					1701	708.1		
1010	4,003.4	309	0.0					1704	307.6		
1102	519.2	402	119.2					1707	492.1		
1103	2,836.8	403	59.1								
1105	2,051.0	404	389.9								
1106	1,367.2	406	25.1								
1203	3,535.8	407	5.7								
1204	1,628.7	501	0.0								
1301	618.7	502	4.8								
1302	669.2	503	18.4								
1303	280.5	504	0.0								
1501	113.6	505	19.8								
1502	162.1	506	65.1								
1503	1,154.9	507	5.0								
1601	902.4	601	10.8								
1602	436.1	602	8.8								
1603	805.4	701	354.2								
1604	264.7	702	185.2								
1703	4,100.5	703	111.1								
1705	524.6	704	138.0								

Table 21: (Continued)

A	Irrigated Land*	B	Irrigated Land*	C	Irrigated Land*	D	Irrigated Land*	E	Irrigated Land*	F	Irrigated Land*
1706	74.4	801	2,855.6								
1802	2,297.5	802	1,422.5								
1803	5,608.0	803	1,052.7								
1804	5,608.0	901	94.7								
1805	471.5	1001	205.8								
1806	806.3	1004	1,286.8								
1807	62.2	1006	389.6								
		1009	200.1								
		1011	120.2								
		1104	138.6								
		1107	80.6								
		1201	155.4								
		1202	629.2								
		1205	654.4								
		1305	755.0								
		1702	2,250.2								
Total acreage (x1000 ac)	45,625.7		14,327.7		1,591.5		332.3		3,065.8		511.7
Total ASA's	35		44		3		4		11		2

*times 1C00 acres

Sum of total acreage (x1000 ac) = 65,454.7
Sum of ASA's = 99

Table 22: Summary of 1985 Artificial Recharge Potential (Need is More Important than Physical Characteristics)

A	Irrigated Land*	B	Irrigated Land*	C	Irrigated Land*	D	Irrigated Land*	E	Irrigated Land*	F	Irrigated Land*
302	96.2	103	21.9	101	3.0	1304	253.1	106	1.4	105	9.1
303	303.9	104	0.7	102	11.1	1403	313.2	204	18.2	1402	502.6
304	1,060.7	201	4.1	401	5.0			405	36.2		
305	1,502.2	202	40.3	1401	856.7			408	13.5		
306	619.5	203	12.0	1801	481.7			705	15.5		
307	75.9	205	100.3					1002	998.3		
1005	399.4	206	18.9					1003	193.3		
1007	2,079.2	301	306.2					1101	281.6		
1008	3,892.2	308	11.7					1701	708.1		
1010	4,003.4	309	0.0					1704	307.6		
1102	519.2	402	119.2					1707	492.1		
1103	2,836.8	403	59.1								
1105	2,051.0	404	389.9								
1106	1,367.2	406	25.1								
1203	3,535.8	407	5.7								
1204	1,628.7	501	0.0								
1301	618.7	502	4.8								
1302	669.2	503	18.4								
1303	280.5	504	0.0								
1501	113.6	505	19.8								
1502	162.1	506	65.1								
1503	1,154.9	507	5.0								
1601	902.4	601	10.8								
1602	436.1	602	8.8								
1603	805.4	701	354.2								
1604	264.7	702	185.2								
1703	4,100.5	703	111.1								
1705	524.6	704	138.0								

Table 22: (Continued)

A	Irrigated Land*	B	Irrigated Land*	C	Irrigated Land*	D	Irrigated Land*	E	Irrigated Land*	F	Irrigated Land*
1706	74.4	801	2,855.6								
1802	2,297.5	802	1,422.5								
1803	5,608.0	803	1,052.7								
1804	301.8	901	94.7								
1805	471.5	1001	205.8								
1806	806.3	1004	1,286.8								
1807	62.2	1006	389.6								
		1009	200.1								
		1011	120.2								
		1104	138.6								
		1107	80.6								
		1201	155.4								
		1202	629.2								
		1205	654.4								
		1305	755.0								
		1702	2,250.2								
Total acreage (x1000 ac)	45,625.7		14,327.7		1,357.5		566.3		3,065.8		511.7
Total ASA's	35		44		5		2		11		2

Sum of total acreage (x1000 ac) = 65,454.7
Sum of ASA's = 99

Table 23: Summary of 1985 Artificial Recharge Potential (Need is Less Important than Physical Characteristics)

A	Irrigated Land*	B	Irrigated Land*	C	Irrigated Land*	D	Irrigated Land*	E	Irrigated Land*	F	Irrigated Land*
302	96.2	1304	253.1	103	21.9	106	1.4	101	3.0	105	9.1
303	303.9	1401	856.7	104	0.7	204	18.2	102	11.1	1402	502.6
304	1,060.7	1801	481.7	201	4.1	405	36.2	401	5.0		
305	1,502.2			202	40.3	408	13.5	1403	313.2		
306	619.5			203	12.0	705	15.5				
307	75.9			205	100.3	1002	998.3				
1005	399.4			206	18.9	1003	193.3				
1007	2,079.2			301	306.2	1101	281.6				
1008	3,892.2			308	11.7	1701	708.1				
1010	4,003.4			309	0.0	1704	307.6				
1102	519.2			402	119.2	1707	492.1				
1103	2,836.8			403	59.1						
1105	2,051.0			404	389.9						
1106	1,367.2			406	25.1						
1203	3,535.8			407	5.7						
1204	1,628.7			501	0.0						
1301	618.7			502	4.8						
1302	669.2			503	18.4						
1303	280.5			504	0.0						
1501	113.6			505	19.8						
1502	162.1			506	65.1						
1503	1,154.9			507	5.0						
1601	902.4			601	10.8						
1602	436.1			602	8.8						
1603	805.4			701	354.2						
1604	264.7			702	185.2						
1703	4,100.5			703	111.1						
1705	524.6			704	138.0						

Table 23: (Continued)

A	Irrigated Land*	B	Irrigated Land*	C	Irrigated Land*	D	Irrigated Land*	E	Irrigated Land*	F	Irrigated Land*
1706	74.4			801	2,855.6						
1802	2,297.5			802	1,422.5						
1803	5,608.0			803	1,052.7						
1804	301.8			901	94.7						
1805	471.5			1001	205.8						
1806	806.3			1004	1,286.8						
1807	62.2			1006	389.6						
				1009	200.1						
				1011	120.2						
				1104	138.6						
				1107	80.6						
				1201	155.4						
				1202	629.2						
				1205	654.4						
				1305	755.0						
				1702	2,250.2						
Total acreage (x100) ac)	45,625.7		1,591.5		14,327.7		3,065.8		332.3		511.7
Total ASA's	35		3		44		11		4		2

Sum of Total acreage (x1000 ac) = 65,454.7

Sum of ASA's = 99

Table 24: Summary of Artificial Recharge Potential for 2000 and 2030 (Need is Equally Important to Physical Characteristics)

A	Irrigated Land*	B	Irrigated Land*	C	Irrigated Land*	D	Irrigated Land*	E	Irrigated Land*	F	Irrigated Land*
302	96.2	202	40.3	103	21.9	705	15.5	101	3.0	105	9.1
303	303.9	203	12.0	104	0.7	1002	998.3	102	11.1	1402	502.6
304	1,060.7	205	100.3	201	4.1	1003	193.3	106	1.4		
306	619.5	206	18.9	301	306.2	1101	281.6	204	18.2		
307	75.9	305	1,502.2	401	5.0	1401	856.7	405	36.2		
1005	399.4	308	11.7	407	5.7	1701	708.1	408	13.5		
1010	4,003.4	309	0.0	501	0.0	1704	307.6				
1103	2,836.8	402	119.2	1202	629.2	1707	492.1				
1105	2,051.0	403	59.1	1205	654.4						
1203	3,535.8	404	389.9	1304	253.1						
1204	1,628.7	406	25.1	1305	755.0						
1302	669.2	502	4.8	1403	313.2						
1501	113.6	503	18.4	1801	481.7						
1503	1,154.9	504	0.0								
1601	902.4	505	19.8								
1703	805.4	506	65.1								
1705	524.6	507	5.0								
1706	74.4	601	10.8								
		602	8.8								
		701	354.2								
		702	185.2								
		703	111.1								
		704	138.0								
		801	2,855.6								
		802	1,422.5								
		803	1,052.7								
		901	94.7								

Table 2: (Continued)

A	Irrigated Land*	B	Irrigated Land*	C	Irrigated Land*	D	Irrigated Land*	E	Irrigated Land*	F	Irrigated Land*
		1001	205.8								
		1004	1,286.8								
		1006	389.6								
		1007	2,079.2								
		1008	3,892.2								
		1009	200.1								
		1011	120.2								
		1102	519.2								
		1104	138.6								
		1106	1,367.2								
		1107	80.6								
		1201	155.4								
		1301	618.7								
		1303	280.5								
		1502	162.1								
		1602	436.1								
		1603	805.4								
		1604	264.7								
		1702	2,250.2								
		1802	2,297.5								
		1803	5,608.0								
		1804	301.8								
		1805	471.5								
		1806	806.3								
		1807	62.2								

Table 24: (Continued)

A	Irrigated Land*	B	Irrigated Land*	C	Irrigated Land*	D	Irrigated Land*	E	Irrigated Land*	F	Irrigated Land*
Total acreage (x1000 ac)	20,855.8		36,720.4		3,430.2		3,853.2		83.4		511.7
Total ASA's	18		52		13		8		6		2

Sum of total acreage (x1000 ac) = 65,454.7
Sum of ASA's = 99

Table 25: Summary of Artificial Recharge Potential for 2000 and 2030 (Need is More Important than Physical Characteristics)

A	Irrigated Land*	B	Irrigated Land*	C	Irrigated Land*	D	Irrigated Land*	E	Irrigated Land*	F	Irrigated Land*
302	96.2	202	40.3	103	21.9	705	15.0	101	3.0	105	9.1
303	303.9	203	12.0	104	0.7	1002	998.3	102	11.1	1402	502.6
304	1,060.7	205	100.3	201	4.1	1003	193.3	106	1.4		
306	619.5	206	18.9	301	306.2	1101	281.6	204	18.2		
307	75.9	305	1,502.2	407	5.7	1701	708.1	405	36.2		
1005	399.4	308	11.7	501	0.0	1704	307.6	408	13.5		
1010	4,003.4	309	0.0	1202	629.2	1707	492.1	1401	856.7		
1103	2,836.8	401	5.0	1205	654.4						
1105	2,051.0	402	119.2	1304	253.1						
1203	3,535.8	403	59.1	1305	755.0						
1204	1,628.7	404	389.9	1801	481.7						
1302	669.2	406	25.1								
1501	113.6	502	4.8								
1503	1,154.9	503	18.4								
1601	902.4	504	0.0								
1703	805.4	505	19.8								
1705	524.6	506	65.1								
1706	74.4	507	5.0								

Table 25: (Continued)

A	Irrigated Land*	B	Irrigated Land*	C	Irrigated Land*	D	Irrigated Land*	E	Irrigated Land*	F	Irrigated Land*
		601	10.8								
		602	8.8								
		701	354.2								
		702	185.2								
		703	111.1								
		704	138.0								
		801	2,855.6								
		802	1,422.5								
		803	1,052.7								
		901	94.7								
		1001	205.8								
		1004	1,286.8								
		1006	389.6								
		1007	2,079.2								
		1008	3,892.2								
		1009	200.1								
		1011	120.2								
		1102	519.2								
		1104	138.6								
		1106	1,367.2								
		1107	80.6								
		1201	155.4								
		1301	618.7								
		1303	280.5								

Table 25: (Continued)

A	Irrigated Land*	B	Irrigated Land*	C	Irrigated Land*	D	Irrigated Land*	E	Irrigated Land*	F	Irrigated Land*
		1403	313.2								
		1502	162.1								
		1602	436.1								
		1603	805.4								
		1604	264.7								
		1702	2,250.2								
		1802	2,297.5								
		1803	5,608.0								
		1804	301.8								
		1805	471.5								
		1806	806.3								
		1807	62.2								
Total acreage (x1000 ac)	20,855.8		37,038.6		3,112.0		2,996.5		940.1		511.7
Total ASA's	18		54		11		7		7		2

Sum of total acreage (x1000 ac) = 65,454.7
Sum of ASA's = 99

Table 26: Summary of Artificial Recharge Potential for 2000 and 2030 (Need is Less Important than Physical Characteristics)

A	Irrigated Land*	B	Irrigated Land*	C	Irrigated Land*	D	Irrigated Land*	E	Irrigated Land*	F	Irrigated Land*
302	96.2	305	1,502.2	202	40.3	103	21.9	101	3.0	105	9.1
303	303.9	1007	2,079.2	203	12.0	104	0.7	102	11.1	1402	502.6
304	1,060.7	1008	3,892.2	205	100.3	201	4.1	106	1.4		
306	619.5	1102	519.2	206	18.9	301	306.2	204	18.2		
307	75.9	1106	1,367.2	308	11.7	401	5.0	405	36.2		
1005	399.4	1301	618.7	309	0.0	407	5.7	408	13.5		
1010	4,003.4	1303	280.5	402	119.2	501	0.0				
1103	2,836.8	1304	253.1	403	59.1	705	15.5				
1105	2,051.0	1502	162.1	404	389.9	1002	988.3				
1203	3,535.8	1602	436.1	406	25.1	1003	193.3				
1204	1,628.7	1603	805.4	502	4.8	1101	281.6				
1302	669.2	1604	264.7	503	18.4	1202	629.2				
1501	113.6	1801	481.7	504	0.0	1205	654.4				
1503	1,154.9	1802	2,297.5	505	19.8	1305	755.0				
1601	902.4	1803	5,608.0	506	65.1	1403	313.2				
1703	805.4	1804	301.8	507	5.0	1701	708.1				
1705	524.6	1805	471.5	601	10.8	1704	307.6				
1706	74.4	1806	806.3	602	8.8	1707	492.1				
		1807	62.2	701	354.2						
				702	185.2						
				703	111.1						
				704	138.0						
				801	2,855.6						
				802	1,422.5						
				803	1,052.7						
				901	94.7						

Table 26: (Continued)

A	Irrigated Land*	B	Irrigated Land*	C	Irrigated Land*	D	Irrigated Land*	E	Irrigated Land*	F	Irrigated Land*
				1001	205.8						
				1004	1,286.8						
				1006	389.6						
				1009	200.1						
				1011	120.2						
				1104	138.6						
				1107	80.6						
				1201	155.4						
				1401	856.7						
				1702	2,250.2						
Total acreage (x1000 ac)	20,855.8		22,209.6		16,102.3		5,691.9		83.4		511.7
Total ASA's	18		19		36		18		6		2

Sum of total acreage (x1000 ac) = 65,454.7
Sum of ASA's = 99

Table 27: Number of ASA's and Acres of Irrigated Lands by Artificial Recharge Classification

	Need vs Physical Characteristics											
	1985 Status						Future (2000 and 2030 Status)					
Classification	=	Acres*		Acres*		Acres*	=	Acres*		Acres*		Acres*
A	35	45.6	35	45.6	35	45.6	18	20.9	18	20.9	18	20.9
B	44	14.3	44	14.3	3	1.6	52	36.7	54	37.0	19	22.2
C	3	1.6	5	1.4	44	14.3	13	3.4	11	3.1	36	16.1
D	4	0.3	2	0.6	11	3.0	8	3.8	7	3.0	18	5.7
E	11	3.0	11	3.0	4	0.3	6	0.1	7	0.9	6	0.1
F	2	0.5	2	0.5	2	0.5	2	0.5	2	0.5	2	0.5

*Approximate figures in millions of acres.

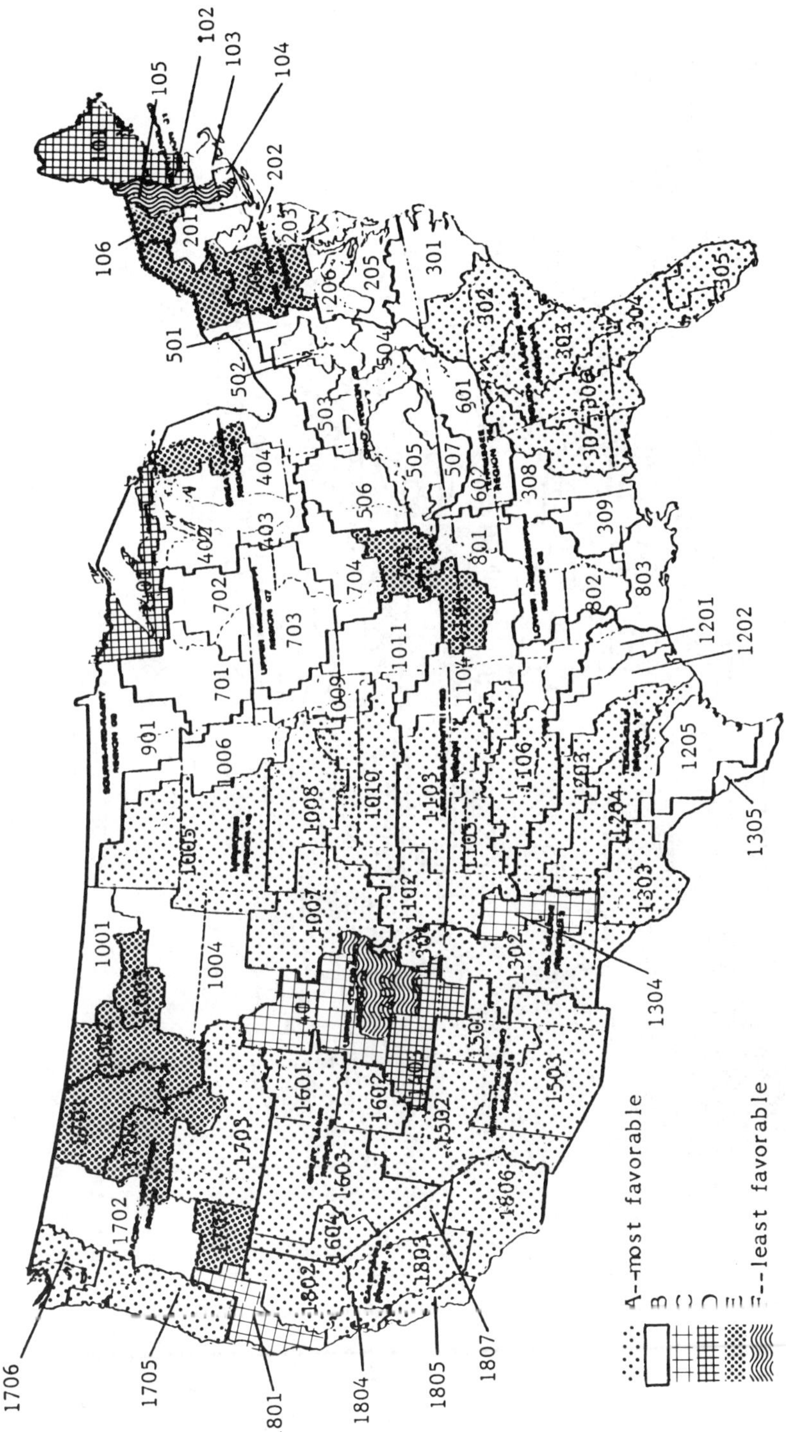

Figure 9: Classification of Artificial Recharge Potential for 1985 (Need is Equally Important to Physical Characteristics)

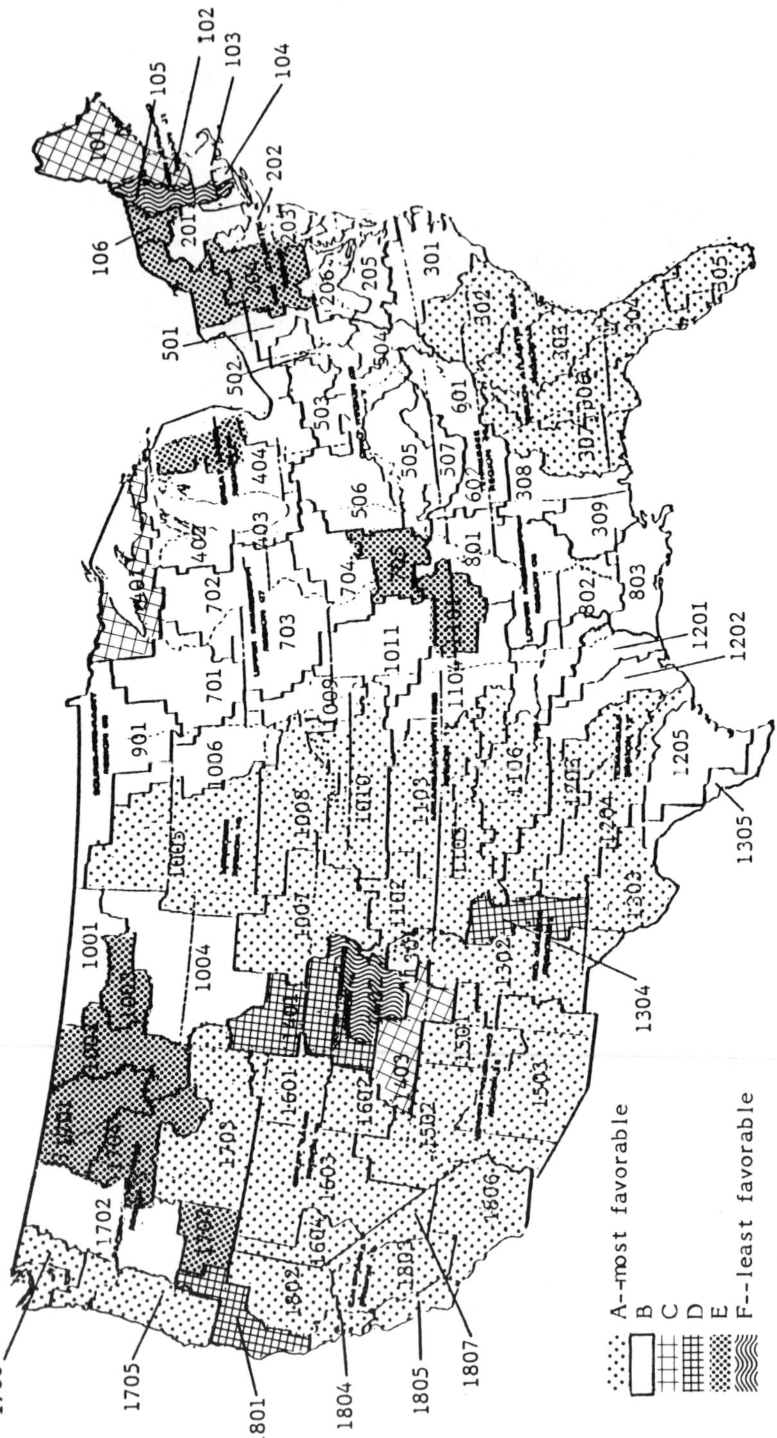

Figure 10: Classification of Artificial Recharge Potential for 1985
(Need is More Important than Physical Characteristics)

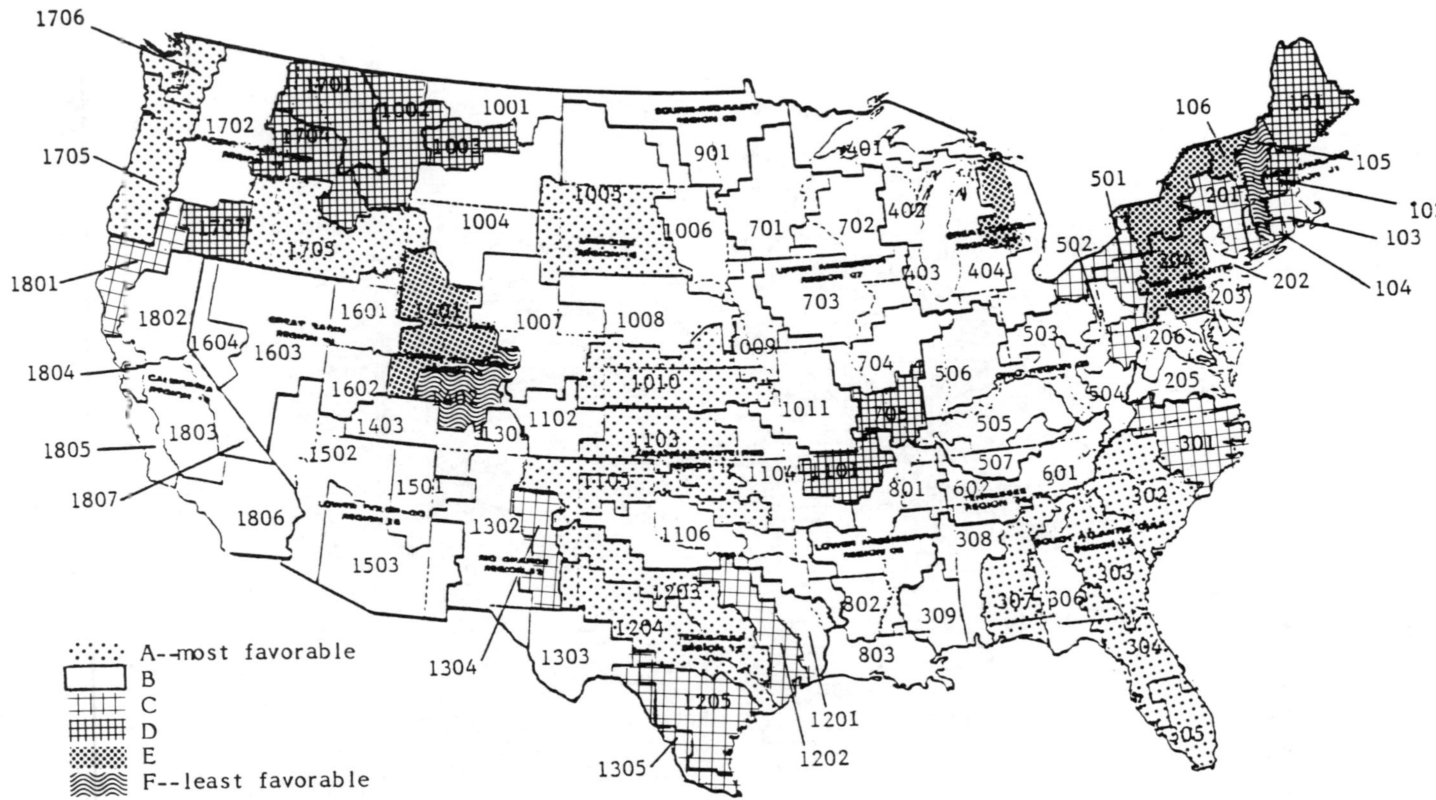

Figure 11: Classification of Artificial Recharge Potential for 1985
(Need is Less Important than Physical Characteristics)

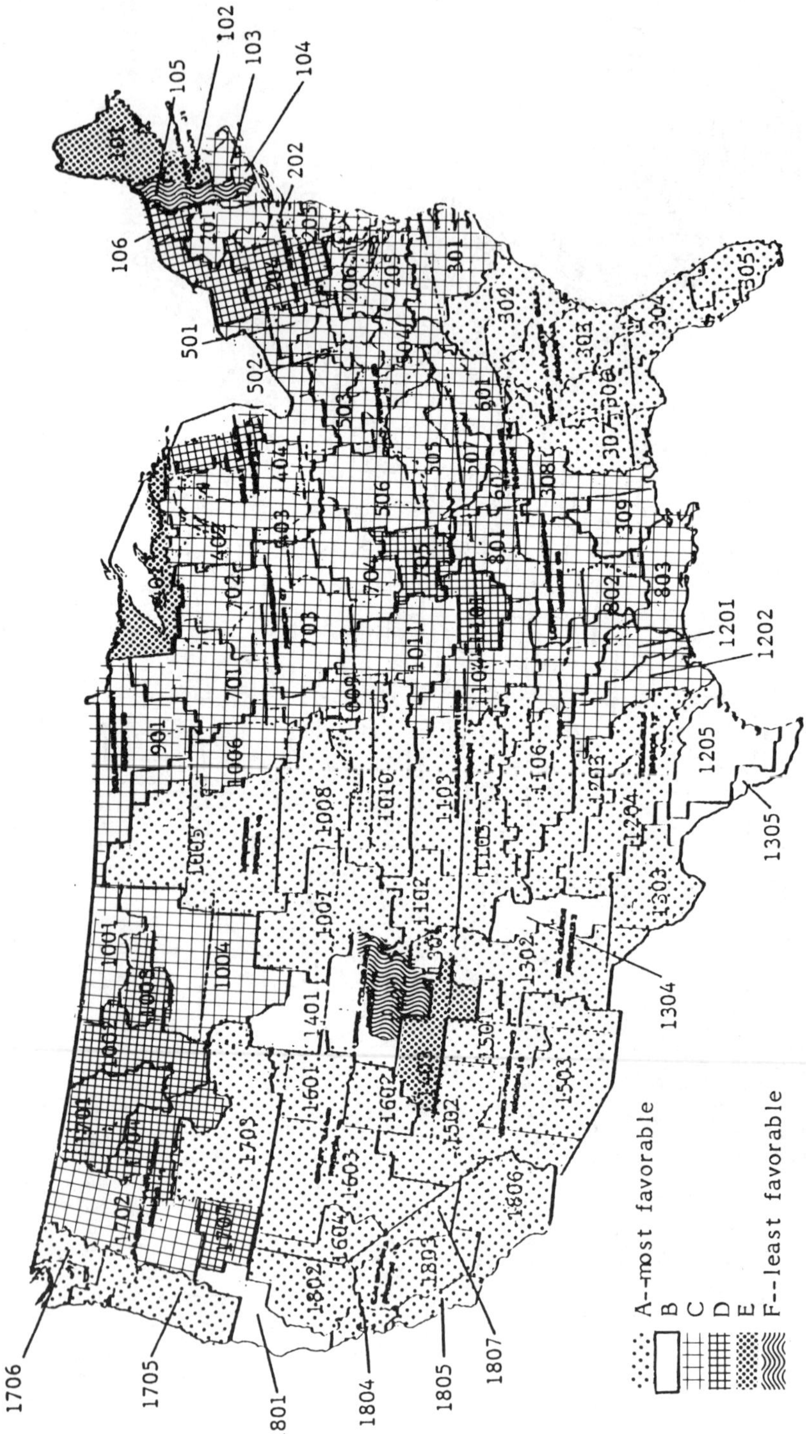

Figure 12: Classification of Artificial Recharge Potential for 2000 and 2030 (Need is Equally Important to Physical Characteristics)

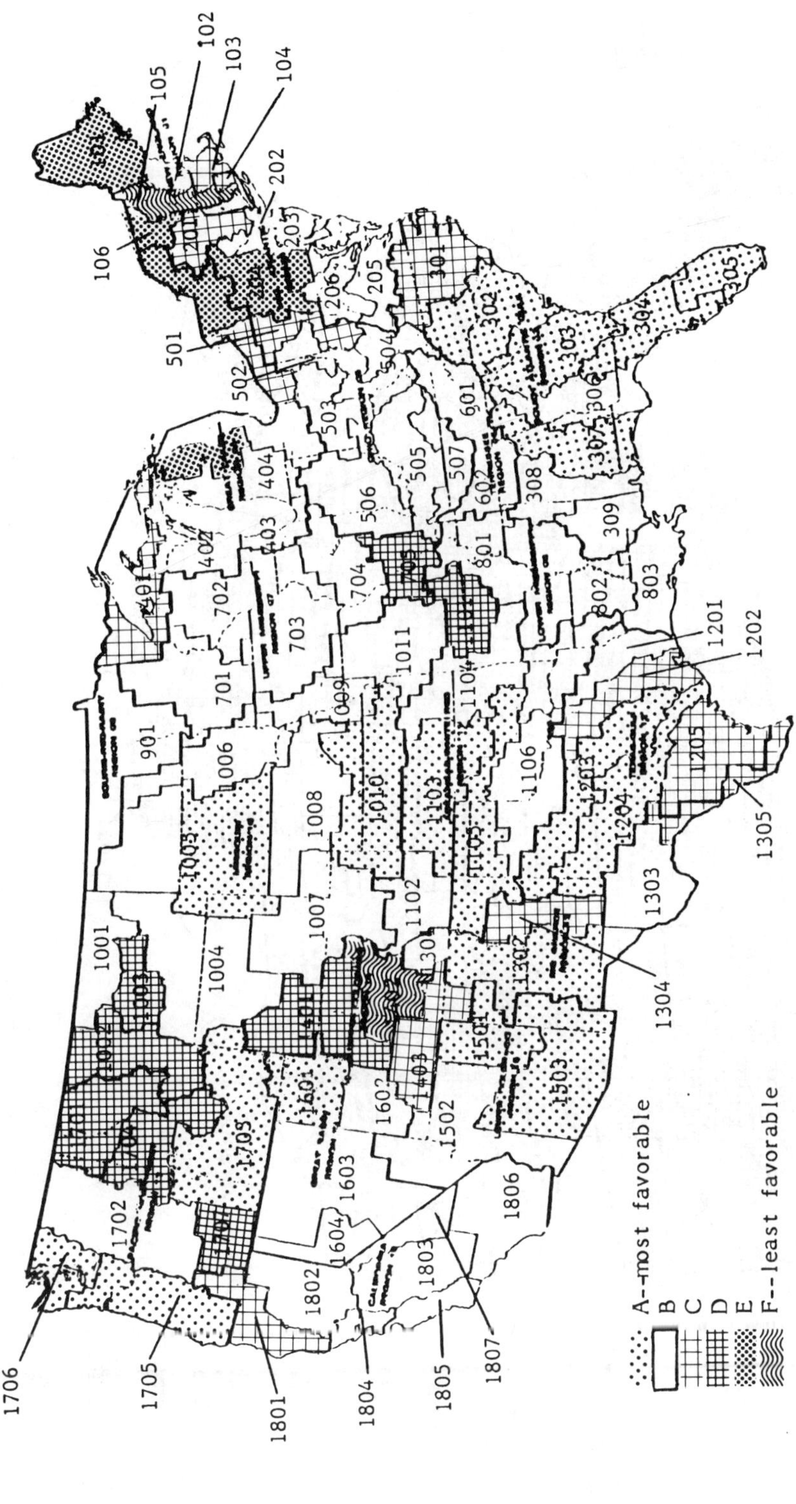

Figure 13: Classification of Artificial Recharge Potential for 2000 and 2030 (Need is More Important than Physical Characteristics)

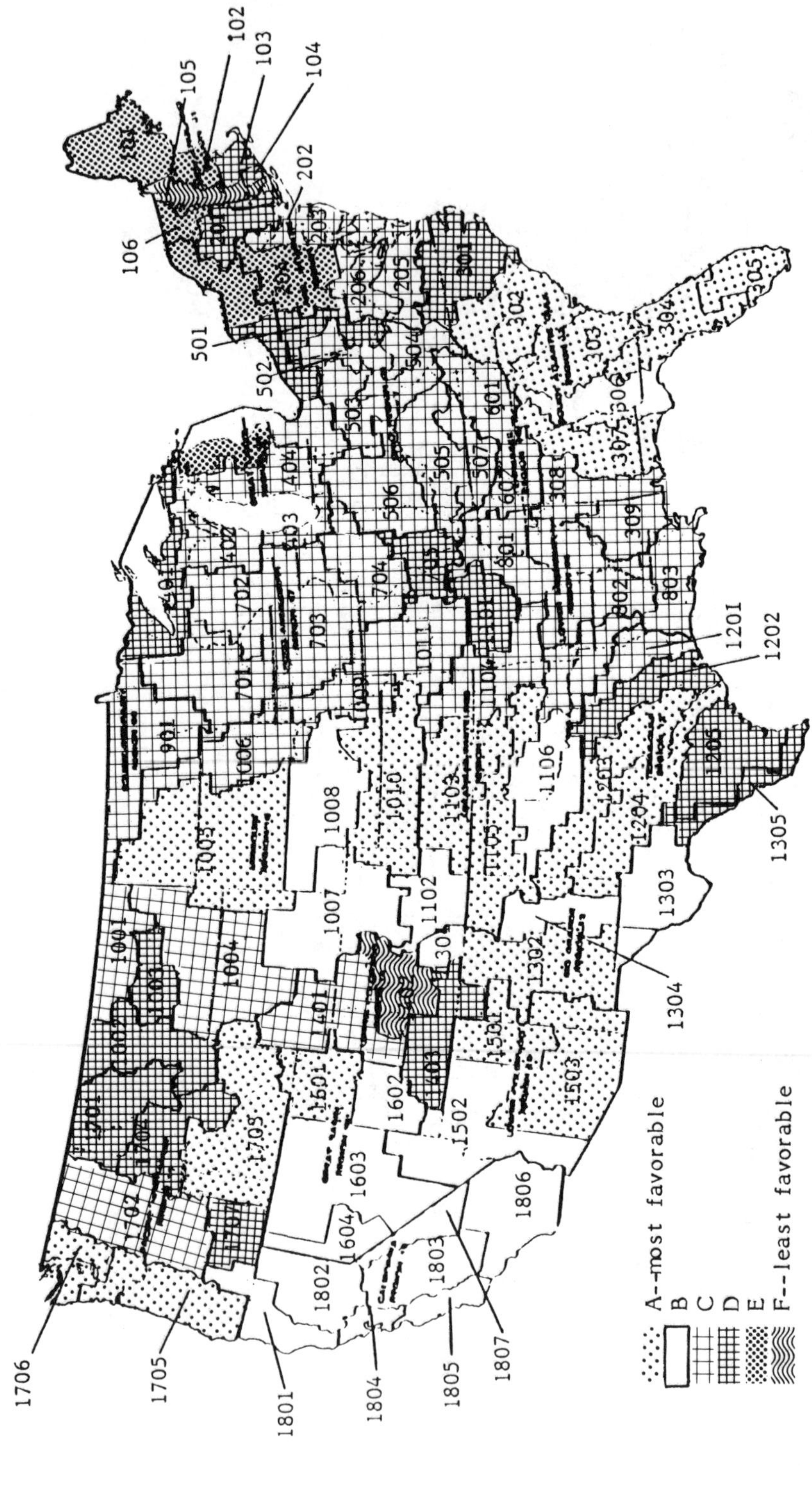

Figure 14: Classification of Artificial Recharge Potential for 2000 and 2030 (Need is Less Important than Physical Characteristics)

Approximation involved detailed hydrogeological studies in the towns of Pittsfield, Fairhaven, Holden, and Amherst (site-specific). The potential use of artificial recharge as means of aquifer contamination control, recharging a lower confined aquifer from an upper unconfined aquifer, and increasing water supplies are also included in the study on the basis of site-specific areas.

First Level Approximation

Before considering artificial recharge as the final method to increase water supply, one must first determine the appropriateness and feasibility of such a method in relation to other water supply alternatives. The feasibility of artificial recharge was determined for the area around each municipal well in Massachusetts that could be located from existing records. The procedures included a plot of the municipal wells on topographic maps, and for each well, the geologic, hydrologic, and soil factors influencing artificial recharge operations were considered and analyzed. These factors are shown in Table 28, including the method used for the evaluation of each factor.

For two reasons, the feasibility study at this level represents a conservative evaluation of artificial recharge potentiality. The reasons are that only three major artificial recharge methods were considered out of a total of ten, and secondly, only the feasibility of artificial recharge around existing wells and not everywhere in the area was considered.

The results of the First Level Approximation method indicate that 52.8 percent of the town wells, and 38.3 percent of the towns, show high and potentially favorable recharge conditions. This indicates that artificial recharge would be an effective way of increasing water supplies in Massachusetts.

Second Level Approximation

The Second Level Approximation method involves more detailed geologic and hydrologic studies which are used in the survey of artificial recharge potentiality of municipal ground water supplies. This method was applied in the Ipswich River Basin north of Boston (Motts et al., 1983). The purpose was to focus on artificial recharge feasibility and not all the alternative methods of obtaining water supplies in the Ipswich River Basin. For example, artificial recharge may be a feasible method in a town; however, detailed study of the water resources of the town may show that drilling new wells along a major stream may be the favorable environmental and cost-effective method.

The success of artificial recharge, as reported by Motts et al., depends on meeting the following requirements (Motts et al., 1983):

(1) a recharge water source of good drinking water quality and low sediment;

(2) good hydraulic connection to water-bearing deposits at the recharge site;

(3) transmissive subsurface media from recharge site to reclamation well;

Table 28: Factors Conducive to Favorable Artificial Recharge Operations by First-Level Approximation Methods (Motts, et al., 1983)

Hydrogeologic and Soils Factors Conducive to Favorable Artificial Recharge	Method Used for Determination of Artificial Recharge Factor
1. Soils and sediments of sufficient permeability to transmit recharged water to the zone of saturation and to prevent or retard clogging on a long-term basis	1. Study of geological and soils maps, generally coarse-grained stratified drift is covered by permeable soils such as the Hinkley Series and similar types.
2. A perennial source of surface water that is relatively silt-free and free from severe contamination, and provides sufficient surface flows in the dry weather months to prevent serious environmental problems resulting from excessive depletion of stream flows, down-stream loss of water supply, or drying up of important wetlands.	2. Determine size and area covered by stratified drift in recharge area.
3. Permeable sediments providing good hydrologic connection between artificial recharge site and well.	3. Character of stratified drift, coarse-grained drift generally has high permeability.
4. No confining beds between the ground surface and main zone of saturation.	4. Study of surficial geology to determine if lake beds or other confining beds are present. Generally, coarse-grained stratified drift has good hydrologic connections between surface and subsurface.

Table 28: (Continued)

Hydrogeologic and Soils Factors Conducive to Favorable Artificial Recharge	Method Used for Determination of Artificial Recharge Factor
5. A zone of ground-water storage in the aquifer (such as a cone of depression) into which recharged water may be discharged. The thickness, extent, and hydrologic characteristics of the storage area are significant.	5. Determined from character of sediment penetrated by wells and analyses of existing geological maps. From aquifer characteristics, compute depth and extent of cone of depression from pumping wells.
6. Suitable geologic and geographic areas relatively free of development where artificial recharge methods could be utilized--such necessary structures as small diversion or check dams could be constructed.	6. Inspection of the most recent topographic maps for evidence of human development activities.

(4) storage capacity; and

(5) reclamation well.

The Second Level Approximation was conducted in three phases. The first phase tried to answer artificial recharge requirements numbers 1,4, and 5 above by using available information sources. Such information included well locations, depth, yield, saturated thickness, hydraulic conductivity, and drainage area and size. The results of this phase identified sites having favorable artificial recharge characteristics and are shown in Figure 15 (Motts et al., 1983).

The purpose of the second phase was to refine the results of the first phase. Site criteria were expanded to include (Motts et al., 1983):

(1) 1990 projected municipal water need indicating type and size of artificial recharge structure. This is shown in Figure 16 (Motts et al., 1983).

(2) Well parameters:

  a. Water level--with depth of well indicate saturated thickness and storage capacity.

  b. Depth of well--with surface geology and water level indicate aquifer dimensions.

  c. Pumping rate--identify transmissivity and average hydraulic conductivity.

  d. Sustained yield--indicator of stress on aquifer.

(3) Recharge characteristics based on 1990 projected water need--is there enough water to provide the artificial recharge structure with a sufficient amount of recharge?

(4) Land use characteristics:

  a. Type of area--forest, agricultural, open, urban, and topographic configuration.

  b. Areal constraints--space limitations in the vicinity of the well site.

(5) Artificial recharge method--choice of the appropriate method is based on the above information.

The results of the second phase evaluation indicated the artificial recharge methods that could be applied at the favorable sites in the Ipswich River Basin. The third phase involved on-site examination of the well location, surficial geology, surface hydrology, soil types, land use, and areal constraints for constructing artificial recharge structures at municipal well sites with favorable artificial recharge characteristics.

Figure 15: Artificial Recharge Phase 1 Results in Ipswich River Basin (Motts, et al., 1983)

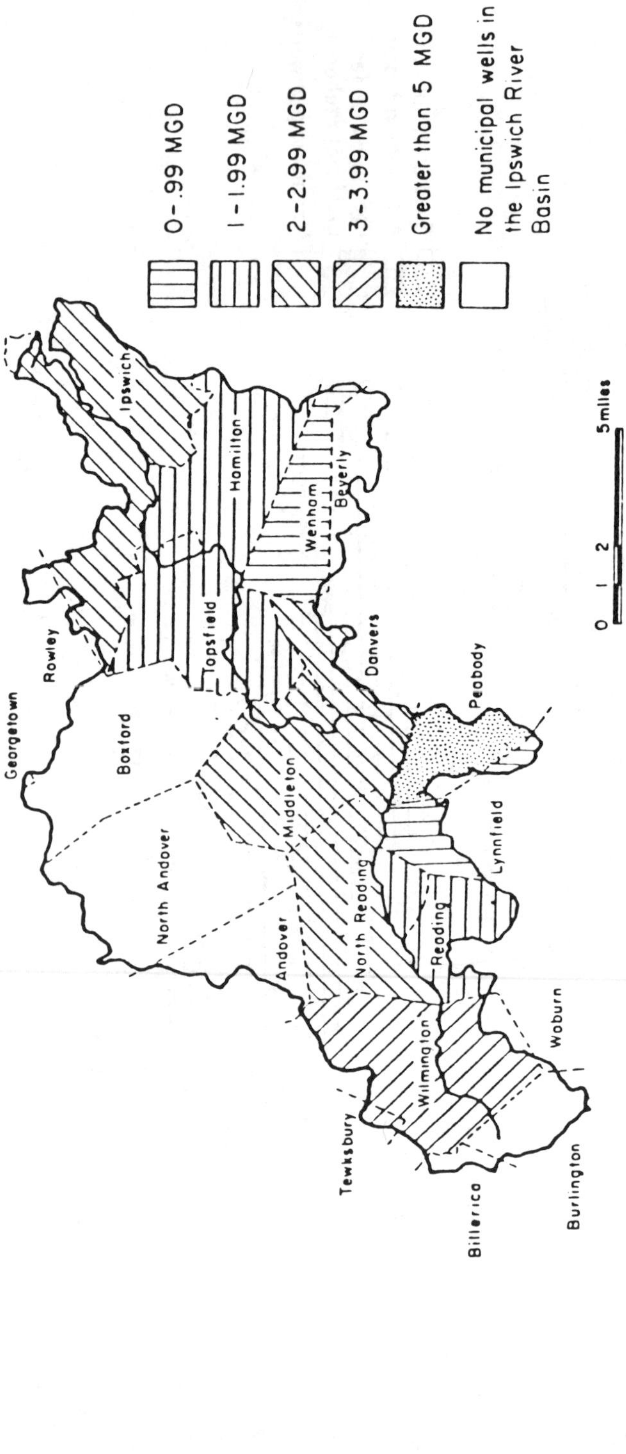

Figure 16: Projected 1990 Water Needs of Municipalities in Ipswich River Basin (Motts, et al., 1983)

Third Level Approximation

The Third Level Approximation consisted of detailed hydrogeological studies made in the towns of Pittsfield, Fairhaven, Holden, and Amherst. The Pittsfield case illustrated the potential use of artificial recharge as a means of preventing aquifer contamination. In Fairhaven the case consisted of recharging a lower confined aquifer from an upper unconfined aquifer. The potential use of artificial recharge as a means of increasing water supplies was illustrated for Holden and Amherst. The Third Level Approximation for the four towns involved detailed and site-specific investigations. The investigations include the hydrogeology of the sites, hydrochemical characteristics of the aquifer systems, water needs, conjunctive water use, water sources and quality of the water, artificial recharge structures, economical factors, and consideration of other alternatives.

Perspectives on State Approach

The study of the Ipswich River Basin utilizing the more detailed Second Level Approximation methods supports the conclusions of the First Level Approximation; namely, that artificial recharge is an appealing alternative in Massachusetts. However, to determine the quantity of water available for municipal development and artificial recharge throughout the Basin, a basin-wide study (site specific) must be done.

The Third Level Approximation method indicated that artificial recharge methods have high potentials to increase water supplies in the towns of Amherst, Holden, Pittsfield, and Fairhaven. For the towns of Amherst, Holden, and Pittsfield, the recommended methods of artificial recharge should be evaluated in relation to other surface and ground water alternatives, short and long term needs, and water problems unique to the town.

In summary, using the First Level Approximation (broad reconnaissance), Second Level Approximation (detailed reconnaissance in the Ipswich Basin), and the Third Level Approximation (detailed town studies), Motts et al. (1983) concluded that artificial recharge is a viable method that is cost effective and environmentally advantageous at many localities in Massachusetts.

SITE SELECTION PROTOCOL

The previous discussion has indicated the need for detailed site-specific information in order to make a valid, informed decision regarding the implementation of artificial recharge as a ground water management option. The initial section of this chapter generalized artificial recharge potential on a national scale to a level for which data was available. When deciding whether a particular site will be receptive to artificial recharge, and whether the particular needs of that area will be met by the artificial recharge possible in that area, the generalized approach is only the starting point.

Consideration of artificial recharge may be borne out of the formation of various goals. Severe depletion of important ground water supplies may lead decision-makers toward artificial recharge as an option to halt, or even reverse, the diminution of the supply. In this case, need is high so the site selection protocol may focus less on economic considerations and more on the rate of recharge attainable and/or the resulting water quality. On the other hand, in a situation not of dire need but of prevention, and of thinking ahead toward inevitable depletion through future increased needs, site selection should focus more on convenience and practicality over perhaps an imported supply, and less on technological considerations. If time is available, research may provide an efficient and effective system in an area that otherwise may appear only moderately favorable for artificial recharge.

Essentially, site-specific artificial recharge evaluation involves a trade-off analysis as shown in Table 29. The relative importance weight of each factor discussed in Chapter 3 (see Table 30 a summary) will vary with the needs and goals for the ground water supply. Not only are the information requirements for determining artificial recharge potential site-specific, so too is the development of a site selection protocol. Perhaps this is the reason why no universal site selection protocol has been previously developed. It is very difficult to delineate a process for site selection when there are so many variables.

Table 29: Steps for a Site Specific Selection Process

1. Outline the needs of the area and define the goals which will meet these needs.
2. Identify site/sites available in the area.
3. Identify factors pertinent to artificial recharge as a management option in reference to the area's needs.
4. Perform importance weighting of factors relative to their significance in meeting the needs of the area.
5. Define the actions necessary to evaluate each factor (pilot studies, laboratory tests, literature research).
6. Arrange actions consecutively (some may be concurrent) in order of significance of the factors they evaluate.
7. Perform actions to determine viability of artificial recharge at each site and eliminate sites as factors are encountered which indicate the site cannot support artificial recharge at the necessary level.
8. Decide whether or not artificial recharge is applicable to meet the needs of the area at a minimum of one identified site.

What is suggested herein is a process by which individual site selection protocols may be developed; Table 29 summarizes the steps. The fourth step involves importance weighting of the factors which determine the potential for artificial recharge. In this way, for each case for which artificial recharge is an option, the factors discussed in Chapter 3 can be examined in a stepwise fashion, beginning with those most crucial to the defined goals. If the crucial factors are found to be contradictory to meeting the needs, then an alternative to artificial recharge (an alternative site in the same area) should be identified before too much time and effort is spent on researching less crucial factors.

Different importance weighting techniques are available. One commonly used in decision-making is the paired comparison technique as described by Canter and Knox (1985). Applied to artificial recharge, the paired comparison technique would simply involve taking two factors at a time and deciding which

Table 30: Some Factors to Consider for Artificial Recharge

1. Availability of waste water
2. Quantity of source water available
3. Quality of source water available
4. Resulting water quality (reactions with native water and aquifer materials)
5. Clogging potential
6. Underground storage space available
7. Depth to underground storage space
8. Transmission characteristics
9. Topography/applicable methods (injection or infiltration)
10. Legal/institutional constraints
11. Costs
12. Cultural/social considerations

of the two is most significant for the defined goals. For example, when the ground water supply is for drinking water, the source water quality would be considered more important than, say, topography. Each factor would be considered relative to every other factor and a dummy factor which is always the least significant. In each comparison pair, the most significant factor is assigned a one (1) and the least significant of that pair is assigned a zero (0). A factor may be assigned a (1) in comparison to one factor and a zero in comparison to another. If two factors are considered of equal importance, each is assigned 0.5. When the numbers assigned to each factor are added, the resulting number, when divided by the total numbers of all factors, indicates the relative importance of that factor when compared to the others. This number is termed the factor importance coefficient (FIC). See Table 31 for a partial hypothetical example. In this example, identification of transmission and storage characteristics in reference to artificial recharge potential would take precedence once these are determined to be acceptable, then clogging potential can be examined, and so on. If they are not acceptable, there is no need to examine this site further.

Table 31: Example--Importance Weighting for Area of High Need for Increasing Agricultural and Industrial Ground Water Supply

Factor	Significance or Importance Weights						SUM	FIC
transmission and storage characteristics	1	1	1				3	0.50
resulting water quality	0			0	1		1	0.17
clogging potential		0		1		1	2	0.33
dummy factor			0		0	0	0	0
							6	1.00

SELECTED REFERENCES

Canter, L.W. and Knox, R.C., Ground Water Pollution Control, Lewis Publishers, Inc., Chelsea, Michigan, 1985.

Dean, B.V. and Nishry, M.J., "Scoring and Profitability Models for Evaluating and Selecting Engineering Projects", Journal Operations Research Society of America, Vol. 13, No. 4, July-Aug. 1965, pp. 550-569.

Heath, R.C., "Ground Water Regions of the United States", Water-Supply Paper 2242, 1984, United States Geological Survey, Washington, D.C.

Motts, W.S., et al., "Feasibility of Increasing Water Supplies and Preventing Environmental Damage by Artificial Recharge in Massachusetts", Publication No. 132, Jan. 1983, Water Resources Research Center, University of Massachusetts, Amherst, Massachusetts.

APPENDIX A

ANNOTATED BIBLIOGRAPHY ON ARTIFICIAL RECHARGE

This annotated bibliography is based on a computer-based search of the literature using key descriptor words. A total of 10 bibliographic bases were searched via the Dialog system: Agricola (1970-1978); Agricola (1979-1984); Compendex; Conference Papers Index; CRIS/USDA; Dissertation Abstracts; National Technical Information Service; Pollution Abstracts; Smithsonian Scientific Information Exchange Current Research; and Water Resources Abstracts. A total of 1022 abstracts were identified using key words such as artificial recharge, ground water, and enhanced natural recharge. The total number of abstracts were screened for relevancy and to eliminate duplication. As a result, 342 abstracts are included in this appendix. Each included abstract is categorized into one of the following categories: bibliographic materials (6 abstracts); techniques of artificial recharge (69 abstracts); wastewater recharge (44 abstracts); water quality impacts of artificial recharge (32 abstracts); modeling of artificial recharge (55 abstracts); economic feasibility (7 abstracts); site selection and system design (31 abstracts); case studies (64 abstracts); miscellaneous (13 abstracts); and international examples (21 abstracts). It should be noted that a given abstract could possibly have been placed in more than one category; however, only one was used. Therefore, abstracts with information pertaining to a particular category could be found in other categories.

BIBLIOGRAPHIC MATERIALS

Keith, S.J. et al., "Bibliography on Ground-Water Recharge in Arid and Semiarid Areas", Completion Report, July 1982, Arizona Water Resources Research Center, Tucson, Arizona.

> This bibliography provides references on natural and man-caused recharge in arid and semiarid areas. The literature on natural recharge includes citations on stream channel recharge, mountain recharge, mountain-front recharge, and precipitation-infiltration recharge. The literature on man-caused recharge includes citations on recharge from activities such as irrigation, mining, and waste-water treatment and disposal, but does not include citations on artificial recharge. Most citations are abstracted; all are keyworded.

Knapp, G.L., "Artificial Recharge of Groundwater, A Bibliography", WRSIC-73-202, Feb. 1973, Office of Water Resources Research, U.S. Department of the Interior, Washington, D.C.

> This report, containing 210 abstracts, is another in a series of planned bibliographies in water resources to be produced from the information base comprising Selected Water Resources Abstracts (SWRA). At the time of search for this bibliography, the data base had 50,631 abstracts covering SWRA through December 15, 1972.

National Technical Information Service, "Artificial Recharge of Aquifers, 1970-1983 (Citations from the NTIS Data Base)", PB84-855154, Dec. 1983, U.S. Department of Commerce, Springfield, Virginia.

> This bibliography contains 57 citations concerning the utilization of artificial recharge methods to augment water supplies and improve water quality. Site studies and evaluations, the use of computer models, hydrodynamic and economic feasibility studies, and the use of artificial

recharge as a ground water resource management tool are presented. Engineering aspects of water recovery are also considered.

National Technical Information Service, "Artificial Recharge of Aquifers, 1977-1983 (Citations from the Selected Water Resources Abstracts Data Base)", PB84-859370, Feb. 1984, U.S. Department of Commerce, Springfield, Virginia.

This bibliography contains 149 citations concerning the utilization of artificial recharge methods to improve water availability and water quality. Topics include site studies and evaluation, the use of impounded fresh water and treated waste water, the role of artificial recharge techniques in ground water resource management, individual state programs, contaminant behavior, and the engineering aspects of utilizing spreading ponds and injection wells. Legal and environmental considerations and the use of computer simulations are also treated.

National Technical Information Service, "Floridan Aquifer, 1977-August, 1983 (Citations from the Selected Water Resources Abstracts Data Base)", PB83-871400, Aug. 1983, U.S. Department of Commerce, Springfield, Virginia.

This bibliography contains 89 citations concerning the Floridan aquifer, a major source of ground water in the State of Florida. Topics include formation descriptions, water resource management, water quality and yields, water balance, water level declines due to pumping, artificial recharge using connector wells, and computer simulations for hydrologic evaluation. Sinkhole development and saltwater intrusion are also investigated.

National Technical Information Service, "Ogallala Aquifer, 1970-August, 1983 (Citations from the NTIS Data Base)", PB83-871418, Aug. 1983, U.S. Department of Commerce, Springfield, Virginia.

This bibliography contains 60 citations concerning the Ogallala aquifer, a major source of irrigation water for the great plains states. Topics include formation descriptions, water resource management, natural recharge, well yields, and the impact of declining water levels due to ground water pumping for irrigation. Artificial recharge and water quality analyses are also presented.

TECHNIQUES OF ARTIFICIAL RECHARGE

Agricultural Research Service, "Percolation Rates with Surface Layers of Gravel or Clay Flocs on Soil Fed Clear and Turbid Waters", Water Resources Research, Vol. 14, No. 3, June 1978, pp. 539-541.

Suspended solids in water can plug the surfaces of aquifer recharge basins, and they are often chemically flocculated and removed in costly separate floc-settling basins. This work evaluated floc and gravel filters on recharge surfaces as substitutes for settling basins. Disturbed soils in permeameters were treated with layers of small gravel and constant feedwater heads of either clear tapwater, untreated turbid runoff water, or turbid runoff water treated to form and retain flocs. Untreated turbid water reduced percolation rates more than the other feedwaters, and gravel layers markedly reduced percolation of all feedwaters. The percolation rate of water with flocs was only slightly less than that of tapwater. These results suggested that untreated turbid water or gravel layers are unacceptable for recharge surfaces. However, combinations of soil, gravel, and clay may have considerable potential for reducing percolation in water retention basins. Flocculation and settling seldom remove all fine particles from turbid water; thus a filter layer of flocs may extend the effective life of a recharge surface and eliminate costly separate floc-settling basins.

Aronovici, V.S., Schneider, A.D. and Jones, O.R., "Basin Recharge of the Ogallala Aquifer", Journal of the Irrigation and Drainage Division, American Society of Civil Engineers, Vol. 98, No. IR 1, Paper 8760, Mar. 1972, pp. 65-76.

Ground water recharge basins were tested to determine the feasibility of percolating surface water to the Ogallala formation in the Texas High Plains. The basins were excavated into porous sediments which separate the ground water formation and the slowly permeable surface soil. Initially, two 1/10-acre basins were tested, one with clear well water and the other with turbid water from a playa (wet weather) lake. Percolation rates with clear well water were as high as 7 ft per day, and the sustained percolation rate with turbid water exceeded 1 ft per day. Results of the initial study were used to design and construct a 1.0-acre basin that can be filled by gravity flow from a playa lake. The maximum percolation rate during a 2-week test with clear well water was 2.0 ft per day. The Ogallala formation can be recharged through basins if the slowly permeable surface soil is removed.

Aronovici, V.S., Schneider, A.D. and Jones, O.R., "Basin Recharging the Ogallala Aquifer through Pleistocene Sediments", The Ogallala Aquifer--A Symposium, Special Report No. 39, 1970, International Center for Arid and Semi-arid Land Studies, Texas Tech University, Lubbock, Texas, pp. 182-192.

Recharge in permeable hardland soils of the High Plains of Texas is negligible, and the Ogallala aquifer is cut off from any other appreciable water source. The relief of the hardlands is characterized by numerous playas or depressions in which most storm runoff collects. The playa bottoms are usually blanketed with clay, which retards downward percolation. Between 1.8 and 5.7 million ac-ft annually are wasted by evaporation from playas. A site 600 ft from a playa at Bushland, Texas, was used for basin recharge tests. Two 66-ft square basins with 1:1 side slopes were excavated to expose the unconsolidated Caliche layer approximately 4 ft below ground surface. Neutron moisture meter access tubes and several piezometers were installed. Playa water containing suspended solids was applied. Percolation rates rose until 0.92 pound of sediment per square foot of basin was added. Suspended solids definitely reduce percolation rates, but zooplankton may have had an equally important influence. The main advantage of basin recharge is the relative low cost and ease of sediment removal should it be necessary.

Aronson, D.A. and Seaburn, G.E., "Appraisal of Operating Efficiency of Recharge Basins on Long Island, New York, in 1969", Water Supply Paper 2001-D, 1974, U.S. Geological Survey, Washington, D.C.

Recharge basins on Long Island are unlined pits of various shapes and sizes excavated in surficial deposits of mainly glacial origin. Of the

2,124 recharge basins on Long Island in 1969, approximately 9% (194) contained water 5 or more days after a 1-inch rainfall. Basins on Long Island contained water because (1) they intersect the regional water table or a perched water table, (2) they are excavated in material of low hydraulic conductivity, (3) layers of sediment and debris of low hydraulic conductivity accumulate on the basin floor, or (4) a combination of these factors. A larger percentage of the basins excavated in the Harbor Hill and the Ronkonkoma Morainal deposits contain water than basins excavated in the Outwash deposits. A larger percentage of the basins that drain industrial and commercial areas contain water than basins that drain highways and residential areas. Storm runoff from commercial and industrial areas and highways generally contains high concentrations of asphalt, grease, oil, tar, and rubber particles, whereas runoff from residential areas mainly contain leaves, grass cuttings, and other plant material. Differences in composition of the soils within the drainage areas of the basins on Long Island apparently are not major factors in causing water retention. Evaporation does not significantly concentrate the chemical constituents and, therefore, evaporation is not a major mechanism of water disposal from these basins.

Aronson, D.A., "Artificial Recharge on Long Island, New York", Long Island Water Resources Bulletin LIWR-9, 1978, U.S. Geological Survey, Syosset, New York.

In eastern and central Long Island, New York, the ground water supply has been diminished by the demands of continued population growth and the loss of replenishment since the advent of sewering. In addition, the quality of ground water is threatened locally by the discharge of effluents from septic tanks, cesspools, and industries, and by saltwater intrusion near the shorelines as a result of heavy pumping. The use of highly treated wastewater (reclaimed water) to replenish the ground water supply on Long Island seems to be both feasible and economical. Two methods that have been tested are injection of reclaimed water into the ground through wells, and ponding it in basins for infiltration to the water table. Major hindrances to these methods are well clogging by suspended particles in the injectant and clogging of the infiltration surface of the basin. Both can be controlled by adequate treatment of reclaimed water prior to recharge and by proper maintenance of infiltration surfaces.

Baffa, J.J., "Injection Well Experience at Riverhead, New York", Journal of the American Water Works Association, Vol. 62, No. 1, Jan. 1970, pp. 41-46.

Injection of fresh potable water was studied in the field and laboratory to compare the hydraulic characteristics of the injection mound with the pumping drawdown characteristics. Injection requires establishment of a boundary pressure for entrance of the injection water into the aquifer pores. Injection was accomplished in two ground wells of different design and in two wells contained in a ground simulator tank so constructed as to permit the sampling of the aquifer material opposite the well screen after pumping, injection and redevelopment. The clogging rate of the double-cased well was less than that of the single-cased well by 0.06 ft per day as measured by the water level inside the casing. The initial specific capacity of the double-cased well was about half of the well with the gravel pack, but after fresh water injection and reclaimed waste water injection it was possible to restore its initial capacity by surging and pumping. After reclaimed waste water injection into the single-cased well its specific capacity was markedly reduced and could not be restored.

Bassett, R.L. et al., "Preliminary Data from a Series of Artificial Recharge Experiments at Stanton, Texas", Open File Report 81-149, 1981, U.S. Geological Survey, Denver, Colorado.

A series of artificial recharge experiments was conducted by the U.S. Geological Survey at an experimental site located in Stanton, Texas. Five tests were performed from March 1977 through December 1978 to: (1) evaluate the hydraulic properties of the aquifer; (2) test sampling and monitoring equipment; (3) compare tracers for future use in hydrologic investigations; and (4) determine the radial and vertical distribution of hydraulic properties at the site. Suites of inorganic, and both volatile and nonvolatile organic tracers were used in the tests, and comparative data were obtained from sampling points at several radial distances and depths from the injection well. Hydraulic data from aquifer tests and geologic data from core material also were obtained during the investigation.

Bianchi, W.C., Nightingale, H.I. and McCormick, R.L., "A Case History to Evaluate the Performance of Water-Spreading Projects", Journal of the American Water Works Association, Vol. 70, No. 3, Mar. 1978, pp. 176-180.

Studies on ways to predict the performance, costs, and benefits of artificial ground water recharge were made for the Fresno, California, Leaky Acres Project. Prediction errors of site recharge rates were as much as 20 times greater than the final operational project rate when soil cores were used, while pilot spreading with proper instrumentation for sublayer perching decreased estimates to within 1.5 times. Recharge costs are difficult to evaluate on a user-benefit basis. Using a zonal response and political boundary approach, 79% of the benefit of recharge benefited city users. The use of the recharge water, however, requires an energy-consuming process, and energy costs soon will equal the operational costs of recharge.

Bianchi, W.C., Nightingale, J.J. and McCormick, R.L., "Fresno, California, Subsurface Drain Collector-Deep Well Recharge System", Journal of the American Water Works Association, Vol. 70, No. 8, Aug. 1978, pp. 427-435.

A system was described that greatly expands the geological profiles that can be used as recharge areas. Water was filtered through sandy soil, collected in an agricultural drain placed in the shallow-perched water table, and gravity-fed into a specially designed recharge well. Recharge volume was increased by 75 percent. The collected water was low in dissolved gases and sufficiently low in suspended solids to allow gravity well recharge. The recharge well was designed to allow controlled mining of clogged aquifer sands. A system employing this technology could greatly expand the soil profiles that can be used as recharge areas by providing a way to bypass low-permeability subsurface horizons limiting vertical water movement into the ground water table. The successful performance of the Fresno drain collector-recharge well system, as indicated by 600 ac-ft of recharge water, results from preconditioning surface canal water by soil filtration before injection.

Bouwer, H., "Groundwater Recharge Design for Renovating Waste Water", Journal of the Sanitary Engineering Division, American Society of Civil Engineers, Vol. 95, No. SA 1, Paper 7096, Feb. 1970, pp. 59-74.

Soil and hydrogeologic conditions permitting, ground water recharge by surface spreading is an economic and esthetic way for further treatment or renovation of secondary sewage effluent, cannery wastes, or similar low-quality water. The design of a system of spreading areas with wells or drains for collecting the reclaimed water should generally be based on: (1) a maximum limit for the elevation of the water table mound beneath the spreading areas; (2) a minimum limit for the underground detention time and travel distance as the water moves to the wells or drains; and (3) minimum contamination of the ground water in the aquifer outside the recharge system. Hydraulic properties of aquifers, including anisotropy, can be determined by analog and by field measurements. These data are used to evaluate the effective transmissibility for recharge, and to calculate water-table positions and underground detention times for a system of long, parallel spreading strips with wells midway between the strips.

Brock, R.R., "Hydrodynamics of Artificial Groundwater Recharge", Completion Report, Feb. 1974, School of Engineering, University of California, Irvine, California.

Ground water recharge to a horizontal water table and to a semipervious layer was investigated. The saturated flow equations for a homogeneous isotropic medium were used to find theoretical solutions. Experiments were conducted in a laboratory sand model. Field observations of recharge mounds developed beneath square ponds near Fresno, California, were also utilized. For the horizontal water table case the classical Dupuit-Forchheimer (D-F) theory and the more exact potential nonlinear (NL) version was used. Solutions for a dimensionless water table rise were obtained using the LDF, NLDF, LP and NLP theories for a strip basin. Most of these were obtained by a finite-difference solution of the equations. For a square basin the LDF and NLDF theory were used. In general it was found that for a shallow flow (which occurs when the ratio of initial depth to basin width is sufficiently small) and small water table rise the LDGF theory is valid, whereas for larger rises the nonlinear effects are important and the NLDF theory must be used. The laboratory results for shallow flows were in close agreement with the NLDF theory. For deep flows the D-F theories are inadequate and the potential theories must be used. The NLP theory was used to obtain detailed flow patterns for both a shallow and deep flow case. For perched mounds on a semi-pervious layer the NLDF theory was used to develop solutions for the mound shape for strip and square basins. Laboratory experiments for steady state perched mounds beneath a strip basin were conducted in a sand model. For small leakage rates the NLDF theory compares favorably with the experimental results. For large leakage rates the NLDF theory is not adequate, probably because of relatively large vertical velocities. A method was developed to determine the permeabilities of the semipervious layer and sand above the layer from observed steady-state mound shapes and recharge rates. The method is illustrated with field data from a square pond.

Brown, R.F. and Signor, D.C., "Artificial-Recharge Experiments and Operations on the Southern High Plains of Texas and New Mexico", Water Resources Investigations 10-73, May 1973, U.S. Geological Survey, Austin, Texas.

The southern High Plains of Texas and New Mexico is an area where ground water is being mined for irrigation, municipal supply, and industrial use. Artificial recharge experiments and operations have been undertaken to develop methods to supplement the declining ground water supplies in the Ogallala formation, a sand and gravel aquifer of tertiary age. Experiments using highly turbid water from playa lakes for injection into the Ogallala formation have resulted in greatly decreased yield and limited life for the recharge wells. Recharge of ground water and surface water of good quality have indicated, however, that injection through wells is an effective method of recharging the aquifer. Water that is slightly turbid can be injected into wells successfully for a period of time, but in most experiments, the use of turbid water resulted in constantly declining yields of the wells and constantly declining capacities for recharge. Redevelopment of wells through pumping and surging will move some of the sediment under certain conditions and significantly prolong the life of the recharge wells. Surface spreading is little practiced, but locally may be a feasible means of artificial recharge.

Brown, D.L. and Silvey, W.D., "Artificial Recharge to a Freshwater-Sensitive Brackish-Water Sand Aquifer, Norfolk, Virginia", Professional Paper 939, 1977, U.S. Geological Survey, Reston, Virginia.

Fresh water was injected into a brackish-water sand for storage and retrieval. The initial injection rate of 400 gpm decreased to 70 gpm during test 3. The specific capacity of the well decreased also, from 15.4 to 0.93 gpm. Current-meter surveys indicated uniform reduction in hydraulic conductivity of all contributing zones in the aquifer. Hydraulic and chemical data indicate this was caused by dispersion of the interstitial clay upon introduction of the calcium bicarbonate water into the sodium chloride bearing sand aquifer. The clay dispersion also caused particulate rearrangement and clogging of the well screen. A pre-flush of 0.2 N calcium chloride solution injected in front of the fresh water at the start of test 4 stabilized the clay. However, it did not reverse the particulate clogging that permanently reduced permeability and caused sanding during redevelopment. Clogging can be prevented by stabilization of the clay using commercially available trivalent aluminum compounds. Test 1 and test 2 showed that 85 percent of the water injected can be recovered, and the water meets U.S. Public Health Standards. Storage of fresh water in a brackish-water aquifer appears feasible provided proper control measures are used.

Bush, P.W., "Connector Well Experiment to Recharge the Floridan Aquifer, East Orange County, Florida", Water Resources Investigations 78-73, 1979, U.S. Geological Survey, Tallahassee, Florida.

An experimental connector well, screened in the shallow sand aquifer, finished with an open hole in the Floridan aquifer, and cased through the confining layer between the two aquifers, was drilled in east Orange County, Florida, to obtain information on the nature and function of the shallow aquifer as related to connector well operation. The potentiometric surface of the shallow aquifer is about 45 ft higher than the potentiometric surface of the Floridan aquifer; hence water flows by gravity from the shallow aquifer to the Floridan aquifer through the well "connecting" the two aquifers. Continuous flow measurement over 10 months shows the well discharge varies seasonally and averages slightly more than 50 gallons per minute. Observation wells show that except for seasonal variation water levels within the area of influence have reached steady state within measurable limits. Vertical anisotropy in the shallow aquifer is apparently caused by the shape and/or arrangement of the sand grains that comprise the shallow aquifer, rather than distinct confining layers of different lithology. Transmissivity of the shallow aquifer at the site is about 600 square feet per day. Extensive dewatering of wetlands in east Orange County by connector wells alone is probably not feasible. Nevertheless, large amounts of water could be channeled to the Floridan aquifer by connector wells. The results of the connector well experiment imply that water is being captured from evapotranspiration and runoff in the vicinity of the connector well.

Cluff, C.B. et al., "Development and Evaluation of Water Harvesting Systems", Report No. OWRR-A-001-ARIZ(6), Dec. 1969, Arizona Water Resources Research Center, University of Arizona, Tucson, Arizona.

Water harvesting systems were evaluated in this study. These systems consisted of gravel-covered plastic and sodium-treated catchments with plastic-lined, butyl-covered storage tanks and underground storage into unused aquifers through artificial recharge. A plastic-laying gravel spreader that attaches to a dump truck was developed and used to install a one-half acre gravel covered plastic catchment. The catchment supplies water to a 100,000-gallon capacity storage tank which is lined with 6-mil polyethylene plastic and covered with a butyl rubber cover. Field tests using sodium chloride were conducted. These tests indicate that the sodium chloride treatment does reduce the infiltration rate on sandy loam soils. The overall increase for the year following treatment was approximately 35 percent. Field tests made using a recharge pit and a recharge well indicated that for small quantities of harvested, sediment free water, the recharge well would be recommended for injection into underground storage, substantial water quality changes could occur in percolating water when using the recharge pit. The same injection well could be used to recover the recharged waste.

Conselman, F.B., Johnson, P. and Crawford, D.A., "Underground Storage of Texas Playa Lake Waters by Injection into the Ogallala Formation under Moderate Pump Pressure", <u>International Symposium on Water Resources Planning</u>, Dec. 1972, Mexico City, Mexico.

An update is presented on an investigation being conducted to determine the feasibility of recharging the Ogallala formation by injection of surface playa lake waters in the semi-arid high plains of west Texas. Ogallala ground water is now being mined, since there are no rechargeable outcrops to the aquifer. Runoff collects in shallow playas since there is no developed drainage in the area. By injection of such unused waters into the Ogallala, it is hoped to establish a method of recharging the aquifer with good quality water that would help end the need for mining the aquifer. Past studies have shown that aquifer clogging by suspended solids and colloids in playa water is a significant problem. In this study, effects of injection are monitored by observation wells, with quality determinations showing playa water better than that of the Ogallala. The major result of this study is that the injection well clogging has been overcome by wells which apply moderate pump pressure and maintain sufficient fluid turbulence at aquifer entry. Flow rates have been attained near 2000 gpm and one well has successfully injected 64 ac-ft of playa water at 1500 gpm at 55 surface psi. Low cost injection wells using good quality waste playa water may increase the aquifer storage in the area to a large extent while offering flood control and new drained agricultural lands.

Deutsch, M., "Artificial Recharge by Induced Interaquifer Leakage", Symposium of Haifa, Artificial Recharge and Management of Aquifers, Mar. 19-26, 1967, Publication No. 72, 1967, International Association of Scientific Hydrology, pp. 159-172.

The U.S. Geological Survey and the City of Kalamazoo, Michigan, conducted tests to demonstrate the feasibility of recharging an artesian glacial-drift aquifer by inducing leakage to it from a surficial sand aquifer. The surficial sand is under water table conditions and in direct hydraulic connection with a creek draining it. Coefficients of leakage for the aquiclude ranged from 0.042 to 1.3 gpd/sq ft/ft. The piezometric surface in the lower aquifer was higher than the water table in the upper sand, and leakage through the aquiclude was upward. Withdrawal of water from the lower aquifer reversed the direction of leakage, but sustained pumping would have resulted in objectionable declines in the piezometric surface and probable inflow of poor-quality water from below. Tests were run by pumping from the lower aquifer under different conditions of artificial recharge of the upper aquifer. Each test resulted in increased head differences between the upper and lower aquifers, which increased the rate of leakage and raised water levels in the lower aquifer. The rates of leakage were 1,600-1,900 gpm, respectively. After six years of pumping at the well field with artificial recharge, water levels have not declined appreciably.

Dvoracek, M.J. and Peterson, S.H., "Artificial Recharge in Water Resources Management", Report No. OWRR-B-041-TEX(4), 1970, Department of Agricultural Engineering, Texas Tech University, Lubbock, Texas.

Water resources for arid and semiarid regions of the United States are only marginal at the present time and new sources of water are no longer available. Artificial recharge techniques are presented as one method of partially alleviating the problem. Artificial recharge is presented in the context of the water depletion of the Ogallala formation on the High Plains of Texas. There, as in the United States as a whole, irrigated agriculture is the major water user. Several recharge mechanisms are discussed, such as recharge wells, shafts, holes, pits, trenches, rubble cones, and water spreading. All of the methods share the problem of sediment content in water used for recharge, but the problem is not insurmountable. In some cases, the economic value of artificially recharged water has already outweighted the limitations.

Dvoracek, M.J. and Peterson, S.H., "Recharging the Ogallala Formation Using Shallow Holes", Proceedings of Hydrology and Water Resources in Arizona and the Southwest, Vol. 1, Apr. 22-23, 1971, Arizona Section-American Water

Resources Association and the Hydrology Section-Arizona Academy of Science, Tempe, Arizona, pp. 245-259.

The southern bed of the Ogallala aquifer is hydrologically isolated from all outside areas of recharge, requiring local precipitation for all natural recharge. Current withdrawals are so much greater than natural recharge that it appears that artificial recharge affords the only means of establishing at least a pseudo-balance. A number of observation wells were drilled at Texas Tech University, and subsequently capped until recharge water became available. The initial recharge was 2.5 AF over 12 days, at a rate of 120 gpm for about the first day, after which 60 gpm was relatively constant. Approximately 1 month later, 1.2 AF were recharged over 3 days at rates ranging over 140-90 gpm. It became evident that a cavity was present at the bottom of the hole being recharged. On a later recharge occasion, the cavity seemed to have enlarged. During a period of 2 years, more than 28 AF of surface runoff water have been recharged through the shallow hole with increases in recharge rates for each subsequent recharge period. The nature of this phenomenon and the cavities are not understood. This may represent the long sought after answer to recharge of the aquifer, but much more extensive research needs to be done.

Dvoracek, M.J. and Peterson, S.H., "Development of Systems for Ground Water Recharge into the Ogallala Formation", OWRR-B-041-TEX(6), June 1971, Water Resources Center, Texas Tech University, Lubbock, Texas.

Water is the most important limiting factor to agricultural production and economic growth in the southern High Plains of Texas. The use of ground water from the Ogallala Formation has helped to remove these water limitations but the remaining water in the Ogallala aquifer is rapidly being depleted. Research was concentrated on the removal of sedimentation as a restriction to artificial recharge. Gravity flow recharge wells appear to be an effective means of recharge. As particle size of the aquifer decreases, increasing attention was given to the possible effects of the introduction of foreign materials into the formation. Other ground water recharge methods discussed are filtration systems, trenches, shafts, and shallow holes.

Dvoracek, M.J. and Wheaton, R.Z., "Development of Systems for Ground Water Recharge into the Ogallala Formation", OWRR-B-016-TEX(2), May 1969, Water Resources Center, Texas Tech University, Lubbock, Texas.

Constructed pits were evaluated as mechanisms for artificial ground water recharge into the Ogallala formation of the High Plains of Texas. The pits proved to be efficient recharge mechanisms when the sediment problem was nonexistent. Drying and freezing or overwintering of this sediment permitted partial recovery of recharge rates without additional maintenance. Buried tile drains used as filter systems for water to be used in recharge were studied. Filter media of sands, gravels, sand-gravel combinations, and organic material were used with sands and gravels proving most effective. Waters resulting from this system were recharged through a multipurpose well without detrimental effects. Some clarification of water is required for economical and efficient artificial recharge or surface runoff waters through multiple purpose wells and constructed pits.

Dvoracek, M.J. and Wheaton, R.Z., "Does Artificial Ground Water Recharge Contaminate Our Ground Waters", OWRR-B-041-TEX(3), 1970, Department of Agricultural Engineering, Texas Tech University, Lubbock, Texas.

Artificial recharge practices produce the possibility that ground waters may be polluted or contaminated by undesirable substances introduced with the recharge waters. Various methods of artificial recharge and the contamination potential of each method are discussed.

Dvoracek, M.J., "Use of Wells and Pits to Recharge Underground Formations in Semi-Arid Areas", International Conference on Arid Lands in a Changing World, June 3-13, 1969, June 1969, University of Arizona, Tucson, Arizona.

A general discussion is given of the different methods that can be employed in artificial ground water recharge in semi-arid areas. Emphasis is placed on the need to collect and retain large amounts of summer rainfall before it is lost through mass evaporation and surface runoff. The use of wells and specially constructed pits in addition to trenches and rubble cones as collection reservoirs for artificial ground water recharge is discussed and suggested.

Espina, J.M., "The Case for Artificial Recharge", The Johnson Drillers Journal, Vol. 52, No. 1, Jan.-Feb. 1980, pp. 7-11.

Artificial recharge can be used for waste disposal, secondary oil recovery, land subsidence problems, salt-water intrusion control, and ground water storage. Various methods of artificial recharge include: water spreading, pit recharge, excavations, wells and shafts, and pumping. Water spreading can be done in a basin, a series of ditches or furrows, a natural stream channel, or irrigation canals. Alternating wet and dry periods in a basin results in greater total recharge than continuous spreading. Water spreading is not effective in areas where subsurface strata restrict the downward passage of water or where the cost of land is prohibitive. In these areas recharge can be conducted through injection wells. Recharge rates rarely equal pumping rates due to clogging of the aquifer with silt and bacteria. Induced recharge involves pumping of ground water from a well adjacent to a lake or stream and can result in high-quality water. The tangible and intangible benefits of artificial recharge are outlined along with the physical requirements for appropriate storage, movement and use of recharge water. The major alternative, surface storage, costs more to construct and is subject to evaporation losses and seasonal availability of water.

Ganus, W.J., "Project Aquarius--Geohydrologic Investigation", CONF-700101, Vol. 2, Jan. 1970, National Technical Information Service, U.S. Department of Commerce, Springfield, Virginia, pp. 1174-1183.

A general discussion of geological and hydrologic conditions for a canyon site in north-central Arizona is given. Ground water recharge by cratering in alluvial materials is rated as the second most favorable application and a site in southeastern Arizona is discussed. Other applications are presented but hold less promise due to lack of geohydrologic and nuclear explosion data necessary for proper evaluation. Primary consideration was given to distance of site from population centers and to the quantity of water that could be developed and managed. For surface water resource development by nuclear explosions, additional considerations for site selection included canyon shape for landslide, throw-out or rock-fill techniques, canyon meanders necessary for diversion channels, low permeable rock types for minimizing problems of radioisotope migration, and a clear identification of any ground water system that could be affected. In the site selection for recharge craters, considerations included adequate subsurface storage space, aquifer characteristics favorable to recharge and recovery, subsurface lithologic type and distribution that would permit control of radioisotopes by either monitoring or removal, suspended sediment load of water source, and estimated increased efficiency of artificial recharge over natural recharge.

Gass, T.E., "Artificial Recharge: Part II", Water Well Journal, Vol. 33, No. 2, Feb. 1979, pp. 32-33.

Surface spreading, the most common method of artificial recharge, relies on infiltration to be effective. Various factors which influence infiltration rates are listed. The types of surface spreading systems include: basins or ponds, ditch or furrow networks, flooding systems, and stream bed modification. Before construction of a surface spreading system, a survey should be conducted to determine the best type of system for the site, potential effects on water quality and the hydrologic budget and possible changes in plant and animal life. Also, a pilot recharge system should be built to finalize design specifications and develop efficient operation and maintenance procedures. General engineering principles for construction of recharge systems are outlined. Recharge problems are caused by poor quality recharge water, changes in soil structure or soil biology, and mechanical clogging. Clogging can be reduced by constructing a settling basin behind the main basin. Both recharge basins and settling basins should be operated intermittently to allow time for drying. Alternative basins should be available to allow other basins to dry out.

Goss, D.W. et al., "Fate of Suspended Sediment During Basin Recharge", Water Resources Research, Vol. 9, No. 3, June 1973, pp. 668-675.

Suspended sediment was tagged with the radioisotope Cs-134 to determine the extent of its movement into materials underlying recharge basins and to aid in evaluating the effect of its accumulation on infiltration rates and the basin life. When openings of naturally occurring large pores were allowed to remain at the basin surface, 50% of the sediment moved deeper than 18 inches. However, when large-pore openings at the surface were destroyed by cultivation, over 90% of the suspended sediment was retained in the upper 1 inch. Most of this sediment could easily be removed. The suspended sediment that moved deeper than about 0.15 inch (the depth removed after each recharge cycle) did not noticeably reduce infiltration rates in an experimental basin.

Goss, D.W. and Jones, O.R., "Movement and Accumulation of Suspended Sediment During Basin Recharge", Underground Waste Management and Artificial Recharge, Vol. 1, 1973, pp. 468-478.

The movement and accumulation of sediment suspended in water used for recharge were determined by microscopic examination of thin sections from material underlying a recharge basin. The basin had been infiltrated to the depth of 91 m by water averaging 191 ppm suspended sediment in six recharge cycles. Flakelike structures were found on the upper 2.5 cm. Two horizontal sheetlike structures, 0.1 mm thick, were found between the depths of 5-8 and 8-12 cm. Fillings in voids, mostly between 2.5 and 23 cm were filled. The average pore volume lost between the depths of 2.5 and 5 cm was 1.5 percent, and, below 5 cm, less than 1 percent. Porosity was maintained by freezing and thawing and by wetting and drying. Infiltration rates of the basin have not been noticeably reduced. The successful use of excavated basins for recharge of turbid water should not be limited by the movement and accumulation of suspended sediment in the basin material.

Hauser, V.L. and Lotspeich, F.B., "Artificial Groundwater Recharge through Wells", Journal of Soil and Water Conservation, Vol. 22, No. 1, Jan.-Feb. 1967, pp. 11-15.

A study was conducted in the Texas High Plains to determine whether the Ogallala formation could be successfully recharged using local surface runoff. Water was injected into the aquifer through 6-inch wells. A field recharge system effectively removed more than 90 percent of the suspended solids by flocculation and settling after treatment with a cationic polyelectrolyte and alum. The aquifer near the wells was gradually clogged during recharge by the remaining solids, however, the wells were successfully redeveloped by bailing. Data is presented on the flow rates into wells, total water recharged, as well as electrical conductivity, and coliform organisms before and after ground water recharge. In arid regions recharge of playa lake water is an important method for conserving water.

Johnson, A.I. and Sniegocki, R.T., "Comparison of Laboratory and Field Analyses of Aquifer and Well Characteristics at an Artificial Recharge Well Site", Symposium of Haifa, Artificial Recharge and Management of Aquifers, Mar. 19-26, 1967, Publication No. 72, 1967, International Association of Scientific Hydrology, pp. 182-192.

Research on artificial recharge of alluvial deposits by means of recharge wells was carried out in the Grand Prairie Region of Arkansas. Extensive test-drilling and construction of observation wells within the 24 sq mi research site provided detailed information on the geology and hydrology of the deposits of quarternary age; aquifer characteristics were determined by means of laboratory analysis of samples collected from the pilot test holes for the recharge wells and by conventional pumping-tests in the field. On the basis of the laboratory analyses, a transmissibility of 60,000 gpd/ft, and a storage coefficient of 0.34 was predicted for the recharge test site. Later pumping tests and recharge tests in the field confirmed that the values of transmissibility predicted from laboratory analysis were comparable to those obtained by the field test. Slow drainage from the fine-textured portions of the aquifer resulted in a change in the storage coefficient with time. However, the field data showed that the storage coefficient probably would have reached the value predicted by laboratory analyses if the field test had been run for a much longer period of time.

Johnson, P., "Application of Surface Pressure to Assist Water Recharge into the Ogallala Formation", The Ogallala Aquifer--A Symposium, Special Report No. 39, 1970, International Center for Arid and Semi-Arid Land Studies, Texas Tech University, Lubbock, Texas, pp. 193-204.

Field studies were performed in west Texas to develop economical techniques of recharging the Ogallala formation at high rates by application of pump pressure. Porous, dewatered zones in the Ogallala are ideal water-storage sites in this developing area. Sources of projected recharge water are excess rainfall and imported water. Cored and logged test wells, a 20-inch diameter, 107-ft cased and 55-ft torch perforated recharge well, and 3 observation wells were drilled, developed and instrumented. The pressure pump was a double-suction centrifugal, 200 gallon per minute capacity at 100 psig and 1760 rpm, powered by a 6-cylinder, 175-hp gasoline engine. Tests from 1964 to 1966 were performed by gravity flow, where a maximum injection rate was 525 gpm, and decreased to 37.5 gpm after 5 ac-ft were recharged. Pressure recharging permitted 1,920 gpm recharge rates at 85 psig with declining pressure to 55 psig after pumping 3 1/2 hours at maximum rate. A cumulated 69,920 lbs of fine suspended particles were injected. Costs were abnormally high (fuel cost of $4.27/ac-ft) because of poor design. Projected costs of $1.95/ac-ft are reasonable. Estimated total costs to develop a recharge well capable of 1700 gpm at 50 psig is $3,969. Pressure recharge is economical, rapid, and efficient, and permits maximum utilization of land.

Jones, O.R., Goss, D.W. and Schneider, A.D., "Surface Plugging During Basin Recharge of Turbid Water", Transactions of the American Society of Agricultural Engineers, Vol. 17, No. 6, Nov.-Dec. 1974, pp. 1011-1019.

Three experimental recharge basins were excavated to determine if artificial recharging of playa water on the southern high plains to the Ogallala aquifer is practical. To establish guidelines and standards for the operation of such basins, it is necessary to develop relationships between the permeability of the pleistocene sediments that overlie the Ogallala aquifer, suspended solids content of the water supply, and quantity of playa water that can be recharged before plugging limits recharge to critically low rates. Most suspended solids in playa water were deposited on or near the surface during basin recharge, except where cracking of the basin surface allowed sediment deposition at undetermined depths. Although plugging was related to the amount of suspended solids introduced into the basin with the recharge water, the total amount of suspended solids required to cause plugging varied by an order of magnitude between tests. The major factor limiting recharge of turbid water through excavated basins was plugging on or near the surface. While substantial quantities of turbid water can be recharged, pretreatment to remove part of the suspended solids load should improve basin performance.

Keys, W.S. and Brown, R.F., "Role of Borehole Geophysics in Underground Waste Storage and Artificial Recharge", Underground Water Management and Artificial Recharge, Vol. 1, 1973, pp. 147-191.

Borehole geophysics is being used by the U.S. Geological Survey to study geohydrologic parameters related to recharging the Ogallala formation on the High Plains of Texas. The optimum utilization of underground space for the emplacement and storage of waste and surface water can be achieved through an understanding of the geohydrologic environment. The emplacement of liquid waste and the artificial recharge of aquifers generally requires the drilling of exploration, injection, and monitoring wells. Some geophysical logs are run on almost all wells drilled for deep disposal. Logs provide data on the location, thickness, and lateral continuity of storage zones and confining beds, amount and distribution of total or effective porosity, and the relative magnitude of permeability. Intergranular and fracture porosity can be discriminated by cross-plotting acoustic-velocity and neutron or gamma-gamma logs. The distribution and orientation of preinjection fractures can be determined by acoustic televiewer logs. Logs provide data on the chemical quality of the native fluids and the mineralogy of the aquifer, which are necessary to predict chemical reactions with injected fluids. The temperature and conductivity of the interstitial fluids may be measured directly and their specific gravity and viscosity may be calculated from log data. After waste injection or artificial recharge has started, logs provide in-situ measurements of changes in the system.

Keys, W.S. and Brown, R.F., "The Use of Well Logging in Recharge Studies of the Ogallala Formation", <u>The Ogallala Aquifer--A Symposium</u>, Special Report No. 39, 1970, International Center for Arid and Semi-Arid Land Studies, Texas Tech University, Lubbock, Texas, pp. 31-48.

The Geological Survey initiated a major study of artificial recharge at Lubbock, Texas, in fiscal year 1969. This paper deals with the use of geophysical well-logging in investigations of artificial recharge through water spreading on the land surface. One of the major research efforts is to develop methods for measuring or estimating the vertical permeability of the unsaturated zone so that favorable sites for surface spreading can be selected on the basis of drilling, logging, and testing. Logging is used to guide construction of air-permeability holes and may be used to provide porosity and moisture data for the calculations. One of the purposes of well logging and sampling is to describe the unsaturated part of the Ogallala formation with sufficient accuracy to permit development of a correlation between air and water permeabilities and other lithologic parameters.

Kimbler, O.K. and Whitehead, W.R., "Effect of Viscosity Ratio on the Recovery of Fresh Water Stored in Saline Aquifers", PB-270 780, Apr. 1977, National Technical Information Service, U.S. Department of Commerce, Springfield, Virginia.

Differences in viscosity between injected and displaced fluids will always result in a lower recovery efficiency during underground water storage in aquifers than would have been realized had the viscosities been equal. This result arises from (1) increased dispersion and (2) viscous fingering (gross frontal irregularities). Experimental studies involving injection and subsequent production from single wells in two physical flow systems (miniaquifers) suggest the following conclusions: (1) the recovery efficiency is a maximum where the fluid viscosities are equal; (2) from an engineering point of view, the recovery efficiency is approximately the same for a fixed viscosity difference irrespective of whether injected or native water is the more viscous; (3) a computational procedure previously developed for a viscosity of unity is applicable for a non-unity viscosity ratio provided an effective dispersivity is used which has been obtained from an injection/production well test; and (4) dispersivity coefficients obtained from the production profile of an observation well rather than the injection well represent highly unrealistic values and if used in the computational procedure will yield false values of recovery efficiency.

Kincannon, D.F. and Tiederman, W.G., "Water Reclamation for Ground Water Recharge", OWRR-A-034-OKLA(1), June 1972, Oklahoma Water Resources Research Institute, Oklahoma State University, Stillwater, Oklahoma.

The purpose of this study was to develop solid-liquid separation techniques so that surface waters can be used to effectively recharge a ground water aquifer. Specifically, the research effort centered around: (1) the development of an efficient, low pressure drop 2 gpm hydrocyclone, (2) the determination of the appropriate chemical treatment for clay-laden water when hydrocyclones are used, and (3) infiltration studies with a laboratory aquifer. The statistical experimental design techniques did yield a significantly improved hydrocyclone design. Furthermore, the separation performance of the "best" hydrocyclone could be improved from an efficiency of 65 percent to an efficiency of 78 percent when the correct chemical treatment is used. However, it is possible for chemical treatment to actually decrease the removal efficiency. The infiltration studies show that a high degree of separation efficiency is required before clay-laden waters can be used as recharge waters. However, further study is needed before direct application of hydrocyclones is made for recharging aquifers with reclaimed surface waters.

Knochenmus, D.D., "Hydrologic Concepts of Artificially Recharging the Floridan Aquifer in Eastern Orange County, Florida--A Feasibility Study", Report of Investigations No. 72, 1975, Florida Bureau of Geology, Tallahassee, Florida.

In East Orange County, Florida, the hydrologic system cycles about 55 inches of rainfall per year of which 10 to 15 inches run off, 40 to 45 inches evapotranspire, and 0 to 2 inches recharge the Floridan (artesian) aquifer. The water level in the nonartesian aquifer is at, or near, land surface and as much as 30 ft above the potentiometric surface of the Floridan aquifer. Little downward leakage (recharge) takes place because the nonartesian aquifer is separated from the Floridan aquifer by a thick sequence of clayey sands. The supply of water for artificially recharging the Floridan aquifer with a network of optimally designed and spaced connector wells (wells that tap both the nonartesian and Floridan aquifer) would be derived from the capture of potential water losses--evapotranspiration and runoff. Water would drain from the nonartesian aquifer into the Floridan aquifer and lower the water table. Lowering of the water table would reduce evapotranspiration losses and provide storage capacity for rainfall that would otherwise run off.

Krone, R.B., "Borehole Recharge: The Compatability of Recharge Water with the Aquifer", Artificial Groundwater Recharge Conference, University of Reading, England, Sept. 21-24, 1970, Vol. 2, Paper No. 10, June 1971, The Water Research Association, Marlow, England, pp. 263-277.

For waters to be recharged through wells, the dissolved gas concentrations should be less than the saturation concentrations at temperatures and pressures that will occur in the aquifer. Aeration should be avoided and the recharge water temperature should be the same or greater than that of the aquifer. The dissolved solids concentration should be such that a sodium absorption ratio less than 3 and as low as is feasible be maintained, and the concentrations of dissolved substances should be suitable for the most sensitive use at the discharge sites. Precipitation of compounds resulting from a combination of recharge water and aquifer water should be avoided. The suspended solids concentration should be as low as is attainable. The recharge waters should be chlorinated to minimize biological growths. While these requirements are restrictive, they are justified in most cases by the high permeability necessary for a satisfactory well, by the high costs of constructing recharge wells, and by the uncertain results of clogged well rehabilitation.

Lichtler, W.F., Stannard, D.I. and Kouma, E., "Investigation of Artificial Recharge of Aquifers in Nebraska", USGS/WRI-80-93, Sept. 1980, U.S. Geological Survey, Lincoln, Nebraska.

Large withdrawals of ground water for irrigation are causing progressive declines of ground water levels in some areas of Nebraska. An investigation was conducted to determine the technical feasibility of artificially recharging aquifers through wells and by surface spreading. The experiments showed that large quantities of water can be recharged through wells and by surface spreading if conditions are favorable. In the well experiments, about 0.5 billion gallons of water was withdrawn from a well near the Platte River and recharged through a well into a Pleistocene sand and gravel aquifer near Aurora where ground water levels are declining. The recharge rate was about 730 gallons per minute during two tests of 6 and 8 months duration. There was no evidence of clogging due to chemical reaction, air entrainment, or bacterial growth. Some clogging of the aquifer did occur in the immediate vicinity of the recharge well due to a small amount of fine sediment in the recharge water. Analysis of water-level buildup in the recharge well during the 6-month test indicated that recharge could have continued at a rate of about 700 gallons per minute for several years before rehabilitation of the recharge well would have been necessary. Infiltration rates from surface-spreading experiments ranged from 0.5 ft per day in the loess-covered plain to 11 ft per day in the sandhills. Perching layers of low permeability existed at each site and could cause problems in recharging.

McCormick, R.L., "Filter-Pack Installation and Redevelopment Techniques for Shallow Recharge Shafts", Ground Water, Vol. 13, No. 5, Sept.-Oct. 1975, pp. 400-405.

The potential for artificial ground water recharge through shallow shafts was conducted at the Leaky Acres Recharge Project in Fresno, California. The research was directed toward developing methods for constructing and placing filter-pack materials and procedures for periodically cleaning filter materials to regenerate initial intake rates. Preliminary results are presented of two differently constructed shallow shafts. Screened canal water was recharged into the shaft. In a combined period of 94 days, 50,800 cubic meters of water were recharged at an average rate of 6.25 liters per second. Shaft No. 1 was regenerated to 109 percent of its original rate, but shaft No. 2 was regenerated to only 78 percent of its original rate. Redevelopment with a submersible pump was more efficient than a centrifugal pump in removing the clogging particles from the shaft-filter-pack materials. Piezometric head losses indicated that most of the clogging that occurred was contained within the filter-pack material.

McWhorter, D.B. and Brookman, J.A., "Pit Recharge Influenced by Subsurface Spreading", Ground Water, Vol. 10, No. 5, Sept.-Oct. 1972, pp. 6-11.

Artificial recharge from pits, trenches, or basins is feasible in situations where thick layer or slowly permeable material exists near the surface provided that a much more permeable stratum exists above the less permeable one. The more highly conductive layer acts as a subsurface spreading basin which greatly enhances the opportunity for recharge. The potential for recharge from pits penetrating a highly conductive layer underlain by a slowly permeable stratum can be calculated from the results of a simple theoretical analysis. A narrow pit or trench 525 ft long and 10 ft deep has the potential for recharging at a rate of 1 million gpd. Subsurface spreading occurs over nearly 10 acres for this case. Evaporation is reduced because the need for spreading the water over large areas on the land surface is eliminated. Since the length of the pit perimeter and not the surface area is the important factor for design, the volume excavated can be minimized by constructing long, narrow pits or trenches.

Nightingale, H.L. and Bianchi, W.C., "Artificial Basin Recharge", Water Engineering and Management, Vol. REF, No. HB, 1981, pp. R85-87, R89-92.

Basin-type artificial recharge and preconstruction field surveys are discussed. These surveys are made to evaluate the magnitude of potential problems as related to both a project's recharge rate and its environmental impact. A multidisciplinary approach is needed to plan and operate a successful recharge facility. Assuming that the size of the ground water basin and amount of storage capacity available are amenable to recharge, seven phases have been identified, in order of priority: precursory site evaluation, soil profile and subsurface geology determinations, recharge water quality determinations, biological considerations, pilot test basin study, construction plans, and the operation, maintenance and performance evaluation plans. In most reports of failures of basin recharge facilities and small experimental sites, soil clogging is to blame. This clogging is caused by sediment in the recharge water. The reason for sediment formation is not always clearly apparent. Sediments can be generated by improper construction techniques. Soil compaction from heavy construction equipment rarely seems to be at fault. The chances of long-term economical basin recharge can be improved by superimposing on the purely engineering aspects of construction and operation the principles of soil chemistry, hydrology, water chemistry, aquatic biology and subsurface geology. Interaction between engineering and environmental factors must be carefully considered to increase the chances for valid decision-making and planning which will produce a successful basin recharge facility.

Nightingale, H.I. and Bianchi, W.C., "Ground Water Chemical Quality Management by Artificial Recharge", Ground Water, Vol. 15, No. 1, Jan.-Feb. 1977, pp. 15-21.

> The effectiveness of basin ground water recharge at the Leaky Acres Recharge Facility in Fresno, California, for improving the regional ground water quality was studied as 65,815,000 cu m of high quality surface water was recharged from 1971 through 1975. Observation wells at the facility showed some variability in chemical parameters associated with each recharge period. The long-term decrease in salinity could be described by a power law decay curve fitted by regression analysis. Without a special network of observation wells outside the facility, scientific evaluation of the enclave of recharged water is not possible. A practical evaluation of water quality changes is possible from producing water wells around the facility. However, the pumping well discharge time variations, well depth, aquifer sequence, and prior use of surrounding land must be considered, since all of these factors affect the pumped-water quality and its seasonal variability. Recharge at Leaky Acres had noticeably decreased the ground water salinity for a distance of up to 1.6 km in the direction of the regional ground water movement.

Pettyjohn, W.A., "Flow Analysis of Hydraulic Connectors in Artificial Recharge Systems, A Model Study", OWRR A-015-OHIO(1), July 1974, Water Resources Center, Ohio State University, Columbus, Ohio.

> Artificial recharge techniques have long been used to augment ground water supplies and increase the yield of infiltration galleries. A case history of the water-supply situation at Minot, North Dakota, which has long suffered from severe water shortages, indicates the usefulness of artificial recharge. Throughout much of the Minot aquifer, the water level was raised more than 20 ft within six months of artificial recharge operation. The cost of the system (about $193,000) was minute in comparison to an estimated cost of $12 million for a pipeline to the Missouri River. A model study of the movement of a wetting front beneath an artificial recharge site, under both dry and damp conditions, illustrates the profound effect of permeability and stratification on the shape of the front. A model study dealing with the shape of the cone of recharge in the vicinity of a recharge well shows that there is a sharp change in gradient directly under the well. Elsewhere the cone develops a shape that is generally consistent with theoretical studies. Where the cone of recharge crosses a boundary, indicating a change in permeability, the cone is slightly distorted.

Pettyjohn, W.A., "Introduction to Artificial Ground-Water Recharge", EPA-600/2-81-236, Oct. 1981, Robert S. Kerr Environmental Research Laboratory, U.S. Environmental Protection Agency, Ada, Oklahoma.

> Artificial ground water recharge has been practiced for scores of years throughout the world. The purpose of artificial recharge is to increase the rate at which water infiltrates the land surface in order to supplement the quantity of ground water in storage. A variety of recharge techniques are feasible. Examples given in this report include methods that increase well yields, reduce the rate of decline of ground water levels, reduce land subsidence, control seawater intrusion in coastal areas, and renovate wastewaters. Two broad types of artificial recharge

are water spreading and well systems. In the former, large areas of land may be flooded, basins constructed, ditches or furrows excavated, or existing stream channels modified. Water is diverted to these structures where it infiltrates. Examples of successful recharge projects in the United States are given.

Prill, R.C. and Aronson, D.B., "Flow Characteristics of a Subsurface-Controlled Recharge Basin on Long Island, New York", Journal of Research of the U.S. Geological Survey, Vol. 1, No. 6, Nov.-Dec. 1973, pp. 735-744.

At the Woodbury recharge basin on Long Island, New York, the principal zones controlling infiltration are a surface-loam stratum and an intermediate gravelly, sandy loam stratum. The saturated hydraulic conductivities of these strata are 0.9 and 0.1 ft per day, respectively. The surface loam acts as the principal zone controlling infiltration until a perched ground water mound develops above the intermediate gravelly, sandy loam and extends to the bottom of the surface loam; then the intermediate gravelly, sandy loam becomes the principal infiltration controlling zone. Infiltration rates at 15.6 deg C are 1.4 ft per day when the surface loam is acting as the principal controlling zone and 0.5 ft per day when the intermediate gravelly, sandy loam is acting as the principal controlling zone. Projections of the probable infiltration rates associated with both the partial and the complete removal of the surface-loam stratum show that if recharge continues for several days total infiltration would not be greatly different whether or not the loam stratum is completely removed. The principal advantage of completely removing the surface-loam stratum is the resulting greater infiltration capacity during the early stages of ponding. This advantage is offset by potential problems associated with more intensive clogging of the subsurface controlling zone.

Rebhun, M. and Hauser, V.L., "Clarification of Turbid Water with Polyelectrolytes for Recharge through Wells", Artificial Recharge and Management of Aquifers, Symposium of Haifa, Mar. 19-26, 1967, Publication No. 72, 1967, International Association of Scientific Hydrology, pp. 218-228.

Storm runoff water is a good source of additional water supply for arid and semiarid regions; however, it must be stored for later use. Storage of these waters in aquifers by artificial recharge is hindered by suspended solids in the water. For recharge by injection through wells, removal of suspended solids must be nearly complete. Polyelectrolyte flocculants effectively clarify suspensions typical of flood runoff water. A simple field system was built and tested, and all elements of a complete system including treatment with polyelectrolytes, a simple clarification system, and ground water recharge through wells were combined into a successful operating unit.

Ripley, D.P. and Saleem, Z.A., "Clogging in Simulated Glacial Aquifers Due to Artificial Recharge", Water Resources Research, Vol. 9, No. 4, Aug. 1973, pp. 1047-1057.

Clogging due to artificial recharge was tested in laboratory simulated unconsolidated aquifers. Recharge with turbid water containing an effective microbial inhibitor showed clogging throughout the aquifers ranging in length from 48 to 123 cm. The rate of clogging at different depths was dependent on the size distribution of the particles in the water relative to the pore size distribution of the porous media. Clogging tended to be more severe at the influent portion of the sample; however, when the particles in the recharge water were smaller in size or when the pore sizes of the porous media were larger, the result was a more uniform rate of clogging with respect to depth. Recharge with nonturbid water and no effective microbial inhibitor shows clogging only in the top few centimeters. If it is not understood that these two patterns result from different causes, such a pattern resulting from the combined factors could be attributed to only one cause, turbidity.

Rubin, H., Glass, J.P. and Hunt, A.A., "Analysis of Storm Water Seepage Basins in Peninsular Florida", Publication No. 39, Sept. 1976, Florida Water Resources Research Center, University of Florida, Gainesville, Florida.

Rapid urbanization, currently proceeding in Florida, has resulted in significant problems with regard to both flood control and pollution abatement. The objective of this study was to search for the design procedures that may improve efficiency, safety and adequacy of drainage systems in peninsular Florida. Special attention was given to possibilities of recharging ground water aquifers with excess storm water. Through such a system, partial solution to problems of inadequate potable water supply in some areas can be achieved simply as a byproduct of flood control systems. A conceptual design framework is presented. In developing this framework, a variety of disciplines involved in the solutions of urban drainage systems in peninsular Florida are considered. Comparatively high levels of ground water in many locations require consideration of two factors relating to the ability of seepage ponds to divert surface water to the aquifer. The engineer should study the ability of the pond to seep water and he should analyze the response of the ground water table to the seeping water. With respect to these two factors, analyses and solutions adapted for engineering application are presented.

Seaburn, G.E. and Aronson, D.A., "Catalog of Recharge Basins on Long Island, New York, in 1969", Bulletin 70, 1973, New York State Department of Environmental Conservation, Albany, New York.

Basic data are compiled for 2,124 recharge basins in operation in 1969 on Long Island, New York. Data consists of basin location, date of construction, capacity to store water, maximum infiltration area, size, bottom altitude, overflow altitude, land-surface altitude, water-table altitude, drainage area, clogging or lack of clogging, use, geologic environment, and soil environment. Tables summarizing much of the data and a cross reference are included. One of the major methods of disposing of storm runoff from urban and suburban areas on Long Island, New York, is to divert the water into recharge basins. This method of disposal, begun in 1935, helps to augment and conserve water in Long Island's ground water reservoir--the sole source of freshwater for nearly 3 million people in Nassau and Suffolk Counties. Also included are the quantity and the quality of precipitation, inflow, and infiltration rates at selected basins, and evaluation of the regional effects of recharge basins on the hydrologic system of Long Island.

Seaburn, G.E. and Aronson, D.A., "Influence of Recharge Basins on the Hydrology of Nassau and Suffolk Counties, Long Island, New York", Water Supply Paper 2031, 1974, U.S. Geological Survey, Washington, D.C.

The hydrology of three recharge basins on Long Island, New York, (Westbury, Syosset, and Deep Park Basins) was studied. The average percentages of precipitation flowing into each basin were roughly equivalent to the average percentages of impervious areas in the total drainage areas of the basins. Inflow hydrographs for each basin typify the usual urban runoff hydrograph--steeply rising and falling limbs, sharp peaks, and short time bases. Infiltration rates averaged 0.9 fph (feet per hour) at the Westbury recharge basin, 0.8 fph at the Syosset basin, and 0.2 fph at the Deep Park recharge basin. Low infiltration rates at Deep Park resulted mainly from silt, clay, and organic debris washed in from construction sites, and a lack of a well-developed plant-root system on the floor of the basin. The apparent rate of movement of storm water through the unsaturated zone below each basin averaged 5.5 fph at Westbury, 3.7 fph at Syosset, and 3.1 fph at Deep Park. On the average, a 1-inch rainfall resulted in a peak rise of the water table directly below each basin of 0.5 ft; a 2-inch rainfall resulted in a peak rise of about 2 ft. The mound commonly dissipated within 1 to 4 days at Westbury, 7 days to more than 15 days at Syosset, and 1 to 3 days at Deep Park. Average annual ground water recharge was estimated to be 6.4 ac-ft at the Westbury recharge basin, 10.3 ac-ft at the Syosset recharge basin, and 29.6 ac-ft at the Deep Park recharge basin. Ground water recharge in the areas where recharge basins are used is probably equivalent to or may slightly exceed recharge under natural conditions (about 148 mgd).

Schneider, A.D. and Jones, O.R., "Basin Recharge of Playa Water", <u>Journal of Irrigation and Drainage Engineering</u>, Vol. 109, No. 3, Sept. 1983, pp. 309-316.

A 0.20-ha (0.5-acre) recharge basin was tested for conserving storm runoff that collected in a 16-ha (40-acre) playa in the southern High Plains, southwest of Amarillo, Texas. The basin was constructed by removing 1.2 m (4 ft) of the slowly permeable Pullman soil to expose a 0.3 m (1-ft) thick caliche layer overlying permeable sediments. Turbid water was pumped from the playa to the basin during eight tests over a 7-yr interval. The average recharge rate during 187 days of flooding was 0.373 m/d (1.22 ft/day). After three recharge tests during the first year, no maintenance was done to the basin bottom. Research showed that recharge basins could be an effective technique for partially replenishing the depleted Ogallala aquifer.

Schneider, A.D., Jones, O.R. and Signor, D.C., "Recharge of Turbid Water to the Ogallala Aquifer through a Dual-Purpose Well", Miscellaneous Paper MP-1001, Aug. 1971, Texas Agricultural Experiment Station, Texas A&M University, College Station, Texas.

The use of irrigation wells for artificial ground water recharge is proposed as a method for conserving runoff water in playa lakes that is otherwise lost to evaporation. To test this procedure, a gravel-packed, dual-purpose well was completed in an area where the Ogallala formation is thin and heterogeneous and consists mainly of sand and fine gravel with some silt and clay. During a 21-hour recharge test, turbid playa water containing an average of 600 ppm suspended solids was injected at 150 gpm. Immediately after recharging ended, the well was pumped and surged for 2.1 hours in an attempt to remove the injected silt and clay. The silt and clay recovery was only 20 percent, and the short interval of recharging reduced the specific capacity of the well from 6.0 to 3.8 gpm/ft. The specific capacity recovered to 4.8 gpm/ft, partially because natural recharge increased the thickness of the saturated formation by 7 ft. However, pumping for 9 days did not increase the specific capacity. Without clarification, the playa water is best suited for methods of ground water recharge in which sediment-laden water is acceptable.

Scott, V.H., "Optimization of Wells and Recharge Facilities in Conjunctive Use of Groundwater Reservoirs", OWRR-B-006-CALIF(1), Oct. 1969, Department of Water Science and Engineering, University of California, Davis, California.

The investigation of wells and recharge facilities was conducted through a series of related studies dealing with well interference, unsteady recharge flow through injection wells, recharge efficiency of wells and trenches, drainage by wells and the application of dynamic programming. Results of the evaluation, analysis and criteria developed are presented in five listed published papers with recommendations of approaches to optimal multisource-multipurpose water allocation in conjunctively managed surface-groundwater-recharge systems.

Signor, D.C. and Hauser, V.L., "Artificial Groundwater Recharge through Basins in the Texas High Plains", Miscellaneous Publication 895, Sept. 1968, Texas Agricultural Experiment Station, Texas A&M University, College Station, Texas.

Ground water supplies in the 22.4-million acre Texas High Plains are being depleted by irrigation pumping. Artificial ground water recharge is a means of managing water collected in playas. A recharge basin 120 by 240 ft was excavated in a bushland playa by removing clay to expose a 3 to 4-ft thick sand. About 89 percent of the clear well water placed in the recharge basin infiltrated. The final loss rate from the basin was 0.06 in per hour. This low rate resulted from a perched water table. The data indicated that a low infiltration rate, sustained over a large area, could recharge most of the runoff collecting in a playa. Geophysical, geological, particle-size distribution, hydraulic conductivity (double-tube permeameter method), stage-area, stage-volume and sub-basin hydrologic data are considered. Tables of soil texture and hydraulic conductivity, and figures showing geologic profile, aerial view of the basin and instrumentation, geohydrologic section, water loss and water levels versus time, elevation head versus depth, and neutron meter moisture determinations are presented.

Signor, D.C., "Recent Advances in Artificial Recharge", <u>Proceedings of the Conference of the Great Plains Resources Council</u>, Apr. 8-9, 1976, Denver, Colorado.

This paper is concerned with the techniques of artificial recharge and the problems that confront their utilization. The approaches to solving these problems constitute the recent advances in artificial recharge. Two recharge techniques are commonly used: (1) spreading basins, in which water that is concentrated on the surface percolates through overlying materials to the formation; and (2) injection wells or shafts, in which there is a direct connection from the surface to the formation. Suspended sediment in the water that is to be recharged by spreading is a serious problem. The sediment settles and is filtered out of the water at the basin surface and restricts subsequent infiltration as the surface becomes sealed by the fine-grained materials. Recent work on this problem has been directed toward developing surface conditions that minimize the sealing effect and toward removing sediment from the input water by chemical treatment. Recharge through injection wells usually requires stringent water-quality considerations, and recharge operations are generally less successful than in spreading basins because suspended sediment in the recharge water clogs the formation near the well. Once the formation is clogged, it is difficult to remove enough sediment to completely rejuvenate the well. The problem can be avoided only if clear water is used for recharge.

Sinclair, W.C., "Experimental Study of Artificial Recharge Alternatives in Northwest Hillsborough County, Florida", Water Resources Investigations 77-13, 1977, U.S. Geological Survey, Tallahassee, Florida.

Extensive water withdrawal from the Floridan aquifer in the urban Tampa Bay area has induced leakage from the overlying surficial aquifer adversely affecting the water table and lake levels. Artificial recharge could reduce the impact of these effects. Four experiments were conducted to investigate possible recharge alternatives; sinkhole recharge, water-spreading, connector wells, and subsurface-tile drainage to a deep well. Experiments indicate that all four methods can be effective. However, the sink-hole recharge experiment moved the greatest volume of water into the Floridan aquifer. The drain-tile experiment indicated greatest potential for draining the surficial aquifer. Combinations of the four methods could be used where potential exists for downward movement of water and sufficient unsaturated aquifer for water storage.

Smith, A.J., "Design is the Key to a Good Recharge Well", Public Works, Vol. 111, No. 10, Oct. 1980, pp. 66-67.

A recharge well located near Grand Island, Nebraska is described which was pumped continuously at a rate of 750 gpm for 6 months. The use of injection water which was nearly sand free is credited with the long-lasting period of pumping. Designing a sand-free well requires sizing of the screen slot in accord with established standards of natural development or gravel packing. The well must be able to recharge at a rate equal to the production rate. The screen must also take full advantage of the aquifer characteristics. The diameter of the well is an important factor which is often determined by the size of the pump selected to be installed. Accurate placement of the screen in the permeable zones is a vital step in design of the well. In injection wells it is often recommended that large parts of the aquifer be screened as opposed to only the most productive parts. Problems such as plugging due to air entrainment, precipitates, and bacteria and thermal interference are discussed.

Smith, C.S., "An Investigation of Filter Media for Recharge Wells in the Grand Prairie Region of Arkansas", MS Thesis, 1972, Department of Agricultural Engineering, University of Arkansas, Fayetteville, Arkansas.

Field tests were used to determine whether gravity flow recharge wells filled with sand and gravel can be used in the Grand Prairie of Arkansas. The wells are of a new type, filled with a sand and gravel filter. A suitable filter media will allow water to be recharged without any other treatment. Plugging should occur at the top of the well instead of in the aquifer, so that the well can be reconditioned simply by replacing or cleaning the filter. The experimental study was divided into the water supply system, the flow measuring system, and the filter. The filter media was made up of sand and gravel provided by a local supply company. Different depths of sand and gravel were used in the tests as well as different sizes of sand. In the preliminary testing, the sand used was graded concrete-mix sand. In the final tests uniformly graded sand (obtained by sieving the concrete mix sand) was used. The resulting particle size ranges were (1) greater than 1.19 mm, (2) 0.59-1.18 mm, (3) 0.25-0.59 mm, and (4) less than 0.25 mm. The sand filters had little effect on the turbidity of the water. A higher degree of treatment will have to be used on the type of water tested before it can be successfully recharged. The concrete mix sand, fine sand, or very fine sand could not be economically used as a filter in a gravity flow type well because of their low percolation rates and fast plugging rates.

Sniegocki, R.T. and Brown, R.F., "Clogging in Recharge Wells, Causes and Cures", Artificial Groundwater Recharge Conference, University of Reading, England, Sept. 21-24, 1970, Vol. 2, Paper No. 13, June 1971, The Water Research Association, Marlow, England, pp. 337-357.

Artificial recharge of surface water to ground water storage is a valuable water management tool. Injection of water into aquifers through wells has generally resulted in partially clogging the aquifer. The clogging is due to air entrainment, the presence of suspended material, the growth of microorganisms, or the chemical incompatibility of the injected water with the aquifer environment. A clogged recharge well generally can be redeveloped to some degree through the use of chemicals and backflushing. However, pretreatment of the injected water to remove suspended material and make it chemically compatible with the aquifer environment will minimize clogging and promote preservation of the aquifer.

Thomas, H.E., "Cyclic Storage, Where Are You Now", Ground Water, Vol. 16, No. 1, Jan.-Feb. 1978, pp. 12-17.

This paper presents information summarizing: the cyclic pattern of precipitation and runoff in the Central Valley of California and the need for cyclic storage to insure regularity and continuity in water supply; the inadequacy of surface water storage because of evaporation losses and probable interference with ground water recharge; the reduced hopes of Federal and State water agencies for control and management of the ground water reservoir, because of potential conflict with private property rights; the artificial recharge by private and local enterprise; the increasing emphasis in Federal and State programs upon surface reservoirs at increasing cost in money and water; the tendency of Federal policy to focus on the needs and clamors for irrigation water, with subsidies; and the availability of water from Federal projects now under construction for artificial recharge and subsurface storage, a first step to a long-range program of cyclic storage.

Tibbals, C.H. and Frazee, J.M., Jr., "Ground-Water Hydrology of the Cocoa Well-Field Area, Orange County, Florida", Open File Report FL 75-676, Jan. 1976, U.S. Geological Survey, Tallahassee, Florida.

The city of Cocoa, Brevard County, Florida, supplies water for much of central Brevard County including Cape Kennedy and Patrick Air Force Base. The water supply is obtained from a well field in east Orange County. Many of the easternmost wells in that well field yield salty water (chloride concentration greater than 250 mg/l). The interface between the fresh and salty water in the west part of the well field occurs at a depth of about 1,400 ft. An upward hydraulic gradient exists between the lower (salty) zones and the upper, or pumped zones of the Floridan aquifer in the west part of the well field. Secondary artesian aquifers in the well-field area are relatively high-yielding but are of limited areal extent. However, they are suitable as a source of water for supplemental supply or for artificially recharging the Floridan aquifer. Fresh water was transferred by siphon from a secondary artesian aquifer to the Floridan aquifer at 90 gallons per minute. Artificial recharge and recovery experiments show that it is feasible to retrieve fresh water stored in salty zones of the Floridan aquifer.

Vecchioli, J., "Experimental Injection of Tertiary-Treated Sewage in a Deep Well at Bay Park, Long Island, New York--A Summary of Early Results", <u>New England Water Works Association</u>, Vol. 86, No. 2, June 1972, pp. 87-103.

As of May 1971, nine ground water recharge tests ranging in duration from 2 to 33 days had been made at Bay Park, Long Island, New York. The injection rate was generally 350 gpm. The maximum volume injected in any one test was 14 million gallons. After each test, the injection well was pumped to waste, partly to restore lost capacity and partly to extract water previously recharged and, thus, to restore the quality of the water in the Magothy aquifer to near-native conditions. Reclaimed water was injected through a gravel-packed well constructed with an 18-in diameter fiberglass casing above a 16-in diameter stainless steel screen. The experiments have shown that the Magothy aquifer can be recharged with reclaimed water through the use of injection wells. Stringent quality requirements would have to be maintained on the injected water to minimize well clogging. For example, turbidity, iron, aluminum, and phosphate contents of the injection water would have to be kept at near-zero levels, and the water would have to be chlorinated; otherwise, the rate of clogging would require frequent redevelopment of the well.

Walton, W.C., Hills, D.L. and Grundeen, G.M., "Recharge from Induced Streambed Infiltration Under Varying Groundwater-Level and Stream-Stage Conditions", Bulletin 6, June 1967, Water Resources Research Center, University of Minnesota, Minneaplis, Minnesota.

Recharge from induced streambed infiltration can be estimated with data on streambed infiltration rates, surface water temperatures, ground water levels, and stream stages. However, estimates of recharge are not valid unless it is recognized that: the rate of leakage through a streambed increases in direct proportion to declines in the water table until water levels have receded to stages below the streambed; and the rate of leakage remains constant and at a maximum value, provided stream stage and temperature do not change. Incorrect water table declines and incorrect potential yields of well fields will be computed if it is assumed that induced streambed infiltration continuously increases in direct proportion to drawdown beneath the streambed regardless of water table stage. An aquifer-stream system for which hydrogeologic data are available was studied using electric analog computers and analytical methods. Results of these computer studies demonstrate that much greater drawdowns are computed for an aquifer-stream system when maximum infiltration conditions are taken into account than when maximum infiltration conditions are ignored.

Watkins, F.A., Jr., "Effectiveness of Pilot Connector Well in Artificial Recharge of the Floridan Aquifer, Western Orange County, Florida", Water Resources Investigations 77-112, Nov. 1977, U.S. Geological Survey, Tallahassee, Florida.

A connector well pilot installation, in continuous operation in western Orange County since December 4, 1970, was transferring water from the lower of two shallow sand aquifers to the Floridan aquifer at a rate of 13 gallons per minute when measured on September 23, 1971. The recharge water is untreated and analyses show it to be chemically and physically compatible with the water in the Floridan aquifer. The temperatures of the recharging and receiving waters were identical, 23 deg C. The transfer of water from the lower sand aquifer to the Floridan aquifer caused only a small buildup of artesian pressure in the Floridan aquifer but it lowered the artesian head 4 ft in the lower sand aquifer near the well which supplied the recharge water. Water levels in the upper sand aquifer were not affected, probably because of the low permeability of an intervening hardpan layer. However, after six auger holes back-filled with sand connected the two sand aquifers on April 5, 1972, a rise of water levels in the lower sand aquifer was noted. The principal chemical and physical effects on the water in the Floridan aquifer were a general improvement in chemical quality and an increase in color. The color may decrease as more water moves through the sand aquifer and the material responsible for the high color is removed by flushing.

Watson, K.K. and Whisler, F.D., "Profile Desaturation During Sediment Deposition in a Groundwater Recharge Trench", Journal of Hydrology, Vol. 33, No. 3/4, 1977, pp. 397-401.

When off-line recharge trenches are used for the artificial recharge of unconfined aquifers, the intake rate often decreases due to the deposition of a sediment layer on the surface of the recharge profile. The buildup of the impeding layer causes progressive hydraulic changes in the profile. The significance of the changes in relation to the intake rate depends on whether air is able to enter the pores of the saturated porous material beneath the trench when the air entry value of the material is exceeded. Flow rates were calculated for recharge through a profile of No. 17 sand for conditions of "no air access" and "adequate air access".

Watson, R. and Merritt, J., "Consumptive Use of Ground Water", 1970, Eastern Water Law Center, College of Law, University of Florida, Gainesville, Florida.

An increasing incidence of water shortage has prompted measures to preserve the supply of ground water. One approach to preservation is regulating the use of water through a well permit system, however, a more positive solution may be a process known as artificial recharge. Incidental recharge, induced recharge and applied recharge are three artificial means to augment water supplies. Each application of artificial recharge must be evaluated to see if it is feasible both in cost and physical terms. The criteria for evaluation could vary from one district of the state to another. There should be flexibility in the outlook for storage and recharge, and new legislation must contemplate future decision-making in the area of underground storage and artificial recharge. To enable each district to meet its specific problems to be remedied by artificial recharge, the Water Resources Act should provide for a specific delegation of legislative authority authorizing each Water Management District to implement rules and regulations dealing with artificial recharge as each district sees fit subject to the supervisory power of the Water Resources Board.

Williams, R.E. and Allman, D.W., "Factors Affecting Infiltration and Recharge in a Loess Covered Basin", Journal of Hydrology, Vol. 8, No. 3, July 1969, pp. 265-281.

Hydrographs of shallow piezometers installed in a basin overlain primarily by loess reveal that water infiltrating and percolating through the loess recharges ground water. Local gravity ground water flow systems adjacent to some intermittent streams thereby are established. Relief on the water table is at a maximum during late winter or spring, but decreases steadily through the summer when dry weather and maximal evapotranspiration minimize recharge. Cultivation practices and type of crop grown are two important factors that make the loess susceptible to infiltration and percolation. Smooth, heavily worked surfaces (commonly planted in winter wheat) act as a major limitation to infiltration capacity; soil moisture concomitantly is low beneath such surfaces. Root mats developed beneath grassy surfaces provide optimal conditions for infiltration and, consequently, for recharge. Measured infiltration capacities are often influenced by near-vertical tubular openings extending at least 30 ft deep in the loess. Tracer studies indicate that when sufficient water is available and when surficial infiltration capacity is sufficiently high to form a zone of positive pressure above these openings, water is conducted ahead of the saturation front.

Winqvist, G. and Mareliun, K., "The Design of Artificial Recharge Schemes", Artificial Groundwater Recharge Conference, University of Reading, England, Sept. 21-24, 1970, Paper No. 3, The Water Research Association, Marlow, England.

Planning of equipment for artificial recharge includes selecting a series of plant components from the raw water intake to the treated water section. The factors which mean most for the design and localizing of the plants are the quality of the raw water and the geological conditions. Even climatic conditions exercise some influence on the design. The basic principle used in planning is that raw water shall be retained sufficiently long in the ground to be modified into ground water. Quality of the raw water, particularly regarding such substances as occur in industrial pollution, and also the capacity of the ground to receive, transport and modify the infiltrated water must be considered.

WASTEWATER RECHARGE

American Water Works Association, "Use of Reclaimed Waste Waters as a Public Water Supply Source", Journal of the American Water Works Association, Vol. 64, No. 10, Part 2, Oct. 1972, p. 66.

The American Water Works Association expressed its belief that the full potential of reclaimed water as a resource should be exploited as rapidly as scientific knowledge and technology will allow. However, noting the increase in planned direct reuse of reclaimed waters, the Association was of the opinion that current knowledge and technology are not sufficiently advanced to permit direct use of treated waste waters as a source of public water supply. It urged and outlined a plan for intensive research and development by the AWWA Research Foundation and the Water Hygiene Division of the Office of Water Programs in the U.S. Environmental Protection Agency to advance technological capability in waste water reclamation.

Aronson, D.A., "Evaluation of Alternative Methods of Supplemental Recharge by Storm-Water Basins on Long Island, New York", Open File Report 76-470, 1976, U.S. Geological Survey, Albany, New York.

Many of the more than 2,200 storm-water basins on Long Island, New York, having runoff and infiltration characteristics similar to those of a basin studied in North Massapequa, are potential sites for returning large volumes of reclaimed water (highly treated waste water) to the ground water reservoir. By use of a finite-difference method of calculation, formulas were devised to calculate changes in basin storage at various time intervals from the start of storm inflow, with and without addition of reclaimed water. The North Massapequa storm-water basin was the prototype design used to test several methods of water application. Calculations using various infiltration rates, storm durations, and storm-return periods indicate that several methods of applying reclaimed water to storm-water basins on Long Island would be feasible.

Aronson, D.A., "Hydrologic and Water-Quality Effects of Recharging Reclaimed Water", 1975, U.S. Geological Survey, Syosset, New York.

A recent engineering feasibility study, done for Nassau County at the request and funding of the U.S. Environmental Protection Agency, has recommended that advanced wastewater-treatment and groundwater-recharge facilities be constructed and operated in Nassau County to demonstrate the feasibility of reclaiming water and using it to recharge the aquifer through basins and shallow injection wells. Research on factors controlling the degree and causes of clogging of basins and wells, on factors causing improvement or degradation of water quality after recharge, and on the hydraulic response of the aquifer system, is required. From these findings, the optimum recharge procedures and costs would be evaluated.

Aronson, D.A. and Beyth, M., "The Meadowbrook Artificial-Recharge Project in Nassau County, Long Island, New York", Long Island Water Resources Bulletin 14, 1980, U.S. Geological Survey, Syosset, New York.

In Nassau and Suffolk Counties, where the quality and quantity of potable ground water have declined as a result of urbanization, the use of highly treated wastewater (reclaimed wastewater) to replenish the ground water supply is technically feasible. Two methods that are particularly well suited to Long Island are surface ponding of the reclaimed water in basins so that it can infiltrate to the water table, and injecting the reclaimed water into the ground through wells. At the Meadowbrook artificial-recharge project site in East Meadow, a system of 11 basins and 5 injection wells will return to the ground water reservoir 4 million gallons per day of reclaimed wastewater, supplied by the Cedar Creek water-reclamation facilities. This demonstration project is intended to study the effectiveness of using reclaimed wastewater to replenish and improve the chemical quality of Long Island's ground water supply. Knowledge gained in the study will have applicability to other areas where the need to conserve and augment fresh water supplies has become critical.

Aronson, D.A., "Preliminary Selection of Storm-Water Basins Suitable for Infiltration of Reclaimed Water in Nassau County, Long Island, New York", Open File Report 76-668, May 1976, U.S. Geological Survey, Albany, New York.

A survey was made of 205 storm-water basins south of the ground water divide and north of Hempstead Turnpike in Nassau County, Long Island, New York, to determine which would be best suited for infiltration of reclaimed water. Selection depended on the infiltration area, location with respect to the ground water divide and to planned transmission mains, tendency to retain storm runoff, underlying lithology, and depth to the water table. The total maximum infiltration area of 14 selected basins is 60.2 acres, or 2,620,900 sq ft (0.24 square kilometers). If 5 ft (1.5 m) high partitions were constructed in the basins to divide each into approximately equal halves and reclaimed water were applied in half of each basin to a depth of 5 ft (1.5 m), using an application-rest cycle, a total area of 25.2 acres (0.10 square kilometers) would be available for supplemental recharge; the remaining infiltration area could be used for disposal of storm runoff.

Asano, T. and Wassermann, K.L., "Groundwater Recharge Operations in California", Proceedings of the Symposium on Wastewater Reuse for Groundwater Recharge, Sept. 6-7, 1979 at California State Polytechnic University at Pomona, May 1980, Office of Water Recycling, California State Water Resources Control Board, Sacramento, California, pp. 12-26.

Wastewater reclamation and reuse have become important factors in meeting California's water demand. Statewide water demand in the year 2000 could exceed supplies by as much as 6.6 million ac-ft/yr unless new supplies are developed and/or management techniques are improved. There are five major ground water recharge operations using reclaimed water in California. In 1977, they reclaimed about 26,000 ac-ft, or 14.2%, of the total reclaimed water; increased recharge is expected at all five in the future, and a ten-fold increase in volume statewide for ground water is projected. Recharge by surface spreading is most advantageous in metropolitan and agricultural areas where ground water overdraft is severe, and has the added benefits of the filtering effect of soils and transporting facilities of aquifers. Recharge by injection is usually used when the ground water is deep or in urban areas where basin recharge is impractical. It is most effective in creating fresh water barriers in coastal aquifers to prevent sea water intrusion, however, extensive above-ground wastewater treatment is necessary. Recent studies show that 50,000 ac-ft/yr of reclaimed water for salt water intrusion barriers, and 80,000 ac-ft/yr for spreading have been identified in Los Angeles and Orange Counties. However, concern for long-term health effects has prevented expansion and discouraged new projects. Four water quality factors are of particular significance: microbiological, total minerals, mineral toxicants of the heavy metal type, and stable organic substances. A consulting panel was created to study these and other problems and to recommend a program of research and development.

Baffa, J.J., "Artificial Groundwater Recharge and Wastewater Reclamation", Journal of the American Water Works Association, Vol. 67, No. 9, Sept. 1975, pp. 471-476.

Economic, conservation, and research factors pertinent to artificial ground water recharge and waste water reclamation are discussed. Artificial recharge of aquifers has been used successfully for ground water management in preventing losses to stream flow. Research has been conducted on hydraulic aspects, particularly on how to most efficiently use the soil and recharge water interface to allow maximum rates of transfer of recharge water to the aquifer. The capability of these systems to remove nitrogen from waste waters has been of considerable interest. A pilot plant shallow well injection system was tested. Effluent from a municipal single-stage trickling filter plant was injected through wells, which were then used for water quality sampling. After a three-year period, water quality parameters were separated into two categories: those of a soluble stable mineral character such as chlorides and alkalinity; and those with latent biological propensities such as nitrogen compounds, sulfates, phosphates, carbon dioxide, and organic solids. Concentrations found were affected by the amount of water recharged, the amount of natural ground water flow (which causes dilution), changes in aquifer permeability from clogging, and biochemical effects of bacterial action. It was concluded that for all water quality parameters, adsorption and dilution were important factors. To reduce soil clogging and to prevent release of unwanted nitrogenous compounds, it was suggested that organics and nutrients be removed by tertiary treatment, particularly when the objective of artificial recharge is municipal water supply.

Baier, D.C. and Wesner, G.M., "Reclaimed Waste Water for Groundwater Recharge", Water Resources Bulletin, Vol. 7, No. 5, Oct. 1971, pp. 991-1001.

Reclaimed trickling filter effluent is injectable and does not cause excessive well clogging. Multi-casing injection wells performed very satisfactorily. The reclaimed water would be acceptable for domestic use after travel through 500 ft of a confined aquifer in that bacteria, virus and toxic material were consistently absent, but the odor, taste, and high concentrations of dissolved inorganics are undesirable characteristics. Methods to eliminate the odor are being tested.

Bouwer, H., "Design and Operation of Land Treatment Systems for Minimum Contamination of Groundwater", Underground Waste Management and Artificial Recharge, Vol. 1, 1973, pp. 23-33.

The use of land treatment systems for sewage effluent and other liquid wastes, as well as some solid wastes, poses a threat to the quality of the native ground water even though the waste water itself undergoes a marked improvement in quality as it moves through the ground and becomes renovated water. To avoid large-scale spread of the renovated water into the ground water basin, the renovated water should be collected again at some point by wells or drains for reuse or release into the surface water. For the Salt River Valley of Arizona, the effective transmissibility of the aquifer for recharge was evaluated from a pilot project and then used in the design of a full-scale system. This effective transmissibility was less than the aquifer transmissibility.

Bouwer, H., Lance, J.C. and Rice, R.C., "Renovating Sewage Effluent by Ground Water Recharge", Proceedings of the Land Spreading Conference held at Orlando, Florida, July 15, 1971, 1971, East Central Florida Regional Planning Council, Winter Park, Florida, pp. 6-1 to 6-20.

A pilot project was installed in 1967 to determine if the tertiary treatment necessary to permit large-scale reuse of sewage could be obtained effectively and economically by using infiltration basins in the normally dry salt river bed at Phoenix, Arizona. The hydrogeological conditions of the salt river bed (about 3 ft of fine, loamy sand underlain by sand and gravel layers to great depth and a ground water table at about 10 ft depth) are favorable for high-rate waste water reclamation by ground water recharge. The infiltration rate in grass-covered basins is 25% higher, and in a gravel-covered basin 50% lower, than in a bare soil basin. Alternating 2-week inundation periods with 10-day dry-up periods yield an annual infiltration rate of about 400 ft. Reclaimed water, pumped from 30 ft depth in the center of the recharge area, has a biochemical oxygen demand of about 0.5 mg/l (BOD of the sewage effluent is about 15 mg/l) and a median fecal coliform density of 10 per 100 ml. Nitrogen is essentially all converted to the nitrate form in the reclaimed water. More nitrogen was removed under vegetated infiltration basins than under nonvegetated basins. Phosphate concentrations in the reclaimed water are around 5 ppm, as compared to 13 ppm in the effluent.

Bramer, H.C. and Hoak, R.D., "Water Reclamation", Water Reuse, American Institute of Chemical Engineers Program Symposium, Series No. 78, Vol. 63, 1967, pp. 92-95.

Present-day water reclamation in the United States consists principally of the use of treated municipal wastewaters for recreation lakes, irrigation, ground water recharge, and industrial purposes. The treatment usually consists of primary settling, secondary treatment by activated sludge or in oxidation ponds, and effluent chlorination. Although there are many instances of water reclamation being practiced in the United States today, the magnitude of the water use involved is small. Insofar as reclamation is a solution to problems of supply, the greatest potentials are in the arid western states, where an expanding economy and a westward moving population will increasingly press upon already limited supplies. Reclamation also, however, presents an attractive alternative to simple stream disposal of treated effluents, in many cases from an economic standpoint even with adequate fresh water supplies. Of particular significance in the future is the increased economic importance which recreational uses of the surface waters will have.

Buros, O.K., "Wastewater Reclamation Project, St. Croix, U.S. Virgin Islands", EPA-600/2-76/134, June 1976, Municipal Environmental Research Laboratory, U.S. Environmental Protection Agency, Cincinnati, Ohio.

Since 1971 work has been proceeding on a project to use tertiary-treated wastewater effluent for artificial recharge of the ground water on St. Croix. A 0.5 mgd (1,890 cu m/day) advanced wastewater treatment plant and recharge facilities were designed and constructed. Background data on water quality and quantity in the surrounding area were collected for 2 1/2 years before recharging began. Recharge operations were carried out for 8 months during 1974, using both spray irrigation and spreading basins. The best method of recharging proved to be the use of spreading basins in an alluvial valley. The cost for the wastewater treatment, recharge operations, and recovery of ground water was estimated to be about \$2.15/thousand gal (\$0.57/cu m) at 0.5 mgd (1,890 cu m/day), with a reduction in estimated costs to \$1.64/thousand gal (\$0.43/cu m) if the operation is expanded to 1 mgd (3,785 cu m/day).

Chan, H.T. and Sykes, J.F., "Ground-Water Mounding Beneath a Large Leaching Bed", Ground Water, Vol. 22, No. 1, Jan.-Feb. 1984, pp. 86-93.

An 84 m (276 ft) by 64 m (205 ft) experimental leaching bed was constructed to receive the secondary effluent from the Norwood, Ontario sewage treatment plant. The distribution system consisted of 100 mm diameter perforated pipes laid 2.1 m apart in 0.45 m wide trenches. The main soil layer at the test site was sandy silt. Ground water levels at the site were measured approximately weekly at 36 piezometers for a period of about 2 1/2 years. An initial hydraulic loading of 122,700 l/day caused surface ponding. A reduced loading of 40,900 l/day was then used for most of the study period. Ground water level fluctuations beneath and adjacent to the leaching bed were attributed to seasonally varying infiltration and hydraulic loading. The experimental data were simulated using a transient two-dimensional, phreatic integrated depth finite-element model and an analytic solution for ground water mounding. Results from the study indicated that seasonal infiltration was a controlling factor in determining leaching bed design loads. Both mathematical models were found adequate for design.

Consoer, Townsend and Associates and Bechtel Corporation, "Reclamation of Wastewater in Santa Clara County: A Look at the Potential", Oct. 1973, Santa Clara Valley Water District, San Jose, California.

In 1970, in response to the growing concern about water sources and water management, a long range plan for wastewater disposition was begun. One facet was a study to determine the potential benefits or disadvantages of reclaiming and reusing wastewater rather than discharging it into San Francisco Bay. There will be a critical need for augmented water supplies both in Santa Clara County and throughout the large San Felipe service area starting in 1978. In looking at the markets for reclaimed wastewater in the area to be served by the proposed San Felipe project, it was found that by far the major market or use for reclaimed wastewater would be ground water recharge in north Santa Clara County. Another large market for reclaimed water would be farms and orchards in the area. Present wastewater quality was compared to present and projected water supplies. Alternative programs using reclaimed wastewater were studied. Four of these plans use reclaimed water rather than importation, and the fifth alternative combines reclamation and importation. A system combining importing water and reclamation would be cheaper than reclamation and reuse alone.

Ehrlich, G.G., Ehlke, T.A. and Vecchioli, J., "Microbiological Aspects of Ground-Water Recharge-Injection of Purified Chlorinated Sewage Effluent", Professional Paper 800-B, 1972, U.S. Geological Survey, Menlo Park, California, pp. B241-B245.

Chlorinated, purified sewage was injected through a well at Bay Park, New York, at rates of 350 gpm (gallons per minute) and 200 gpm for 33 days. Observed injection-well head buildup was correlated with the content of mostly inorganic particulate matter in the injectant. Bacterial slimes were not present in repumped water or in a sand probe placed near the injection zone during the test. Several species of aerobic and facultative anaerobic bacteria were found in repumped water samples. Microbial growth occurs in the aquifer after chlorine has been inactivated by the aquifer materials.

Ehrlich, G.G. et al., "Microbiological Effects of Recharging the Magothy Aquifer, Bay Park, New York, with Tertiary-Treated Sewage", Professional Paper 751-E, 1979, U.S. Geological Survey, Menlo Park, California.

Injection of highly treated sewage (reclaimed water) into a sand aquifer on Long Island, New York, stimulated microbial growth near the well screen. Chlorination of the injectant to 2.5 mg/l suppressed microbial growth to the extent that it did not contribute significantly to head buildup during injection. In the absence of chlorine, microbial growth caused extensive well clogging in a zone immediately adjacent to the well screen. During a resting period of several days between injection and well redevelopment, the inhibitory effect of chlorine dissipated, and microbial growth ensued. The clogging mat at the well-aquifer interface was loosened during this period, probably as a result of microbial activity. Little microbial activity was noted in the aquifer beyond 20 ft from the well screen; it probably resulted from small amounts of biotransformable substances not completely filtered out of the injectant by the aquifer materials. Movement of bacteria from the injection well into the aquifer was not extensive. In one test, in which injected water had substantial total-coliform, fecal-coliform, and fecal-streptococcal densities, no fecal-coliform or fecal-streptococcal bacteria, and only nominal total-coliform bacteria, were found in water from an observation well 20 ft from the point of injection.

Faust, S.D. and Vecchioli, J., "Injecting Highly Treated Sewage into a Deep-Sand Aquifer", Journal of the American Water Works Association, Vol. 66, No. 6, June 1974, pp. 371-377.

The growth in population in Nassau County, Long Island, over the past twenty years has greatly increased the demand for water. At the same time, urbanization has reduced the recharge capacity of the local aquifer. In an effort to solve the problem, a series of artificial-recharge experiments are being carried out which examine the feasibility of injecting reclaimed water into a network of wells. The effect on the resultant water quality is discussed.

Garcia-Bengochea, J.I., "Recharge of a Carbonaceous Saline Aquifer of South Florida with Treated Sanitary Wastewater", Conference on Artificial Groundwater Recharge, University of Reading, England, Sept. 21-24, 1970, Supporting Paper C, 1970, The Water Research Association, Marlow, England.

Artificial recharge of a carbonaceous saline aquifer of south Florida is used to dispose of treated sanitary waste. The deep open-hole injection well is located immediately south of Miami, Florida. The treated wastewater has a lower specific gravity (approximately 1.025). Mixing will occur at the very beginning of the injection period and thereafter a fresh-water bubble will develop, floating on the saline water and limited above by confining layers. Injection of the fresh-water effluent into the deep artesian aquifer of south Florida will lead to the formation of an enormous fresh-water bubble at the top of the aquifer which would be under normal artesian pressure. Such a volume of water could be used for irrigation or for future water supply when a shortage of present fresh water sources would justify the cost of reclaiming the stored water.

Hanor, J.S., "Aquifers as Processing Plants for the Modification of Injected Water", Bulletin 11, Aug. 1980, Louisiana Water Resources Research Institute, Louisiana State University, Baton Rouge, Louisiana.

Gulf Coast (Baton Rouge) Region aquifer systems offer the potential for infested water storage and water composition modification. A study of natural processes on ground water chemistry provides information on processes potentially influencing water composition during artificial recharge. Baton Rouge area fresh water aquifer processes include fermentive breakdown of organic matter, sulfate reduction, Donnan behavior, and dissolution/formation of new silicates. Although appearing to be in exchange equilibrium with sediment clay minerals, ion-exchange influences native water composition less than previously thought. Potentially-occurring artificial recharge reactions (carbon dioxide formation and reaction with minerals; ion-exchange; diffusion) were studied. Altering montmorillonite appears to control water composition. Clay contents down to 0.1 weight-percent rapidly cause dissolved calcium removal and water softening during simulated injection/storage/production of fresh waters from brackish/saline Gulf Coast aquifers, corroborating field study results. Aquiclude clays soften injected water only if pore water concentrations of calcium and magnesium are less than in injected water. Of a variety of other possible reactions/processes (oxidation-reduction; membrane effects), many are incompletely understood.

Heisel, J.E. and Gonzalez, J.R., "Water Budget and Hydraulic Aspects of Artificial Recharge, South Coast of Puerto Rico", USGS/WRD/WRI-79/048, May 1979, U.S. Geological Survey, Fort Buchanan, Puerto Rico.

An analog model was used to evaluate ground water conditions on the south coast of Puerto Rico. Water levels during a normal period and during an extended drought were simulated. Recharge and discharge values are reported. The model was also used to evaluate the possibilities of using treated waste water to recharge the aquifer. Three methods were considered: infiltration basins, injection, and irrigation. The tests were planned to determine what changes in water levels would result if certain rates of application were used. Because of the limited vertical hydraulic conductivity, irrigation is suggested as the most practical method of waste-water use. Injection, though practical from the mechanical standpoint, may be objectionable from health and aesthetic standpoints.

Jenks and Adamson, "A Program for Water Reclamation and Groundwater Recharge, Environmental Impact Statement", Apr. 1973, Santa Clara County Flood Control and Water District, San Jose, California.

A water reclamation plant is to be located at the existing Palo Alto Subregional Water Quality Control Plant, together with a ground water well injection and monitoring system. The results of an environmental impact study are reported. The project will not in any significant way alter the existing terrestrial environment. However, the project will impact to a significant extent both the ground water and surface water environment in the affected area. The primary purpose of the proposed project is to prevent further salt water intrusion to the underground water resource supply. In addition, the project will reduce pollution to the receiving waters of San Francisco Bay. Thus, the significant impact of the project will be positive both in respect to quality and quantity of the natural water resource. The adverse effects from the plant will be noise and air pollution. The project was adopted for implementation after consideration and evaluation of five basic alternatives to meet the water supply needs of the affected area. Two alternatives involve importation of water, one alternative involves desalination of San Francisco Bay water, and the remaining two alternatives involve use of reclaimed wastewater to either: (1) percolate into the ground water basin, or (2) inject into the ground water basin.

Katz, B.G. and Mallard, G.E., "Chemical and Microbiological Monitoring of a Sole-Source Aquifer Intended for Artificial Recharge, Nassau County, New York", Open File Report 80-567, 1980, U.S. Geological Survey, Syosset, New York.

In late 1980, approximately 4 million gallons per day of highly treated wastewater will be used to recharge the ground water reservoir in central Nassau County through a system of 10 recharge basins and 5 shallow injection wells. To evaluate the impact of large-scale recharge with reclaimed water on ground water quality, the U.S. Geological Survey has collected hydrologic and water-quality data from a 1-square mile area around the recharge site to provide a basis for future comparison. Extensive chemical and microbiological analyses are being made on samples from 48 wells screened in the upper glacial (water-table) aquifer and the upper part of the underlying Magothy (public-supply) aquifer. Preliminary results indicate that water from the upper glacial aquifer contains significant concentrations of nitrate and low-molecular-weight chlorinated hydrocarbons and detectable concentrations of organochlorine insecticides and polychlorinated biphenyls. At present, no fecal contamination is evident in either aquifer in the area studied. In the few samples containing fecal indicator bacteria, the numbers were low. Nonpoint sources provide significant loads of organic and inorganic compounds; major sources include cesspool and septic-tank effluent, cesspool and septic-tank cleaners and other over-the-counter domestic organic solvents, fertilizers, insecticides for termite and other pest control, and stormwater runoff to recharge basins. The water-table aquifer is composed mainly of stratified, well-sorted sand and gravel and, as a result, is highly permeable. In the 1-square mile area studied, some contaminants seem to have traveled 200 ft downward to the bottom of the water table aquifer and into the upper part of the public supply aquifer.

Ku, H.F.H., Vecchioli, J. and Ragone, S.E., "Changes in Concentration of Certain Constituents of Treated Waste Water During Movement Through the Magothy Aquifer, Bay Park, New York", Journal of Research of the U.S. Geological Survey, Vol. 3, No. 1, Jan.-Feb. 1975, pp. 89-92.

Approximately 7 million gallons of tertiary-treated sewage (reclaimed water) was injected by well into the Magothy aquifer, Long Island, New York, and was subsequently pumped out. As the reclaimed water moved through the aquifer, concentrations of certain dissolved constituents decreased: total nitrogen, 7 percent; methlene blue active substances, 49 percent; chemical oxygen demand, 50 percent; and phosphate, more than 93 percent.

Miller, D.G. and Hunter, B.A., "The Principles and Practice of Pretreatment for Artificial Recharge", Conference on Artificial Groundwater Recharge, University of Reading, England, Sept. 21-24, 1970, Paper No. 4, 1970, The Water Research Association, Marlow, England.

In selecting suitable pretreatment methods for basin recharge, account must be taken not only of the achievement of economic recharge rates by minimizing clogging but also of the necessary conditioning of lower quality waters to protect underground supplies and aquifers. To assess pretreatment requirements for water quality purposes it is necessary to have adequate data on the reliable purification afforded by infiltration. Pilot scale investigations are needed to understand the generalized mechanism involved. Principal requirements for most recharge waters are for suspended solids reduction and aeration. Considerable care is needed when recharging polluted waters, especially those containing industrial wastes, to prevent aquifer contamination. Where potable water is concerned, particular care must be taken with contaminants which are not affected by percolation or where removal cannot be assured. These include iron, manganese, aluminum, heavy metals, inorganic salts, ammonia, and persistent insecticides. Of the pretreatment methods available for suspended solids removal the various forms of deep bed filtration would seem to have an acceptable combination of removal efficiency and cost. Other methods are likely to be too efficient and too expensive.

Prill, R.C., Oaksford, E.T. and Potorti, J.E., "A Facility Designed to Monitor the Unsaturated Zone During Infiltration of Tertiary-Treated Sewage, Long Island, New York", Water Resources Investigations 79-48, 1979, U.S. Geological Survey, Syosset, New York.

A facility consisting of a circular recharge basin 6.10 meters in diameter with a central observation manhole was developed on Long Island, New York, to study the role of the unsaturated zone during aquifer recharge with tertiary-treated sewage. The manhole extends through most of the 7.5-meter thick unsaturated zone, which is composed of glacial outwash sand and gravel, and enables collection of water samples and monitoring of dynamic characteristics of the unsaturated zone during recharge experiments. The system contains instrumentation for monitoring infiltration rate, pressure-head distribution, soil-moisture content, ground water levels, and soil gases. The 24.55-square meter recharge basin has operated in all seasons intermittently since April 1975 and, as of April 1978, has transmitted 62 million liters of tertiary-treated effluent to the water-table aquifer. Overall performance of the facility indicates that it is suitably designed for monitoring the unsaturated zone during artificial recharge experiments.

Ragone, S.E., "Geochemical Effects of Recharging the Magothy Aquifer, Bay Park, New York, with Tertiary-Treated Sewage", Professional Paper 751-D, 1977, U.S. Geological Survey, Mineola, New York.

Thirteen recharge tests using tertiary-treated sewage (reclaimed water) and six tests using city (drinking) water were completed between 1968 and 1973 as part of an experimental recharge program at Bay Park, New York. In the longest test, reclaimed water was injected during 84.5 days over a 199-day period. Although the iron concentration of reclaimed water and native water in the recharge zone is less than 0.5 mg/l, the iron concentration in samples taken 20, 100, and 200 ft from the recharge well approached 3 mg/l at times. Within a 20-ft radius of the recharge well, dissolved oxygen in the reclaimed water oxidized pyrite and caused iron to be released to solution. Because the amount of iron in water increases with distance from the recharge well even though dissolved oxygen is no longer present in water reaching the 20-ft radius, the exact dissolution mechanism is not understood. Some cation exchange occurs. Ammonium and potassium cations exchange with the proton in clay, resulting in a decrease in pH of over 1 pH unit at times. Phosphorus and organic compounds can react chemically and biologically to degrade water quality. Reclaimed water contains an average of 0.17, 0.07, and 9 mg/l of phosphate, MBAS (methylene blue active substances), and COD (chemical oxygen demand), respectively. During recharge, phosphate concentrations remain at native-water levels at the 20-, 100-, and 200-ft observation wells, which indicates retention by the aquifer; MBAS and COD are partly retained at the 100- and 200-ft wells, presumably by adsorption reactions.

Ragone, S.E. and Vecchioli, J., "Chemical Interaction During Deep Well Recharge, Bay Park, New York", Ground Water, Vol. 13, No. 1, Jan.-Feb. 1975, pp. 17-24.

Tertiary-treated sewage (reclaimed water) was recharged into the Magothy aquifer in 13 recharge experiments in Long Island, New York, between 1968 and 1973. The recharge resulted in a degradation in water quality with respect to iron concentration and pH. Iron concentration increased from the range 0.14 to 0.30 mg/l to as much as 3 mg/l at the 20, 100, and 200 ft or 6.1, 30, and 61 meter observation wells as the reclaimed water displaced native water. The increase was probably a result of pyrite dissolution. The pH of the water decreased from the range 5.22 to 5.72 to a low of about 4.50, predominantly as a result of cation-exchange reactions.

Ragone, S.E., Ku, H.F.H. and Vecchioli, J., "Mobilization of Iron in Water in the Magothy Aquifer During Long-Term Recharge with Tertiary-Treated Sewage, Bay Park, New York", Journal of Research of the U.S. Geological Survey, Vol. 3, No. 1, Jan.-Feb. 1975, pp. 93-98.

Tertiary-treated sewage (reclaimed water) has been recharged by well into the Magothy aquifer at Bay Park, New York, intermittently since 1968. The longest of 13 recharge tests lasted 84.5 days. This was sufficient time for the reclaimed water to reach an observation well 200 ft from the recharge well. Although the iron concentrations of the reclaimed water and the native water were less than 0.4 mg/l, the iron concentrations of samples from observation wells 20, 100, and 200 ft from the recharge well at times approached 3 mg/l. The source of the iron is pyrite that is native to the aquifer.

Ragone, S.E., Vecchioli, J. and Ku, H.F.H., "Short-Term Effect of Injection of Tertiary-Treated Sewage on Iron Concentration of Water In Magothy Aquifer, Bay Park, New York", Underground Waste Management and Artificial Recharge, Vol. 1, 1973, pp. 273-290.

Tertiary-treated sewage (reclaimed water) is recharged by a deep well into the Magothy aquifer, the primary water supply source for Nassau County, Long Island, New York. As of September 1972, 12 recharge tests have been run since the inception of the recharge program in September 1968. Although the iron concentrations of reclaimed and native water averaged 0.44 mg/l and 0.24 mg/l, respectively, the iron concentration of the mixed native and reclaimed water at times exceeded 3 mg/l. The most probable source of iron is the pyrite in the Magothy aquifer. During recharge, the natural reducing condition in the aquifer is replaced by a progressively more oxidizing environment. The initial response to this change is the oxidation of pyrite, which releases Fe into solution. Eventually, ferric hydroxide precipitates, and the Fe concentration decreases. The oxidation mechanism involves inorganic and organic constituents in the reclaimed water.

Rice, R.C. and Raats, P., "Underground Travel of Renovated Waste Water", Journal of the Environmental Engineering Division, American Society of Civil Engineers, Vol. 106, No. EE6, Dec. 1980, pp. 1079-1098.

Equations are presented to predict the minimum underground detention time for ground water recharge facilities utilized in land application of sewage effluents. Shallow aquifers in which the reclaimed water is collected in drains were analyzed using horizontal flow theory. Flow nets were used to determine the detention time for deep aquifers. The nitrate peak associated with newly infiltrated waste water in deep aquifers would be considerably flattened by the time the renovated water would be collected for reuse by wells. It would not be practical to separate the portion of renovated water that has a high nitrate content either for further treatment or for recycling through the land use system. The flow system analysis of shallow aquifers determined that more than 60 percent of the nitrate could be recovered in 25 percent of the renovated water when the infiltration rate was greater than 1 ft/day.

Roberts, P.V., McCarty, P.L. and Roman, W.M., "Direct Injection of Reclaimed Water into an Aquifer", Journal of the Environmental Engineering Division, American Society of Civil Engineers, Vol. 104, No. EE5, Oct. 1978, pp. 933-949.

Renovating and recycling wastewater are alternate means of meeting future water demands. The Environmental and Water Studies program at Stanford University has embarked on a 3-year study concerning the changes in water quality and aquifer characteristics resulting from the injection of reclaimed water. The ultimate goal is to ascertain whether direct injection of reclaimed water is a feasible and reliable strategy for producing potable water. An advanced wastewater treatment and injection-extraction system built by the Santa Clara Valley Water District (SCVWD) not only provides reclaimed water for landscape irrigation but also inhibits saline water intrusion and subsidence. An injection pilot study at the SCVWD facility found that during passage of reclaimed water through an aquifer most contaminants were removed by various mechanisms. These include: adsorption, ion exchange, precipitation and dissolution, chemical oxidation, biological nitrification and denitrification, aerobic or anaerobic degradation of organic substrates, mechanical dispersion, and filtration. The dominant removal mechanism is believed to be ion exchange on clays. This pilot study will provide important information regarding the retention capacity of an aquifer for attenuating the movement of a wide spectrum of inorganic, organic, and microbial pollutants. These data will be used in monitoring a full-scale well field operation.

Roberts, P.V., "Water Reuse for Groundwater Recharge: An Overview", Journal of the American Water Works Association, Vol. 72, No. 7, July 1980, pp. 375-379.

Using reclaimed water to artificially recharge ground water aquifers is one of the more promising approaches to replenishing dwindling water resources in water-short areas. When this reclaimed water will be used for domestic consumption, however, many questions arise that range from the health effects to the appropriate water quality standards for the points of recharge and extraction to the available alternative methods of water reclamation technology. This summary reports the progress made toward defining and answering these questions.

Russell, L.L., "Chemical Aspects of Ground Water Recharge with Wastewaters", Ph.D. Dissertation, 1976, University of California, Berkeley, California.

Studies on waste water-soil systems and recharge water quality indicated that nitrification in the soil rather than in a waste water treatment plant could substantially reduce percolant water quality. This was attributed to low carbon dioxide concentrations in aerated effluents which could also dissolve calcareous soils, resulting in high calcium concentrations of up to 200 mg/l. Field studies at the 14-mgd Whittier Narrows Water Reclamation Plant and the 26-mgd San Jose Creek Water Renovation Plant indicated that 90-95% of the CO_2 produced during treatment was stripped during aeration. A systematic, mathematical model for the chemistry of a waste water-soil system is presented. The effects of minor buffering systems, complex ion formation, temperature, and ionic strength are incorporated into the model. The model is useful in simulating the effects of waste water treatment on effluent quality and the effects of land application on percolant quality, mixing, dilution, and subsequent chemical reactions during percolation.

Schicht, R.J., "Feasibility of Recharging Treated Sewage Effluent into a Deep Sandstone Aquifer", Ground Water, Vol. 9, No. 6, Nov.-Dec. 1971, pp. 29-34.

Artificial recharge with tertiary treated sewage effluent is suggested as one remedial measure for projected ground water deficits in the Chicago region. A deep sandstone aquifer, and improvement of the source of ground water in the region, offers the best opportunity for artificial recharge. Recharge will be through wells because the aquifer is deeply buried. Expected problems in maintaining well injection capacity were studied by recharging treated effluent through formation cores of the sandstone. Recharge rates were maintained at constant heads for several days.

Schmidt, C.J. and Clements, E.V., "Reuse of Municipal Wastewater for Groundwater Recharge", EPA-600/2-77-183, Sept. 1977, U.S. Environmental Protection Agency, Washington, D.C.

This report concerns only formal ground water recharge operations with municipal waste water effluents. As of 1977, this type of activity was underway at 10 sites with an estimated total capacity of 77 mgd; however, the inadvertent uncontrolled recharge of treated effluent discharged to land (e.g., by small rural facilities) and continuous release of septic tank waste amounts to far greater volumes than for the formal programs discussed. Seven of the sites studied employ percolation-type programs, the most successful practice being alternate flooding and drying. The other three sites practice well injection, which can be successful only if rigorous control of injected water quality is maintained. Sufficient data to define the movement of salts, trace organics, or pathogens as a function of soil and ground water characteristics and hydraulics have not yet been developed to permit quantitative water quality requirements to insure successful recharge over long periods. Intensive monitoring of present and other new sites is therefore recommended as a means of gathering data upon which rational design criteria can be based.

Schmidt, C.J., Clements, E.J. and Skelton, S.P., "A Survey of Practices and Regulations for Reuse of Water by Groundwater Recharge", Journal of the American Water Works Association, Vol. 70, No. 3, Mar. 1978, pp. 140-147.

Present interest in formal, regulated wastewater reuse projects is simply bringing focus upon an informal practice which has existed for a long time. To date, ground water recharge by direct injection of wastewater is practiced at one site (Orange Co., California), under construction at another (Nassau Co., New York), and in planning stages at three. Formal ground water recharge by landspreading and percolation is practiced at six sites, five located in California. By 1976, approximately 26 states had prepared regulations and guidelines for ground water recharge; proposed U.S. EPA regulations for underground injection programs would generally allow the states to exercise maximum flexibility to prevent ground water pollution. Direct injection of treated wastewater is regulated on an individual project basis in California and New York, the two states where such systems are in use. Individual case regulation is also practiced for spreading and percolation but with less stringent requirements for the quality data for each recharge site discussed, as well as procedural details of the various operations.

Smith, H.H., "Sequential Use of Reclaimed Wastewater Destined for Aquifer Recharge", PB80-199680, June 1979, National Technical Information Service, U.S. Department of Commerce, Springfield, Virginia.

Effluent from the advanced wastewater treatment plant in St. Croix, U.S. Virgin Islands, a water-deficient island, was used to recharge aquifers by use of spreading basins. Wells in the recharge area were monitored to determine changes in static levels and in quality. Results indicate an increase in the quantity of water in the aquifer and an upgrading in the quality of the effluent. An economic analysis showed the cost of recharge/recovery operations to be significantly less than the cost of desalination of sea water which supplies the major portion of water to the island.

Smith, H.F., Schicht, R.J. and Humphreys, H.W., "Pilot Scale Investigations of Well Recharge Using Cored Samples", Conference on Artificial Groundwater Recharge, University of Reading, England, Sept. 21-24, 1970, Vol. 2, Paper No. 11, June 1971, The Water Research Association, Marlow, England, pp. 279-296.

Laboratory studies were conducted in Illinois to determine if artificial recharge of sandstone aquifers with sewage effluent is feasible. The apparatus used in this research consisted of four main components: (1) permeameter; (2) filters; (3) chlorinator; and (4) constant head tank. The Ironton-Galesville sandstone, since it is the most productive formation in the deep sandstone aquifer, was selected for investigation. The core samples are described by the Illinois State Geological Survey as fine to medium grained sandstone, moderately well sorted, and friable. The median grain size is 0.30 mm, with the maximum grain size about 0.75 to 1.0 mm. After discharge of deionized water through the sandstone core has stabilized, filtered and chlorinated sewage effluent was recharged. A pretreated effluent of the type used could be successfully recharged through a sandstone well having similar characteristics to the cores used. The pretreatment would consist of removal of solids, a disinfectant, and control of pH. Sand filters are the most reliable type of filter with fewer problems of maintenance.

Todd, D.K., "Future Prospects of Artificial Groundwater Recharge", Conference on Artificial Groundwater Recharge, University of Reading, England, Sept. 21-24, 1970, Vol. 2, Paper No. 15, June 1971, The Water Research Association, Marlow, England, pp. 385-399.

The traditional basis for artificial recharge has been to augment ground water storage using water obtained from controlled releases of upstream reservoirs or from importations. The method provides for storage of large volumes of water in natural reservoirs without the cost of building a surface reservoir. Moreover, in the western United States few sites remain where it would be economically feasible to construct dams. Storing water underground has the advantages of minimizing evapotranspiration losses, equalizing water temperatures, providing natural filtration, and insuring a protected local water supply during times of emergency. More than 400 cities in the United States now use their treated effluent for agricultural irrigation or ground water recharge. Water reclamation systems may be classified into infiltration basins, ridge-and-furrow, spray-irrigation, spray-runoff, and recharge wells. Renovation is most successful when wastewater is applied to a soil system which incorporates vegetation, such as a woodlot, pasture, or cropped land. Removal of nutrients by the vegetation produces a better quality of water than if the vegetation were absent. The major problems of water reclamation systems are maintenance of high infiltration rates and removal of odor and taste. Recharge rates can be maximized by alternate wetting and drying cycles of surface installations and by periodic redevelopment of recharge wells. The injection of liquid wastes into deep underground formations by means of wells is in reality a storage method rather than a disposal method because the wastes injected into the stratum remain there indefinitely. It is often not recognized that most potential injection zones are filled naturally with brine or another unusable fluid, hence the pore space only becomes available by displacement or extraction of the natural fluid, which merely relocates the disposal problem.

Vecchioli, J., Ku, H.F.H. and Sulam, D.J., "Hydraulic Effects of Recharging the Magothy Aquifer, Bay Park, New York, with Tertiary Treated Sewage", Professional Paper 751-F, 1980, U.S. Geological Survey, Syosset, New York.

From 1968-73, public supply water and water reclaimed from sewage were injected into the Magothy aquifer, Long Island, New York, in a series of 19 recharge experiments. Injection was through a fiberglass-cased well with a 16-in diameter stainless steel screen set from 418 to 480 ft below land surface. Most tests ranged from 2 to 33 days; in the final test a total of 42 million gallons of water was injected intermittently for 6 months. In some tests, chlorination or other chemical treatments were applied to evaluate their effect on well clogging and to observe geochemical reactions within the aquifer. After each test, the injected water was pumped from the aquifer. Clogging developed to varying degrees in the recharge well and contiguous parts of the aquifer and resulted in excessive head buildup in the well. Clogging was due primarily to suspended solids in the injected water. Microbial growth was an insignificant factor in clogging as long as a total residual chlorine level of 2 mg/l was maintained. Specific capacity was restored by pumping and surging the well. Clogging materials underwent biodegradation when the well was idle. Redevelopment by pumping and surging was improved after idle periods of several weeks.

Vecchioli, J., "Water Quality Aspects of Well Recharge with Reclaimed Water, Bay Park, New York", Proceedings of the International Conference on Biological Water Quality Alternatives, Mar. 3-5, 1975, 1976, University of Pennsylvania, Philadelphia, Pennsylvania and American Society of Civil Engineers, pp. 295-299.

From 1968 to 1973, water reclaimed from sewage was used in a series of deep-well artificial-recharge experiments at Bay Park, New York, by the U.S. Geological Survey in cooperation with the Nassau County Department of Public Works. This paper summarizes water quality aspects of the recharge tests. Microbial growth around the well screen caused significant well clogging when unchlorinated water was injected. The higher filter efficiency of the fine to medium sand of the aquifer restricted movement of bacteria away from the point of injection. In tests in which chlorinated water was recharged, clogging by microbial growth was insignificant. However, whenever there was an appreciable pause between the end of injection and the start of repumping, the first water repumped had a strong sulfide odor, was highly turbid, and was laden with bacteria. Chemical quality of the reclaimed water was modified by interaction with the aquifer. The iron concentration of the mixed native and reclaimed waters increased to as much as 3 mg/l from background levels of a few tenths in each; pH decreased to a low of about 4.5 from levels of about 5.5 for native water and 6.1 for reclaimed water. Residual phosphorus was removed almost completely, but nitrogen uptake by the aquifer was minimal.

Vecchioli, J., et al., "Wastewater Reclamation and Recharge, Bay Park, New York", Journal of the Environmental Engineering Division, American Society of Civil Engineers, Vol. 101, No. 2, Apr. 1975, pp. 201-214.

About 600,000 gal/day of effluent from an activated-sludge type sewage treatment plant was further purified by clarification, filtration, activated-carbon adsorption, and chlorination. Significant quality parameters of the reclaimed water and their usual concentrations were: chemical oxygen demand, approximately 10 mg/l; phosphorus, approximately 0.1 mg/l; methylene blue active substances, less than 0.1 mg/l; and turbidity, less than 1 mg/l. Reclaimed water was injected intermittently in a series of tests into the Magothy aquifer through a well screened at a depth of 418 ft to 480 ft (127 m to 146 m) below land surface. The rate of excessive head buildup observed during the recharge of 41,700,000 gal averaged 3 ft (0.9 m)/1,000,000 gal. Operation and maintenance costs (1972) of further purifying the secondary-stage effluent to rechargeable quality, but without nitrogen removal, are estimated to be about $0.27 per 1,000 gal on a 3-mgd scale. Operation and maintenance of recharge wells should be on the order of $0.04 to $0.05/1,000 gal.

Wesner, G.M. and Baier, D.C., "Injection of Reclaimed Wastewater into Confined Aquifers", Journal of the American Water Works Association, Vol. 62, No. 3, Mar. 1970, pp. 203-210.

The Orange County Water District, California has conducted research in wastewater reclamation and subsurface injection since July, 1965. The first phase of the work was intended to determine the feasibility of treating and injecting secondary effluent. The second phase has the following objectives: (1) determine the hydraulic characteristics of the proposed injection barrier system of multi-point injection wells; (2) determine the long-term fate of reclaimed wastewater in the injection system; (3) determine the feasibility of utilizing wastewater for a barrier; and (4) determine the chemical composition of blended reclaimed water and deep ground water. The multiple casing injection wells have performed very satisfactorily. The treated trickling filter effluent is injectable and would not cause excessive well clogging. Coliform bacteria have appeared sporadically 100 ft from the injection well, and have not been found at 245 ft. Many chemical constituents do not move conservatively in the injected water. Hardness and alkalinity increase; ammonia and other oxygen-demanding materials are significantly reduced by travel in the confined aquifer. The odor and taste which persist in the injected reclaimed water is probably the most serious deterrent to utilizing this source for injection in a barrier system.

Willis, R., "Optimal Management of the Subsurface Environment: Parameter Identification", OWRT-A-052-NY(1), May 1976, Office of Water Research and Technology, U.S. Department of the Interior, Washington, D.C.

Mathematical models are presented for assessing the potential impact of the artificial recharge of municipal wastewater effluents on the subsurface environment. Wastewaters are assumed to be recharged to a shallow unconfined ground water aquifer in surface spreading basins. A Galerkin finite element simulation model predicts the time and spatial variations of conservative and nonconservative constituent concentrations within the soil profile. The model incorporates time dependent nonequilibrium absorption and chemical and/or biochemical reactions. The Galerkin procedure transforms the partial differential equations describing mass transport and absorption into systems of first-order ordinary differential equations. The resulting equations are recursively solved by finite difference techniques. The simulation equations form the basis for the parameter identification model which is formulated as a quadratic programming problem using a generalized Newton-Raphson procedure. The model is used to estimate: (1) the kinetic, dispersion, and adsorption parameters characterizing quality changes within the soil system; and (2) the initial distribution of a contaminant within the sorbed or solution phases at the inception of the spreading cycle.

WATER QUALITY IMPACTS OF ARTIFICIAL RECHARGE

Bouwer, H., "Water Quality Aspects of Intermittent Systems Using Secondary Sewage Effluent", Conference on Artificial Groundwater Recharge, University of Reading, England, Sept. 21-24, 1970, Paper No. 8, 1970, The Water Research Association, Marlow, England.

An experimental project on reclaiming water from secondary sewage effluent by ground water recharge with infiltration basins in the dry salt river bed near Phoenix, Arizona, shows that the infiltration rate in grass-covered basins is 25% higher, and in gravel-covered basins 50% lower, than in a bare soil basin. Alternating 2-week inundation periods with 10 day dry-up periods (17 days in winter) yields an annual infiltration rate of about 100 m (330 ft). The hydrogeological conditions of the salt river bed, about 0.9 m (3 ft) of fine, loamy sand underlain by sand and gravel layers to great depth and a ground water table of about 3 m (10 ft) depth, are favorable for high-rate waste water reclamation by ground water recharge. Reclaimed water, pumped from 9 m (30 ft) depth in the center of the recharge area, has a biochemical oxygen demand of about 0.5 mg/l (BOD of the sewage effluent is about 15 mg/l) and a median fecal coliform density of 1 per 100 ml. Nitrogen, which is almost all in the ammonium form at a concentration of 25 ppm N in the sewage effluent, is essentially all converted to the nitrate form in the reclaimed water if sequences of short inundation periods (3 days or less) are held. Phosphate concentration in the reclaimed water pumped from 30 ft depth in the center of the recharge area are around 10 ppm P, as compared to 26 ppm in the effluent. To reclaim the sewage flow of 370 million cu m per year that is expected in the Phoenix area by the year 2000, about 370 ha (950 acres) of infiltration basins would be required. The cost of reclaiming water in this manner would be about $4 per 100 cu m.

Bouwer, H., "Water-Quality Improvement by Ground-Water Recharge", Proceedings of the 2nd Seepage Symposium Phoenix, Arizona, Mar. 25-27, 1968, Report 41-147, 1969, Agricultural Research Service, Phoenix, Arizona, pp. 23-27.

Aerobic percolation of low-quality water through soil to the water table and subsequent lateral movement below the water table can be effective mechanisms for removing biodegradable material, pathogenic organisms, and certain inorganic substances such as metals from the water. Recharge basins should be managed to avoid accumulation of suspended material on the bottom and to allow sufficient oxygen to enter the soil during dryups to satisfy the total oxygen demand of the water that has entered during inundation. Also, the basins should be so laid out and operated to avoid excessive water-table buildups while at the same time achieving maximum recharge per unit basin area. Nitrogen in the low-quality water will appear mostly as nitrates in the renovated water. For renovated sewage effluent, the nitrate content may be two to four times higher than the permissible limit for drinking water. Nitrate reduction by denitrification or other processes is emerging as a major problem in renovating sewage effluent, and research in this area is being intensified.

Chang, A.C. and Page, A.L., "Fate of Inorganic Micro-Contaminants During Groundwater Recharge", Proceedings of the Symposium on Wastewater Reuse for Groundwater Recharge, held September 6-7, 1979 at California State Polytechnical University at Pomona, May 1980, Office of Water Recycling, California State Water Resources Control Board, Sacramento, California, pp. 118-136.

This paper reviews the occurrence of inorganic micro-contaminants in treated wastewater effluents, their potential impact in terms of beneficial use of the water, and the capability of soils in attenuating contaminants in recharge water. The inorganic constituents of wastewater reflect the mineral composition of the source water; additional minerals may be picked up during use in urban areas. Trace elements in treated industrial effluents constitute most of the inorganic micro-contaminants in artificial ground water recharge. In drinking water, the mineral content usually poses no health hazards but could be aesthetically unpleasant; for irrigation purposes, a high dissolved mineral content increases the soil's salinity. Depending on the physical characteristics of the water's impurities, the attenuation mechanisms vary. For constituents in suspended form, these include filtration, trapping and adsorption. Many micro-contaminants are present in recharged water in a dissolved state. A chemical rather than physical reaction is needed to immobilize the dissolved constituents, such as cation exchange, precipitation, surface adsorption, and chelation and complexation. Soils do not have unlimited capability in attenuating micro-contaminants, but they can retain large amounts of trace metal elements. However, long-term, continuous high rate applications of wastewater could cause substantial trace element enrichment of the surface soil, which may render the land unfit for subsequent agricultural development. For recharge systems, the composition of the specific wastewater effluent should be determined, and trace metal inputs for each case should be carefully evaluated.

Collins, H.F., "Effects of Sewage Disposal and Reclamation of Ground Water Quality", Proceedings of the 9th Biennial Conference on Ground Water, Sept. 13-14, 1973, Francisco Torres Conference Center, Goleta, California, Report

No. 26, Dec. 1973, California University Water Resources Center, Berkeley, California, pp. 66-77.

Waste water spreading operations are used at selected sites in California. The city of Oceanside, followed by the County Sanitation Districts of Los Angeles County, were among the first to employ recharge by a planned spreading operation. Other planned surface recharge operations have been carried out at various locations throughout the state but only four which recharge an aquifer supplying domestic water are in operation at the present time. A few projects in California have investigated the use of reclaimed water as a source of injected water. Pilot studies were carried out from 1952 to 1963 at the Hyperion Sewage Treatment Plant, and from 1965 to 1971 at the Orange County Water District. The effects of percolating waste water on the quality of the ground water will vary widely from one location to another depending on: (1) application rates, (2) the characteristics of the waste water, (3) geological considerations, (4) the nature of the soil, and many other factors. Constituents in the waste water that are of primary concern from a health viewpoint are biological pathogens, stable organics, and heavy metals. The soil mantle appears to be quite efficient in the removal of bacteria, viruses and many other constituents. In direct injection, the treatment phenomena which occur with percolation through the soil are not provided, nor are they duplicated with ground water movement. Consequently, the quality of the injected water must be such that it is suitable for the eventual intended uses. The quality must be such that no adverse changes will occur in the underground travel which will impair its use for extraction.

Dvoracek, M.J. and Wheaton, R.Z., "Does Artificial Groundwater Recharge Contaminate Our Groundwaters", Agricultural Waste Management Conference, Jan. 1970, Cornell University, Ithaca, New York.

The authors describe the methods of artificial ground water recharge and raise the question of whether artificial recharge contaminates ground water. They feel that no clearcut answer is available, but they hope to stimulate research and evaluation on planned and operational artificial recharge installations. In the past, severe ground water contamination has generally been localized. However, the potential for widespread contamination is quite real and the authors fear it may already be occurring. Any damage may upset the total ecological system by the phenomena of biological amplification. Augmentation of the water supply by artificial recharge should not be considered so important as to openly invite the use of contaminated recharge water.

Eren, J. and Goldshmid, Y., "Sanitary Significance of Coliform Bacteria in Recharge Wells", Artificial Groundwater Recharge Conference, University of Reading, England, Sept. 21-24, 1970, Supporting Paper B, 1970, The Water Research Association, Marlow, England.

When water from Lake Kinnereth, Israel, is recharged into a sandy aquifer through dual purpose wells, there are often high coliform counts in the water pumped from these wells. The number of coliforms depends usually on the length of the pause period between the end of the recharge and pumping. When the water is pumped immediately at the end of the recharge, the number of coliforms is low. A pause period of a few days or weeks is followed by high counts. After reaching a peak, the coliform count decreases with time. These coliform bacteria are not of fecal origin and there is evidence that they multiply in the well. Tests for enteropathogenic bacteria (Salmonella and Shigella) in recharge wells were negative. Water from recharge wells can be used for public supply after chlorination when it meets the standard set for turbidity and odor, even if the coliform counts are high.

Fetter, C.W., Jr. and Holzmacher, R.G., "Groundwater Recharge with Treated Wastewater", Journal of the Water Pollution Control Federation, Vol. 46, No. 2, Feb. 1974, pp. 260-270.

Artificial recharge of aquifers is a proven method for conservation of ground water resources. The use of open basins, stream channels, and spray irrigation is possible with a typical secondary effluent, but recharge through wells requires wastewater of higher quality. Very slight turbidity and bacterial growths may lead to rapid clogging of recharge wells. Recharge of a fine- or medium-grained aquifer does not seem to lead to problems caused by pathogenic bacteria or viruses if the water travels several hundred feet and the aquifer is not fissured. Water quality improvement may also be affected by physical and biological processes in the soil.

Gass, T.E., "The Health Effects of Ground Water Recharge, Part I", Water Well Journal, Vol. 33, No. 4, Apr. 1979, pp. 34-35.

The potential health effects resulting from the use of surplus wastewater for artificial ground water recharge are discussed. The quality and chemical composition of such wastewater and excess surface runoff must be carefully examined and controlled to avoid long-term contamination of aquifers. Contaminating agents are either biological or chemical. Giardiasis is a common bacteria found in contaminated drinking water which causes an intestinal disease. Viruses, such as hepatitis, are extremely difficult to monitor but have been clearly traced to sewage-contaminated water. Arsenic and barium are two potentially toxic inorganic compounds found in water. Unlike inorganic chemicals, the presence of even trace amounts of organic chemicals in treated, reclaimed water is of great concern. The long-term effects of low doses of potentially toxic compounds in water is not known.

Gass, T.E., "The Health Effects of Ground Water Recharge: Part II", Water Well Journal, Vol. 33, No. 5, May 1979, pp. 38-39.

Depending on its quality, method of recharge, and intended reuse, water may undergo primary, secondary, or advanced treatment processes. Primary treatment mechanically removes solids from water. Secondary treatment removes dissolved or colloidal materials using biological systems. Advanced Wastewater Treatment (AWT) extracts heavy metals or toxic substances through physical or chemical means. After AWT, chlorine or ozone disinfection rids water of bacteria and viruses still present. Storing effluent in open reservoirs is another way to destroy bacteria and viruses which may remain in treated water. Water spreading, an inexpensive and popular recharge method, uses soil systems to filter out bacteria, viruses and other contaminants. Basic criteria used in selecting a recharge site include soil texture, water-transmitting capability, and effects of soil constituents on water. Most biodegradation takes place in the upper 24 to 30 inches of soil. Chemicals are further reduced through absorption by plants in the upper 4 to 5 ft of soil. Chemical adsorption removes pollutants until the soil's total adsorptive capacity is reached. Without adequate contact time for the occurrence of necessary chemical and biological reactions there will be little or no attenuation of pollutants. Contact time is reduced by coarse-textured soil, fractures, and pumping.

Kimrey, J.O. and Fayard, L.D., "Geohydrologic Reconnaissance of Drainage Wells in Florida--An Interim Report", Open File Report 82-860, 1982, U.S. Geological Survey, Tallahassee, Florida.

Drainage wells are used to inject surface waters directly into an aquifer, or shallow ground waters directly into a deeper aquifer, primarily by gravity. Such wells in Florida may be grouped into two broad types: (1) surface water injection wells, and (2) interaquifer connector wells. Surface water injection wells are commonly used to supplement drainage for urban areas in karst terrains of central and north Florida. Data are available for 25 wells in the Ocala, Live Oak, and Orlando areas that allow comparison of the quality of water samples from these Floridan aquifer drainage wells with allowable contaminant levels. Comparison indicates that maximum contaminant levels for turbidity, color, and iron, manganese, and lead concentrations are equaled or exceeded in some drainage well samples, and relatively high counts for coliform bacteria are present in most wells. Interaquifer connector wells are used in the phosphate mining areas of Polk and Hillsborough Counties, to drain mining operations and recharge the Floridan aquifer. Water quality data available from 13 connector wells indicate that samples from most of these wells exceed standards values for iron concentration and turbidity. One well yielded a highly mineralized water, and samples from 6 of the other 12 wells exceed standards values for gross alpha concentrations.

Kimrey, J.O., "Preliminary Appraisal of the Geohydrologic Aspects of Drainage Wells, Orlando Area, Central Florida", Water Resources Investigations 78-37, May 1978, U.S. Geological Survey, Tallahassee, Florida.

The Floridan aquifer contains two highly transmissive cavernous zones in the Orlando area: an upper producing zone about 150 to 600 ft below the land surface and a lower producing zone about 1,100-1,500 ft below the land surface. Natural head differences are downward and there is hydraulic connection between the two producing zones. Drainage wells are finished open-end into the upper producing zone and emplace surface waters directly into that zone by gravity. Quantitatively, their use constitutes an effective method of artificial recharge. Their negative aspects relate to the probably poor, but unknown, quality of the recharge water. Caution is suggested in drawing definite and final conclusions on the overall geohydrologic and environmental effects of drainage wells prior to the collection and interpretation of a considerable quantity of new data. Though few ground water pollution problems have been documented, the potential for pollution should be seriously considered in light of the probable continuing need to use drainage wells; the probable volumes and quality of water involved; and the hydraulic relations between the two producing zones.

Klein, J.M. and Bradford, W.L., "Distribution of Nitrate and Related Nitrogen Species in the Unsaturated Zone, Redlands and Vicinity, San Bernardino County, California", Water Resources Investigations 79-60, July 1979, U.S. Geological Survey, Menlo Park, California.

Nitrogen in the unsaturated soil zone in the vicinity of Redlands in San Bernardino County, California has been suspected as the source of nitrate in water from wells. Plans have been made to recharge the aquifer with imported surface water. If this occurs, the rising water table will intercept soluble nitrate in the unsaturated zone. This study was made to quantify and delineate the distribution of the nitrate in the unsaturated zone. Thirteen test holes were drilled in areas of various land uses, and samples of the unsaturated zone materials were obtained with depth. In 10 of the test holes, the maximum nitrate-nitrogen (NO3-N) concentrations occurred within 10 ft of land surface suggesting NO3-N sources at the surface. Ranking the test holes according to near-surface NO3-N concentrations showed that, of the top six, only two are in irrigated citrus areas. One is in an abandoned feedlot. Of the lowest six, four are in irrigated citrus areas and one is in the Santa Ana River channel. The control test hole ranked fourth suggesting that relatively high NO3-N concentrations can occur in near-surface soils even in the absence of human habitation. Analysis according to one conceptual model suggests that raising the water table altitude by aquifer recharge will intercept NO3-N now in the unsaturated zone causing increases in NO3-N concentrations in water from wells of up to 48 mg/l. The largest increases will occur in the vicinity of irrigated citrus areas.

Klein, J.M. and Bradford, W.L., "Distribution of Nitrate in the Unsaturated Zone, Highland-East Highlands Area, San Bernardino County, California", Water Resources Investigations 80-48, Sept. 1980, U.S. Geological Survey, Menlo Park, California.

In the vicinity of Highland-East Highlands, San Bernardino County, California, plans to recharge local aquifers with imported water would raise the water table and intercept nitrogen in the unsaturated zone. Samples from 13 test holes were analyzed to ascertain the distribution of nitrate and related inorganic nitrogen species and other water quality characteristics. Test holes were ranked according to maximum NO3-N in the top 10 ft, total NO3-N in the top 10 ft, and total NO3-N in the top 40 ft. In all three rankings, the top seven test holes were the same--five in or near past or present agricultural areas (primarily citrus groves), one in a feedlot, and one adjacent to an abandoned sewage treatment plant. Two test holes in uninhabited areas ranked lowest. The control test hole ranked high in geometric mean ammonium nitrogen, suggesting that NH4-N is present in freshly weathered granite. The geometric mean NH4-N concentrations in six of eight test holes in citrus-related areas were significantly lower than in the control hole. High NO3-N concentrations tended to be found with high specific conductance and chloride concentrations. If recharge is carried out as planned, assuming complete mixing of the recharge water and interstitial pore water in the top 20 ft of the unsaturated zone, NO3-N concentrations in water at the top of the saturated zone may exceed 10 mg/l in seven areas studied.

LeGrand, H.E., "Environmental Framework of Ground Water Contamination", Ground Water, Vol. 3, No. 2, Apr. 1965, pp. 11-15.

The volume of usable ground water is shrinking in many places because of dispersion of contaminated water. Ways of contamination and pertinent parts of the physical environment include: waste disposal practices (at or near land surface and in deep formations), artificial recharge (at land surface and in aquifers), accidents, and salt water contamination of aquifers. Evaluation of waste disposal problems calls for appreciation of two opposing tendencies--the tendency of wastes to move with ground water and the tendency to be attenuated near disposal sites by decay or inherent decrease in potency, by chemical and physical sorption, and by dilution through dispersion of ground water.

Moravcova, A., Masinova, L. and Bernatova, V., "Biological and Bacteriological Evaluation of Pilot Plant Artificial Recharge Experiments", Water Research, Vol. 2, No. 4, 1968, pp. 265-276.

The effect of the algicide CA-350 (a combination of copper and silver) was investigated on four sand filter models of an artificial recharge basin. Using 300 microgram/l (ppb) Cu in a preventive treatment, the development of filamentous algal forms was fully controlled with the conservation of a normal bacterial microflora. The processes of mineralization of organic matter continued under aerobic conditions. Copper was gradually eluted from the sand layers of the model filters, never reaching its maximum allowable concentration in the outflowing water.

Nightingale, H.I. and Bianchi, W.C., "Ground-Water Chemical Quality Management by Artificial Recharge", Proceedings of the 3rd National Water Well Association-EPA National Ground Water Quality Symposium, Special Issue Vol. 2, Sept. 1976, pp. 15-22.

The effectiveness of basin ground water recharge at the Leaky Acres Recharge Facility in Fresno, California for improving the regional ground water quality was studied as 65,815,000 m^3 of high-quality surface water was recharged from 1971 through 1975. Observation wells at the facility showed some variability in chemical parameters associated with each recharge period. The long-term decreases in salinity could be described by a power law decay curve fitted by regression analysis.

Nightingale, H.I. and Bianchi, W.C., "Ground-Water Turbidity Resulting from Artificial Recharge", Ground Water, Vol. 15, No, 2, Mar.-Apr. 1977, pp. 146-152.

Turbid ground water is rarely observed in domestic or public supply aquifers. At the Leaky Acres Recharge Facility at Fresno, California, water of low salinity (less than 50 micromhos/cm) and turbidity (less than 5 FTU, Formazin turbidity units) is recharged in the spreading basins. Six months after the start of the third (1973) recharge period, the ground water salinity was decreased to about 100 micromhos/cm from the initial mean of 147 micromhos/cm, and the ground water became visibly turbid (more than 5 FTU). Two months later, some peripheral domestic wells also began to become turbid. After two more recharge periods (1974 and 1975), turbidity at 10 observation wells beneath Leaky Acres averaged 18 FTU, and salinity averaged 74 micromhos/cm. By this time, ground water turbidity in peripheral wells near Leaky Acres had decreased to less than 0.5 FTU. This turbidity was traced to poorly crystallized and extremely fine colloids, which have leached from the surface soils because of the low salinity of the recharge water. Laboratory and field studies showed that gypsum application will reverse the phenomena, but such treatment is uneconomical. This phenomenon is a transient one, and now turbidity outside the recharge area is insignificant from a water quality viewpoint. However, the magnitude of the mass of material in transit through the profile, if stabilized through flocculation or sieving in soil pore space, could greatly change the water transmission and the recharge project performance. However, this effect has not yet been noted at Leaky Acres.

Okubo, T. and Matsumoto, J., "Effect of Infiltration Rate on Biological Clogging and Water Quality Changes During Artificial Recharge", Water Resources Research, Vol. 15, No. 6, Dec. 1979, pp. 1536-1542.

The effect of infiltration rate on biological clogging and water quality changes was investigated experimentally using a sand column. The concentration of soluble COD in the effluent was stable during 3 months of infiltration, and the lower the initial infiltration rate was, the higher the average percentage of the reduction of soluble COD became. The soluble COD was primarily removed in the surface layer. The hydraulic conductivity in the bottom layer decreased at the highest infiltration rate. The variation of infiltration rate was divided into four stages. In addition, the biological-clogging model was proposed, and the mechanism of biological clogging was discussed.

Olsen, R.C., "Pesticides and Ground Water", Water and Oregon's Ecology, Fall Quarter 1970 Seminar Report WR 013.70, Jan. 1971, Oregon Water Resources Research Institute, Oregon State University, Corvallis, Oregon, pp. 79-81.

Problems are discussed in both natural and artificial ground water recharge employing surface water that has been subjected to mild pesticide contamination. Two specific cases in the United States are mentioned and discussed, one in Colorado, and one in Texas. Depending on aquifer conditions, ground water moves at rates ranging from about 5 ft per day to 50 ft per year. Therefore, pollution may not show up until long after the source has ceased polluting. A recharge experiment on the Ogallala aquifer involved addition of DDT, nitrate and a tracer with the recharge water for 10 days of injection. The well was then pumped for 12 days. Of the 3.11 lb of DDT used in recharging, only 0.142 lb was received by the pumping, and this recovery was from adsorption on the sand particles. Standards are especially needed for persistent complex organic pesticides.

Rittmann, B.E., McCarty, P.L. and Roberts, P.V., "Trace-Organics Biodegradation in Aquifer Recharge", Ground Water, Vol. 18, No. 3, May-June 1980, pp. 236-243.

The low organic concentration and high specific surface area that characterize ground water recharge environments allow biofilms to predominate microbial kinetics. Consideration of the kinetics of substrate utilization and growth of biofilms indicated that microbiological activity occurs very near the injection well. The aggregate substrate expressed, for example, as chemical oxygen demand, can be considered the primary substrate; the biofilm's growth is supported through the utilization of the aggregate primary substrate. Individual trace organic compounds, none of which could support biofilm growth alone, were utilized as secondary substrates. Although biodegradable, secondary substrates that have slow utilization kinetics will pass through the biologically active zone undegraded. Field data from the ground water recharge project in Palo Alto, California, illustrated that naphthalene and heptaldehyde are biodegraded, while other compounds, such as chloroform and chlorobenzene, pass through the biologically active zone without biodegradation.

Rizet, M., Mallevialle, J. and Cournarie, J.C., "Pilot Plant Investigation of the Evolution of Various Pollutants During Artificial Recharge of an Aquifer by a Basin", Progress in Water Technology, Vol. 9, No. 1, 1977, pp. 203-215.

An artificial recharge pilot plant (1 cu m/hr) was employed to investigate the evolution of various pollutants in the charging basin and successive ground layers. The compounds were introduced separately at different concentrations: 1 mg/l of lead in the form of nitrate, 1 mg/l of sodium alkyl aryl sulfonate, 2 mg/l of a ziegler synthetic alcohol (nonionic surfactant), 2.5 mg/l of phenol, 12.5 mg/l of a mixture of hydrocarbons and surfactants (soluble oil), 300 mg/l of a refinery effluent, and finally, lindane at concentrations of 1 micrograms/l and 10 micrograms/l. Specific algal and bacterial counts were made on the water in the basin as well as in the sand before and after passage of the polluting flows. A certain delay in penetration was observed for the non-biodegradable compounds, while the passage of the biodegradable compounds into the surface layer was stopped rapidly by acclimatization of the biotope.

Roberts, P.V., Schreiner, J. and Hopkins, G.D., "Field Study of Organic Water Quality Changes During Groundwater Recharge in the Palo Alto Baylands", Water Research, Vol. 16, No. 6, June 1982, pp. 1025-1035.

A field study was performed wherein reclaimed water was injected directly into an aquifer in the Palo Alto Baylands, near San Francisco Bay. The injection rate was 6 liters/second of reclaimed municipal wastewater. The concentrations of organic constituents were affected by processes such as adsorption and biodegradation during recharge even though the recharged water had received advanced treatment including granular activated carbon. The influence of adsorption was observed as retardation of the transport of trace organic compounds, which are retained to varying degrees in the aquifer. The effects of biodegradation are evidenced by decreasing concentrations of total organic carbon, chemical oxygen demand, and dissolved oxygen as the water spread out from the injection point. In general, the concentrations of TOC and COD responded rapidly to fluctuations in input concentration, indicating that the bulk of organic constituents in the reclaimed water did not engage in sorptive interaction with the aquifer material. Halogenated organic compounds showed no evidence of retardation or retention during aquifer passage. Organic pollutants remaining after water reclamation treatment pose a potential health threat if the reclaimed water is to be directed to potable reuse. Ground water recharge by direct injection affords protection by affording opportunities for adsorption and degradation and by smoothing concentration fluctuations.

Roberts, P.V. and Valocchi, J., "Principles of Organic Contaminant Behavior During Artificial Recharge", Journal of the Science of the Total Environment, Vol. 21, 1981, pp. 161-172.

The behavior of a variety of organic contaminants having low molecular weight has been observed during ground water recharge with reclaimed water. The evidence is site specific, but is believed to have broader implications regarding the general behavior of organic contaminants in ground water. The movement of specific contaminants such as chloroform and chlorobenzene is retarded with respect to that of a conservative tracer such as chloride. The measured retardation factors are approximately 3 and 35 for chloroform and chlorobenzene, respectively. The retardation is caused by the sorption of the solute, apparently by the organic constituents of the soil material. The magnitude of the retardation factor of an organic solute can be predicted approximately from knowledge of the organic carbon content of the aquifer material and the octanol-water partition coefficient of the solute. Sorption also contributes to attenuation (damping) of concentration fluctuations. It is shown that the degree of attenuation depends strongly on the retardation factor, the distance traveled, and the hydrogeologic characteristics of the aquifer. There is evidence of biodegradation of organic solutes in the vicinity of the recharge well.

Sasman, R.T., "Thermal Pollution of Ground Water by Artificial Recharge", Water and Sewage Works, Dec. 1972, pp. 52-55.

Of the five industries in northeastern Illinois recharging cooling water, four have reported no undesirable effects. The effect on well water temperatures of one industry recharging high temperature cooling water effluent is described. The water for this industry is primarily obtained from two wells about 1350 ft deep. The recharge well is 50 ft from one supply well and about 600 ft deep. The temperature of farm wells as well as of the supply wells were measured and a temperature gradient away from the recharge well was found. High temperature recharge water has had a significant effect on the ground water temperature for a distance of at least 500 ft and perhaps as far as 1000 ft. Additional consideration should be given to this problem in future operations of this type. Monitoring injection rate, water levels, temperature and quality would be beneficial in attempts to determine long range effects on ground water aquifers.

Scalf, M.R., et al., "Fate of DDT and Nitrate in Ground Water", Apr. 1968, R.S. Kerr Environmental Research Laboratory, U.S. Environmental Protection Agency, Ada, Oklahoma.

Increasing demands on the water resources of the High Plains area of West Texas and Eastern New Mexico have given rise to the possibility of storing water underground. This would be accomplished by recharging the Ogallala aquifer. Artificial recharge practices give rise to concern that ground water may be contaminated by pesticides and nitrates introduced with the recharge water. This report presents findings concerned with the fate of DDT and nitrate under actual recharge conditions in the Ogallala aquifer.

Toups, J.M., "Water Quality and Other Aspects of Ground Water Recharge in Southern California", Journal of the American Water Works Association, Vol. 66, No. 3, Mar. 1974, pp. 149-151.

Ground water recharge became necessary in southern California during the 1950's because ground water levels along the coastline had dropped below sea level in many places and seawater was intruding into the ground water basins. Most of the recharge water is Colorado River water, which is of poorer quality than native ground water or local streamflow. Recharge water, whether it be placed underground by spreading or by injection, can undergo water quality changes through interaction with the soil. The changes are dependent upon the specific character of the recharge water and upon the soil type.

Vecchioli, J., "A Note on Bacterial Growth Around a Recharge Well at Bay Park, Long Island, New York", Water Resources Research, Vol. 6, No. 5, Oct. 1970, pp. 1415-1419.

Highly treated sewage plant effluent is injected into a deep sand aquifer at Bay Park, Long Island, New York. High bacterial counts are observed in water pumped initially from the injection well after recharge termination, even though the injected water generally meets potable standards. Moreover, the number of bacteria recovered during pumping is much larger than the number injected. This suggests a growth of bacteria in the aquifer around the injection well, possibly in part on a filter mat of organic material formed at or near the aquifer-gravel pack interface during injection. Continued pumping results in lower counts until virtually bacteria-free water is produced. However, whenever pumping is stopped and restarted, the initial slug of water pumped shows higher counts than before shutdown. Apparently the surging action of the water dislodges bacteria that otherwise would not enter the well.

Vecchioli, J., Ehrlich, G.G. and Ehlke, J.A., "Travel of Pollution-Indicator Bacteria Through the Magothy Aquifer, Long Island, New York", Professional Paper 800-B, 1972, U.S. Geological Survey, Mineola, New York, pp. B237-B239.

Experimental injection of reclaimed water (highly treated but unchlorinated sewage) into the Magothy aquifer at Bay Park, Long Island, New York, showed very restricted travel of pollution-indicator bacteria. Although the injected water had substantial densities of total-coliform, fecal-coliform, and fecal-streptococci, no fecal-coliform or fecal-streptococcal bacteria and only nominal total-coliform bacteria were found in water from an observation well 20 ft away from the point of injection. Restriction of bacterial travel through the Magothy aquifer is due primarily to the high filter efficiency of the fine to medium sand characteristic of the aquifer and to bacterial capture by a filter mat and slime deposits that form around the injection well during injection.

Warner, D.L. and Doty, L.F., "Chemical Reaction Between Recharge Water and Aquifer Water", Artificial Recharge and Management of Aquifers, Symposium of Haifa (Mar. 19-26, 1967), Publication No. 72, 1967, International Association of Scientific Hydrology, pp. 278-288.

Chemical reactions between recharge water and aquifer water may influence water quality and aquifer permeability, and may cause fouling in discharge wells. However, evidence suggests that the danger of permeability reduction may not be very great. Precipitates of calcium and magnesium carbonate and iron compounds are the most likely reaction products. The amount and distribution of reactive chemicals remaining in solution can be related to the dispersive characteristics of a porous medium. Reactions of various rates are considered. It may be desirable to prevent reaction. It is suggested that, in some cases, this can be accomplished by emplacing a buffer zone of nonreactive water between the recharge water and the aquifer water.

Wilson, L.G., "Management of Artificial Recharge Wells for Groundwater Quality Control", Proceedings of Hydrology and Water Resources in Arizona and the Southwest, Apr. 22-23, 1971, Vol. 1, 1971, Arizona Section-American Water Resources Association and the Hydrology Section-Arizona Academy of Science, Tempe, Arizona, pp. 261-268.

Recharge wells may be used in various problems relating to chemical water quality because of the phenomenon of in-aquifer mixing. This paper reviews specific recharge well-mixing techniques of possible utility in underground mixing operations for nitrate control. Illustrative data from field studies at a recharge site near Tucson, Arizona are presented. Both single-and 2-well types of mixing were investigated. In single-well operations, effluent recharge and pumping of the subsequent mixture occur at the same well. Differences in chlorine ion levels were used to distinguish between recharge effluent and native ground water. Undiluted effluent was discharged in single-well operations until a pumped volume ratio of about 0.4 was attained. Dilution increased steadily with increased pumping and the relative concentration versus pumped volume curve was S-shaped. Seven-day pauses after effluent recharge resulted in immediate pumping of almost completely diluted water, probably because ground water movement swept the effluent beyond the pumping unit during the pause. With 2-well pumping, the chlorine breakthrough curve reached a constant level at about 13 days and was close to that of the pause-type, single-well regime.

Wood, W.W. and Signor, D.C., "Geochemical Factors Affecting Artificial Groundwater Recharge in the Unsaturated Zone", Transactions of the American Society of Agricultural Engineers, Vol. 18, No. 4, July-Aug. 1975, pp. 677-683.

Recent research on artificial ground water recharge has focused on spatial-temporal aspects. The type and magnitude of chemical controls at a site are studied. Chemical considerations are placed into two categories: the change in geometry of the interstitial pore space and the prediction of the quality of the water with time. The observations discussed fall into the second category. The recharge facility used, a 0.4-ha basin near Lubbock, Texas, is typical of the southern High Plains of Texas and New Mexico, and the water was imported from Lake Meredith, near Amarillo. Porous ceramic cups were used to collect water samples at depths of 0.6, 2, 8, 16, 23, and 33 meters, first daily and then weekly. Chemical analyses were performed for pH, bicarbonate and specific conductance. Chloride concentration increased with depth. Ion exchange was one of the most important chemical processes observed in the system. Cation and anion exchange took place, but cation exchange dominated. This could be an important consideration in a predictive model, depending on whether or not the water is intended for human consumption. Some sulfate adsorption takes place and the effects are still being studied. Desorption of silica was the other major mechanism that affected the water quality. Sulfate reduction signaled a hydrologic change but the effect was small because of the rapid loss of hydraulic conductivity. Some mineral solution occurred but also had a very small effect. Studies should be conducted at other sites in the southern High Plains of Texas and New Mexico, including more detailed analyses. With enough pertinent data incorporated into a model, accurate predictions can be made on water quality changes.

Wood, W.W. and Basset, R.L., "Water Quality Changes Related to the Development of Anaerobic Conditions During Artificial Recharge", Water Resources Research, Vol. 11, No. 4, Aug. 1975, pp. 553-558.

Artificial recharge basins or spreading sites commonly exhibit reductions in infiltration rates after prolonged periods of submergence. This loss in infiltration rate has often been shown to be associated with a large population of anaerobic or facultative anaerobic bacteria in the material underlying the basin floor. An artificial recharge experiment in a spreading basin at Lubbock, Texas, showed a significant change in the chemical quality of recharged water that was associated with the growth of anaerobic bacteria.

MODELING OF ARTIFICIAL RECHARGE

Abiodun, A.A., "Analysis of Seepage into Ground-Water System", Journal of the Hydraulics Division, American Society of Civil Engineers, Vol. 99, No. HY7, July 1973, pp. 1203-1208.

Analytical methods were developed to simulate seepage into a ground water system from a range of geometries. The solutions to such problems might find practical application in situations where subsurface lateral distribution of seepage flows is to be restricted and also in cases where a porous medium that is to be replenished is overlain by a fairly thin impermeable layer. The particular case chosen is that of seepage through a channel with impervious side walls and permeable bottom. The driving force of the flow system is the force of gravity and the flow is two-dimensional. The seepage flows are steady and fully saturated, taking place in an isotropic, homogeneous, porous medium, and Darcy's Law is applicable. The solution to this problem was obtained through the application of the hodograph technique.

Amar, A.C., "Hydrodynamics of Artificial Groundwater Recharge (Saturated Flow Theory)", Ph.D. Thesis, 1973, School of Engineering, University of California, Irvine, California.

Use of artificial ground water recharge as an effective means of reclaiming low-quality water has grown steadily since its inception late in the nineteenth century. As a prerequisite to an optimum design of a recharge system the practicing engineer must be able to predict with reasonable accuracy the hydrodynamic behavior of the proposed system for a given geometry and recharge operation. This study describes the hydrodynamic behavior of a (typical) water recharge system under saturated flow conditions and methods, through the use of a sand model and mathematical models for the prediction of the transient and steady-state hydrodynamic behavior of the proposed system. The results of the investigation are considered significant in determining the authenticity and the accuracy of the various theories over a wide range of parameters relating to the hydrodynamics of the recharge operation. A comprehensive literature survey as well as an extensive bibliography are also included.

Bagchi, S. and Goodman, A.S., "Emergency Water Supplies from Ground Water in Humid Regions", Water Resources Bulletin, Vol. 15, No. 2, Apr. 1979, pp. 536-549.

This paper focuses on the development and testing of a mathematical model of an emergency ground water supply operated principally during periods of low streamflow. The process of ground water withdrawal and recharge was simulated taking account of streamflow, water demand, evapotranspiration, natural and artificial recharge, and increased evapotranspiration due to artificial recharge, ground water pumpage, and streamflow contribution to pumped water. The model determined whether natural recharge is possible in less time than the return period of drought and also whether artificial recharge is needed. By simulating operation over a long period of time, the model can examine different droughts of short and long duration and can test the operating rules for ground water storage development in an area. Submodels analyze the components of the operating process, including ground water flow into the stream, seepage losses, stream portion of well discharge due to induced infiltration, and recharge from rainfall or water spreading. The model has been tested for areas in the humid northeastern United States.

Bamgboye, O.A., "Groundwater Basin Modeling Using the Finite Element Method--Yolo County Case Study", MS Thesis, 1979, Department of Agricultural Economics, University of California, Davis, California.

In order to effectively monitor and manage aquifers subjected to varying natural and artificial recharge as well as increased pumping demands, it becomes necessary to analyze their hydrodynamic response. In this study a regional two-dimensional flow analysis is formulated using finite element techniques. A general algorithm is described to handle confined or unconfined flow. The advantage of using finite element analysis for aquifer flows becomes evident when attempting to model irregular boundaries and variable inputs such as natural recharge, pumping and aquifer properties. Numerical solutions are presented which, when compared with known analytical results, verified the technique and showed some limitations of the model. An analysis is presented of flow in the aquifer systems underlying Yolo County, a region where time variable pumping and natural recharge do occur. Unconfined and confined ground water flow analyses were employed to simulate the hydrodynamic response of the Yolo County aquifer systems. The Galerkin time step scheme formulation was found applicable for small time steps. Based on minimization of differences between recorded and computed ground water levels, the confined analysis using reliable constant transmissivity values is regarded as the most appropriate model for the aquifer systems.

Behnke, J.J., "Clogging in Surface Spreading Operations for Artificial Groundwater Recharge", Water Resources Research, Vol. 5, No. 4, Aug. 1969, pp. 870-876.

Clogging is a complex phenomenon dependent upon many variables. For the conditions studied it is essentially a surface-sealing process. When turbid water contains a range of grain sizes, clogging is initiated by the deposition of a surface layer that is graded because of the gravitational segregation of heavier particles. Most of the clogging results from the straining of finer particles from the flowing water by the graded depositional layer. An equation was developed which predicts accumulated water quantity versus time relationships for several concentrations of a specific turbidity. Even coarse-grained porous media clogged at turbid water concentrations as low as 50 ppm.

Brock, R.R., "Dupuit-Forchheimer and Potential Theories for Recharge from Basins", Water Resources Research, Vol. 12, No. 5, Oct. 1976, pp. 909-911.

Artificial ground water recharge from strip and square basins to an unconfined aquifer was investigated. The classical Dupuit-Forchheimer (DF) theory and the most exact potential theory were used to determine the motion of the water table resulting from recharge. For each theory, both a linear and a nonlinear version were considered. Finite difference schemes were employed to solve the nonlinear models. Analytical solutions exist for the linear models. By comparing the results from the various theories and from experimental results in a sand model, the range of validity of the theories was determined. For sufficiently shallow flows, the DF theories gave results that are good approximations to the potential theories. For sufficiently deep flows, the potential theories must be used, particularly beneath the basin and for small times. The velocity field near a strip basin was determined from the nonlinear potential theory for both a deep and a shallow flow. Tabular results were presented that can be used to determine the rise of the water table for strip and square basins.

Brock, R.R. and Amar, A.C., "Ground-Water Recharge Strip Basin-Experiments", Journal of the Hydraulics Division, American Society of Civil Engineers, Vol. 100, No. HY4, Paper 10503, Apr. 1974, pp. 569-592.

The problem of recharging a strip basin to a horizontal water table overlying an impervious base was formulated using the saturated flow equations. From this formulation the number of dimensionless variables that determine a dimensionless water table rise were identified using the exact or potential theory and an approximate (Dupuit-Forchheimer) theory. Experiments were performed in a 5-m long sand model. As a dimensionless recharge rate was increased, an expected systematic deviation from the theory is presumably a result of the invalidity of the assumption of the theory for deep flows. Experimental results are presented in the form of dimensionless graphs.

Brown, G., Jr. and Deacon, R., "Economic Optimization of a Single-Cell Aquifer", Water Resources Research, Vol. 8, No. 3, June 1972, pp. 557-564.

Optimal economic use of an aquifer over time was analyzed under conditions of economic growth, inequality of ground water withdrawal and consumption, and availability of surface water and artificial recharge; the value of an aquifer as a natural water quality treatment facility is derived. These conditions are discussed in depth and are presented explicitly as four subject areas that have hitherto remained unexplored. The stage was set with a simple nonstochastic ground water model using optimal control; equations were plentiful throughout the study. Discussed are optimal lift levels and pumping tax for a ground water basin characterized by simple growth of the net benefits of water. Second, the model makes explicit the important distinction between ground water use and consumption: ground water users with different rates of return flow should pay different pumping tax rates because they borrow different amounts of capital. Third, it is shown that introducing surface waters or artificial recharge replenishment raises the water table and lowers the pumping tax. Fourth, an explicit distinction is made between water quantity and water quality that permits the determination of the value of a ground water basin as a natural water quality treatment facility.

Burns, A.W., "Hydrologic Analysis of the Proposed Badger-Beaver Creeks Artificial Recharge Project, Morgan County, Colorado", Water Resources Investigations 80-46, July 1980, U.S. Geological Survey, Lakewood, Colorado.

A hydrologic analysis of the proposed Badger-Beaver Creeks artificial-recharge project in Morgan County, Colorado, was made with the aid of three digital computer models: a canal-distribution model, a ground water flow model, and a stream-aquifer model. Statistical summaries of probable diversions from the South Platte River based on a 27-year period of historical flows indicate that an average-annual diversion of 96,000 ac-ft and a median-annual diversion of 43,000 ac-ft would be available. Diversions would sustain water in ponds for waterfowl habitat for an average of about five months per year with a maximum pond surface area of about 300 acres with the median diversions, and a maximum pond surface area of about 1,250 acres at least one-half of the years with the historic diversions. If the annual diversion were 43,000 ac-ft, recharge to the two alluvial aquifers would raise water levels sufficiently to create flowing streams in the channels of Beaver and Badger Creeks while allowing an increase in current ground water pumping. The only area of significant waterlogging would be along the proposed delivery canal on the west edge of Badger Creek Valley. If the total water available were diverted, the aquifer system could not transmit the water fast enough to the irrigation areas to avoid considerable waterlogging in the recharge areas. The impact of the proposed project on the South Platte River basin would be minimal once the ground water system attained steady-state conditions, but that may take decades with a uniform diversion of the 43,000 ac-ft annually.

Charbeneau, R.J. and Street, R.L., "Modeling Groundwater Flow Fields Containing Point Singularities: A Technique for Singularity Removal", <u>Water Resources Research</u>, Vol. 15, No. 3, June 1979, pp. 583-594.

Artificial recharge of ground water resources through wells leads to a need for computer simulation of the hydraulics and transport of contaminants in a flow field. The basic numerical technique used is the Galerkin finite element method. Greatly improved accuracy of the simulation near wells is achieved first by removing the singular behavior of the flow with an analytic solution for the near-well zone, second by solving the remaining nonsingular problem, and third by combining the results. The resulting solutions compare favorably with known analytic solutions for confined and leaky aquifers (both isotropic and anisotropic). No additional computation time is required for this method compared to standard finite element methods programs. The resulting combined velocity fields are accurate enough throughout the flow field for direct use in contaminant transport simulations.

Chowdhury, P.K., Shakya, S.K. and Anjaneyulu, B., "Drainage by Recharge Wells in a Leaky Aquifer", Journal of Hydrology, Vol. 36, No. 1/2, Jan. 1978, pp. 87-93.

Theoretical solutions were developed for predicting the rate of recharge through a well in a leaky artesian aquifer overlain by an unconfined aquifer of limited areal extent. The ground surface was considered to be waterlogged. A solution to the basic steady-state differential equation was presented. The solution can be used to compute the recharge rate on the basis of aquifer characteristics and the steady infiltration rate of the soil.

Cole, J.A., Hirsch, H. and Thorn, B.M., "Finding Natural Recharge Patterns and Abstraction Phases to Maximise Yield of Aquifers", Proceedings of the International Symposium on Modelling Techniques in Water Resources Systems, May 9-12, 1972, Vol. 2, 1972, The Water Research Association, Marlow, England, pp. 240-249.

Identification of a unit hydrograph response of well water level is presented as a means of detecting sequences of natural recharge to aquifers. This method yields a key to the quantity of infiltration arriving at the water table itself, as distinct from its earlier departure from the root zone of soil. A simple electrical analogue is then used to simulate a block of unconfined aquifers. Various seasonal patterns of natural and artificial recharge can be represented. For each set of boundary conditions the annual ground water yield is found as a function of the pumping phase, i.e., the timing of pumping relative to the recharge pattern. Results are presented as families of graphs and reveal the amount of recharge that may be captured for any pumping phase.

Dowden, J.E. et al., "Artificial Recharge of Surface Runoff in Cold Spring Valley, Nevada", OWRT-B-109-NEV(1), Dec. 1982, Office of Water Research and Technology, U.S. Department of the Interior, Washington, D.C.

Cold Spring Valley, Nevada was examined with the intent of developing a methodology for studying and implementing artificial recharge--surface runoff harvesting schemes that could be employed in other basins of Nevada. A steady-state, finite element model was used to simulate the Cold Spring Valley ground water flow system. The primary limiting factor to any artificial recharge scenario in Cold Spring Valley is the response of the water table in the Reno Park subdivision caused by an increase in the septic field load. The existing ground water and waste management practices mitigate against the implementation of artificial recharge in the valley at this time. However, with increasing growth in the valley, artificial recharge will undoubtedly become viable in the future.

Durbin, T.J. and Morgan, C.O., "Well-Response Model of the Confined Area, Bunker Hill Ground-Water Basin, San Bernardino County, California", Water Resources Investigations 77-129, July 1978, U.S. Geological Survey, Menlo Park, California.

The Bunker Hill ground water basin, in the vicinity of San Bernardino, California, is being artificially recharged with imported water. Current and future artificial recharge of the basin may cause the potentiometric surface in an area of confined ground water to rise above the land surface and water to flow from uncapped and unplugged wells. This could cause damage to structures where the soil becomes waterlogged and where buried wells begin to flow beneath the structures. A well-response model was used to generate a series of water-level hydrographs representing the response of the ground water basin to six possible combinations of conditions for each well; one pumping rate, two artificial-recharge rates, and three natural-recharge rates. Inflow to the ground water basin exceeds outflow for all tested combinations. According to model predictions, the accumulation of stored ground water resulting from the excess of inflow is sufficient to cause the water level in the selected wells to rise above the land surface for all but one of the combinations of conditions tested. Water levels in wells are predicted to rise above the land surface as early as 1981 for the combination with the greatest excess of inflow.

Fogel, M.M., Thames, J.L. and Duckstein, L., "Modeling of Hydrologic Processes and Water Salvage Procedures in Semiarid Regions", PB-206 121, Dec. 1971, National Technical Information Service, U.S. Department of Commerce, Springfield, Virginia.

A methodology is set forth to model and analyze rainfall and runoff events utilizing the basic concepts of probability theory to the general modeling problem. The procedure looks at an event such as thunderstorm rainfall in light of the number of occurrences in a given time period and the magnitude and spatial distribution of each event. A poisson distribution was found to adequately describe the number of events per season and a geometric distribution for the depth of rainfall at a point per storm. These distributions were then combined to produce an extreme-value distribution for thunderstorm rainfall. Random variables representing rainfall at several points were convoluted to yield mean areal values of rainfall whose distribution was determined to be a negative binomial. Chicago and New Orleans data seemed to validate the model developed in Tucson. The above event-based procedure is then used to describe seasonal rainfall combinations of rainfall depths per event and number of events per season. Corresponding likely sets of runoff amounts can then be obtained using watershed models or other appropriate rainfall-runoff relationships. A Monte Carlo simulation can be set up to generate an unordered succession of yearly combinations of events. This synthetic rainfall set is useful for determining the annual water yield, for evaluating runoff modification due to vegetative manipulation and urbanization and determining the effectiveness of water conservation practices such as artificial recharge and water harvesting.

Griffis, C.L., "Groundwater-Surface Water Integration Study in the Grand Prairie of Arkansas", OWRR-A-010-ARK(4), Dec. 1972, Office of Water Resources Research, University of Arkansas, Fayetteville, Arkansas.

A mathematical model of the Quaternary aquifer of the Grand Prairie, Arkansas was developed and used to evaluate a variety of methods of artificially recharging this aquifer. In addition, the model was used to evaluate the impact of various levels of water management and the probable movement of artificially recharged water in the aquifer. Improved water management and the use of recharge wells were the most promising alternatives. The rate of movement of recharged water was determined by the model to be 300 ft/yr under a gradient of 16 ft/mile.

Griffis, C.L., "Modelling a Groundwater Aquifer in the Grand Prairie of Arkansas", Transactions of the American Society of Agricultural Engineers, Vol. 15, No. 2, Mar.-Apr. 1972, pp. 261-263.

Most of the irrigation water for the Grand Prairie of Arkansas comes from the Quaternary aquifer lying beneath the area. Constant pumping has lowered the water table and shortened the water supply. A model was developed to investigate the possibility of artificial recharging of the aquifer. The area was divided into sections of one square mile. The flow equations were then solved for each section. Various methods of recharge were then evaluated using the model. Recharge wells were found to provide the best solution.

Haimes, Y.Y., "Integrated System Identification and Optimization for Conjunctive Use of Ground and Surface Water, Phase II", SED-WRG-74-1, Sept. 1974, Ohio Water Resources Center, Ohio State University, Columbus, Ohio.

The ground water parameter identification model developed in Phase I has been further modified and improved as part of Phase II work. A hierarchical management model was developed and tested. The coupling of a complex multicell aquifer system was studied via an extension of the algebraic technological function approach. Functions were developed to relate the infiltration from a stream into an aquifer due to pumpage at a well. A new approach was proposed to incorporate time varying boundary conditions in a ground water system. A predictive model was developed to demonstrate the usefulness of these response oriented models, where the decision variables include pumpage, artificial recharge, and imported water. A model for the planning and management of a ground water system was developed, incorporating a tax-quota system. The research results were applied to the area of the lower Great Miami River with data provided by the Miami Conservancy District (MCD), Dayton, Ohio.

Hanson, H.J. and Brock, R.R., "Strip Basin Recharge to Inclined Water Table", Journal of the Hydraulic Engineering Division, American Society of Civil Engineers, Vol. 110, No. 4, Apr. 1984, pp. 436-449.

The hydrodynamics of recharge mounds that develop on an inclined water table beneath a strip recharge basin are considered. The nonlinear potential (NLP) theory is used to predict the motion of the water table resulting from the recharge operation. The theoretical derivation of the NLP theory uses the saturated flow equations for a homogeneous isotropic medium. Solutions to the NLP theory use the saturated flow equations for a homogeneous isotropic medium. Solutions to the NLP theory for recharge to an inclined water table are obtained by using a finite difference method. A few examples of flow patterns showing the motion of the recharged and native water are given. The mathematical model can be utilized to predict the response and movement of recharged water in the design and operation of recharge facilities.

Hardt, W.F. and Hutchinson, C.B., "Development and Use of a Mathematical Model of the San Bernardino Valley Ground-Water Basin, California", Open File Report 80-576 (WRI), Sept. 1980, U.S. Geological Survey, Menlo Park, California.

Part of the San Bernardino urbanized area in California overlies formerly swampy lands with a history of flowing wells. This area, upgradient from and adjacent to the San Jacinto fault, contains a zone in an alluvial ground water basin that is under artesian pressure. Since 1945, withdrawals have exceeded recharge and caused head declines of more than 100 ft. Artificial recharge of imported water in the upgradient areas may cause ground water levels to rise, which could cause abandoned but unplugged wells to resume flowing. If so, structures could be damaged. A two-layer Galerkin finite-element digital model was used for predicting the rate and extent of the rise in water levels from 1975 to 2000. Six hydrologic conditions were modeled for the basin. Artificial recharge of one-half entitlement and full entitlement from the California aqueduct were each coupled with low, average, and high natural recharge to the basin. According to model predictions, the greatest water level rises will be along the San Bernardino front. This area encompasses the artificial recharge sites and also has a thick section of unsaturated sediments for storing ground water. The formerly swampy lands between Warm Creek and the Santa Ana River adjacent to the San Jacinto fault have little additional storage capacity, and water levels could rise to the land surface as early as 1983 under maximum recharge conditions and 1970-74 average pumping conditions. If pumping rates are reduced in the Warm Creek area, water levels may rise to land surface prior to the dates predicted by the model, regardless of the artificial recharge program.

Hardt, W.F. and Hutchinson, C.B., "Model Aids Planners in Predicting Rising Ground-Water Levels in San Bernardino, California", <u>Ground Water</u>, Vol. 16, No. 6, Nov.-Dec. 1978, pp. 424-431.

Problems in the orderly development of the current water-management program in the San Bernardino Valley could result from imported water from Northern California recharging the ground water system. Intensive agricultural development in this area led to a decline in artesian water levels from 50 ft or more above land surface to 50-150 ft below land surface. If the artesian head were to again extend above the land surface due to artificial recharge, buildings, public works and utilities could be damaged. A mathematical well-response model was developed by Durbin and Morgan to predict water-level changes that would occur in selected wells due to artificial recharge. In this study, a conceptual model of the flow mechanics and geologic framework of the basin was first developed. Then, using the data from the analysis of the conceptual model, a mathematical model was devised to represent the prototype. The solution to the problem is based on the combined influence of natural recharge, artificial recharge, and ground water pumping. The greatest unknown is the quantity and distribution of natural recharge. Artificial recharge and pumpage can be controlled. In addition to the predictive mathematical model, key monitoring wells are located between the recharge sites and the former swampy area to reflect the effects of the recharge program.

Harp, J.F. and Laguros, J.G., "A Dimensionless Parameter Study of Groundwater Recharge, Phase II", June 1972, Research Institute, University of Oklahoma, Norman, Oklahoma.

The method of coefficients has been used to predict ground water recharge for several years. A new approach was attempted using a "dimensionless parameter" concept to relate recharge to other known parameters, i.e., pumpage, permeability, rainfall, recharge area, etc. Data from a total of fifteen observation wells from two locations in Oklahoma and two locations in Kansas were used. The high-use municipal wells in southwestern Oklahoma show periodic "mining" which can be avoided if pumpage rates are modified. The wells in Kansas are located very far from other pumping locations, thus rendering the recharge area excessively large. Regression analysis was performed encompassing recharge periods of one month, six months, and twelve months. The resulting linear equations are multiterm, wherein positive coefficients imply no overuse while negative coefficients substantiate "water mining"; these equations predict ground water recharge rates more accurately than heretofore.

Kisiel, C.C. and Duckstein, L., "Ground-Water Models", Systems Approach to Water Management, ed. by Asit K. Biswas, Ch. 4, 1976, McGraw Hill Book Company, Inc., New York, New York, pp. 80-155.

A comprehensive analysis of ground water problems is presented. Identified is the current state of modeling, including its deficiencies. The focus is upon emerging approaches to describing and managing the ground water system, not on the many theoretical and empirical solutions to regional and specialized ground water problems. Examples of specific methodologies are given. Various ground water models are summarized, along with statements about their properties which should enable one to select a suitable methodology. A significant part of this chapter is relevant to problems other than ground water; these are some of the important techniques and considerations, i.e., model choice, multilevel optimization, hybrid computations, adjustment algorithms, multiobjective features, and fundamental problems such as uncertainties, error growth, worth of data, and economic losses. It is suggested that planners know enough about uncertainties to interpret their possible consequences. The chapter considers: (1) general systems framework; (2) general systems theory and ground water hydrology, including a continuous systems model of a lake-aquifer system; (3) classical ground water euations and linear systems theory; (4) aquifers as finite state machines (FSM), including an example of FSM of well systems; (5) inverse problems; (6) extraction of water from an aquifer; (6) natural and artificial recharge; (7) chemicals and pollutants in an aquifer; and (8) conjunctive operation of surface and ground water resources.

Klemt, W.B., et al., "Ground-Water Resources and Model Applications for the Edwards (Balcones Fault Zone) Aquifer in the San Antonio Region, Texas", Report 239, Oct. 1979, Texas Department of Water Resources, Austin, Texas.

The scope of this study was primarily directed toward the

quantitative aspects of ground water withdrawals, spring flows, and aquifer characteristics. A digital computer model of the Edwards (Balcones Fault Zone) aquifer was successfully verified by taking historical input data (initial heads, pumpage, recharge, and other data) and computing water-level changes and spring flows which agree well with water-level declines and spring flows observed from actual field measurements. The digital computer simulation for the period 1972 through 2049 of the Edwards (Balcones Fault Zone) aquifer in the San Antonio region indicates the following: (1) ignoring any water-quality constraints, the aquifer is capable of meeting projected demands through the year 2049; (2) natural flow of Comal and San Marcos Springs will cease before the year 2020, if projected pumpage in the region occurs at predicted rates; (3) the addition of currently proposed artificial recharge does not result in an appreciable increase in the aquifer's available water; (4) the drought-flood sequence of the 1950's was introduced at various points in the simulation and produced only minor effects when compared to the aquifer's long-term simulations; (5) spring flows at Comal Springs and San Marcos Springs can be maintained through ground water management plans; and (6) there is sufficient storage in the aquifer to allow spring flow of Comal and San Marcos Springs to be replaced by augmentation pumpage through the year 2020.

Latinopoulos, P., "The Effect of Infiltration Distribution on Artificial Recharge Schemes", Journal of Hydrology, Vol. 49, No. 3/4, Feb. 1981, pp. 279-286.

An analytical solution of transient ground water flow is presented in this study so that quantitative relations between the time-dependent input to an aquifer and the corresponding ground water response may be derived. The effect of a varying natural infiltration on the ground water response was determined to be of relative importance, with ground water response also being influenced by the characteristics of the aquifer. Artificial recharge offers several advantages which are still dependent on many factors, and design criteria should therefore be examined carefully in every case of application. Using a simple recharge scheme taken as an example to obtain a long-term approximation, it was concluded that existing data for the aquifer properties and the infiltration distribution over time played a decisive role when planning for such schemes. Aquifers with low transmissivity and high storage values together with a moderate throughout-the-year infiltration distribution seem the most efficient for artificial recharge, from the viewpoint of long-term regulation of the ground water output.

Latinopoulos, P., "Response of Groundwater to Artificial Recharge Schemes", Water Resources Research, Vol. 17, No. 6, Dec. 1981, pp. 1712-1714.

Analytic solutions are presented for the ground water flow in an unconfined aquifer under seasonal artificial recharge schemes of variable duration. The results can be used for a preliminary assessment of the response of ground water to artificial recharge. Variations in storage and outflow from the aquifer, apart from the siting of recharge, are also highly dependent on both the aquifer characteristics and the duration of recharge.

Latinopoulos, P., "Well Recharge in Idealized Rectangular Aquifers", Advances in Water Resources, Vol. 5, No. 4, Dec. 1982, pp. 233-235.

Analytical solutions for ground water flow in rectangular aquifers are presented in the case of a single well recharge. The problem concerns the impact of a seasonal recharge scheme of variable duration on aquifers with various boundary conditions. The results obtained from these idealized aquifers can be used in a preliminary assessment of the ground water response to artificial recharge schemes. Among the goals of an artificial recharge scheme is storage augmentation and river regulation, for example, low flow supplementation. Since the natural inputs and outputs of an aquifer vary seasonally, the efficient way of using artificial recharge should be such that deficiencies during the dry period of a year could be beneficially supplemented from surplus water added artificially during the wet period. However, the dry and wet periods are repeated almost consecutively year after year, so a reasonable approach to operating any recharge system is to run it under an annual periodicity. The flow is assumed to take place through an idealized aquifer, rectangular in shape, which is nonleaky, homogeneous with respect to storage, and homogeneous and isotopic with respect to transmissivity. The analytical solutions for the two-dimensional aquifer recharge are of great help in deriving preliminary estimates of the future consequences from artificial recharge schemes.

Manbeck, D.M., "Recharge Simulation Model", Completion Report, June 1975, Department of Agricultural Engineering, University of Nebraska, Lincoln, Nebraska.

A mathematical model simulating the movement of water from the soil surface to ground water was developed and programmed for the IBM 360/65 computer in Fortran IV language. Components of the model include precipitation, irrigation amounts, air temperature, saturation soil moisture content and hydraulic conductivity, volumetric moisture vs. moisture potential relationship, and thickness and number of strata. The model generates the amount of water from precipitation and irrigation sources moving vertically downward from one soil layer to the next on a monthly basis with given initial conditions. The input values and mathematical relationships are used to simulate the following processes: (1) interception by vegetation; (2) surface runoff; (3) evapotranspiration; and (4) percolation. The percolation or moisture movement from the last layer just above the water table is considered recharge to the ground water system. Flow of moisture in the soil is assumed to be one dimensional and occurs under saturated and unsaturated conditions. The concept of one-dimensional flow restricts the accuracy of the model because soil moisture movement is very complicated and more complex relationships are not amenable to easy solution.

Marino, M.A., "Artificial Groundwater Recharge, I--Circular Recharging Area", Journal of Hydrology, Vol. 25, No. 3/4, May 1975, pp. 201-208.

Artificial ground water recharge is of increasing importance as an essential component of water management programs. The fluctuation of the water table in response to recharge depends on the size and shape of the recharging area, on the duration and type of variation of the recharge rate, and on the hydraulic and geometric parameters of the underlying aquifer. Numerical solutions describing the formation of ground water mounds were obtained by applying the Douglas-Jones predictor-corrector methods to the second order, nonlinear, partial-differential equation characterizing the unsteady flow in an unconfined aquifer beneath a circular uniformly recharging area. An advantage of using the method is that the predictor-corrector equations result in a tridiagonal set of simultaneous equations which can be solved rapidly in a digital computer. Furthermore, the method is computationally stable.

Marino, M.A., "Artificial Groundwater Recharge, II--Rectangular Recharging Area", Journal of Hydrology, Vol. 26, No. 1/2, July 1975, pp. 29-37.

The nonlinear, partial-differential equation characterizing the unsteady flow in an unconfined aquifer receiving uniform vertical recharge from a rectangular spreading basin was solved numerically. The method of solution consisted of the alternating-direction implicit scheme with an inhomogeneous grid spacing. The numerical solution was compared with an available linearized solution for a flow system in which the rate of recharge was much smaller than the hydraulic conductivity of the underlying aquifer. The comparison showed that the values of the maximum height of the water table obtained from the numerical scheme are always higher than those calculated from the linearized solution.

Marino, M.A., "Mathematical Models of Artificial Recharge Systems", Paper No. 2005, July 1975, Department of Water Science and Engineering, University of California, Davis, California.

The nonlinear, partial differential equation characterizing the unsteady flow in an unconfined aquifer beneath a circular recharging area was solved numerically using a predictor-corrector method. The results obtained were compared with available linearized solutions. The predictor-corrector method resulted in a tridiagonal set of simultaneous equations which could be solved rapidly by computers. Furthermore, the method was computationally stable. A rectangular artificial ground water recharge system was analyzed by use of the alternating-direction implicit scheme with a nonuniform grid spacing. Values of the maximum height of the water table obtained from the numerical scheme were always higher than those calculated from an available linearized solution. A recharge well-unconfined aquifer system was analyzed by coupling the fourth-order Runge-Kutta method with the LaGrange interpolation method. The method was computationally inherently stable for any times. Dimensionless graphical solutions of water table profiles at different times were presented for various conditions. A closed-form solution based on the linearized equation was developed that closely approximates the flow system under consideration. Dimensionless graphical solutions for volume and average rate of recharge were presented for conditions commonly encountered in practice.

Marino, M.A. and Yeh, W.W.G., "Nonsteady Flow in a Recharge Well-Unconfined Aquifer System", Journal of Hydrology, Vol. 16, No. 2, June 1972, pp. 159-176.

A recharge well-unconfined aquifer system is considered in which the well completely penetrates a homogeneous, isotropic, extensive aquifer. The system is characterized also by nonuniformity of the recharge rate, constancy of the injection head, hydraulic conductivity, and specific yield of the aquifer, and the validity of the Dupuit-Forchheimer assumptions. Dimensional analysis and similarity approximation are used to transform the governing second-order, nonlinear, partial differential equation into a second-order, nonlinear, ordinary differential equation. The differential equation is solved by coupling the fourth-order Runge-Kutta method with the LaGrange Interpolation method. The method is inherently stable for any times. Dimensionless graphical solutions of water table profiles at different times are presented for various conditions. A closed-form analytical solution based on the linearized equation is developed that closely approximates the flow system under consideration. Dimensionless graphical solutions for volume and average rate of recharge are presented for conditions commonly encountered in practice. The dimensionless form of the solutions makes possible their application to any well-aquifer system with similar characteristics regardless of the values of the hydraulic properties of the aquifer.

Molden, D., Sunada, D.K. and Warner, J.W., "Microcomputer Model of Artificial Recharge Using Glover's Solutions", Ground Water, Vol. 22, No. 1, Jan.-Feb. 1984, pp. 73-79.

An interactive program written for an Apple II+ 48K microcomputer provides an analytical solution for recharge from a rectangular basin. R.E. Glover published in 1960 through the USDA Agricultural Research Service in Ft. Collins, Colorado, a paper offering a solution to mathematical derivations in ground water recharge. This microcomputer model written in BASIC adapts Glover's solution to computerized application. The model is capable of graphically displaying the rise and decline of the recharge mound for either an infinite homogeneous medium or for a stream aquifer system. Using this program a large number of recharge strategies can be graphically worked out by hydrologists, including soil parameters and characteristics, plus unusual boundary conditions.

Nutbrown, D.A., "A Model Study of the Effects of Artificial Recharge", Journal of Hydrology, Vol. 31, No. 1/2, Sept. 1976, pp. 57-65.

Some simple analytic and numerical results are obtained which may be used to derive preliminary estimates of the effects of artificial recharge on the storage and flow of ground water in an unconfined aquifer. In particular, two situations are considered. In the first, a steady-state approach is adopted to describe the long-term effects of recharge and an analytic expression is obtained giving values of the resulting increase in storage. These values compare well with model recharge experiments performed with a detailed description of ground water flow in the Chalk of the South Downs, U.K. In the second situation, some numerical results are obtained from a transient analysis which describes the seasonal effects of recharge, both on aquifer outflow and storage. These results are relevant to the use of artificial recharge to supplement ground water storage for river regulation and direct-supply purposes.

Ortiz, N.V. et al., "Artificial Ground-Water Recharge with Capillarity", Journal of the Irrigation and Drainage Division, American Society of Civil Engineers, Vol. 104, No. IR1, Proceedings Paper 13627, Mar. 1978.

The effect of capillarity on the transient response of the water table to recharge was evaluated by a numerical and a porous media model. The numerical model was developed to simulate the growth and spread of ground water mounds, taking into consideration the flow and storage in the capillary regions. The contribution from the capillary region was described analytically in terms of recharge rates and the measurable soil properties of bubbling pressure, pore-size distribution index, and hydraulic conductivity. The numerical model was verified by comparing its predictions for various flow conditions with the result obtained from the porous media model. The effects of bubbling-pressure head, pore-size distribution index, initial saturated depth, depth to water table, and recharge rate on the predicted mound height were determined. It was shown that the effect of the capillary region significantly influences the growth and spread of ground water mounds.

Prickett, T.A. and Lonnquist, C.G., "Selected Digital Computer Techniques for Groundwater Resource Evaluation", Bulletin 55, 1971, Illinois State Water Survey, Urbana, Illinois.

Generalized digital computer program listings are given that can simulate one-, two-, and three-dimensional nonsteady flow of ground water in heterogeneous aquifers under water table, nonleaky, and leaky artesian conditions. Also included are programming techniques involving time varying pumpage from wells, natural or artificial recharge rates, the relationships of water exchange between surface waters and the ground water reservoir, the process of ground water evapotranspiration, and the mechanism of converting from artesian to water table conditions. The discussion of the digital techniques includes the necessary mathematical background, documented program listings, theoretical versus computer comparisons, and field examples. A finite difference approach is used to formulate the equations of ground water flow. A modified alternating direction implicit method is used to solve the set of resulting finite difference equations. The programs are written in Fortran IV and will operate with any consistent set of units.

Prickett, T.A. and Lonnquist, C.G., "Aquifer Simulation Model for Use on Disk Supported Small Computer Systems", Circular 114, 1973, Illinois State Water Survey, Urbana, Illinois.

A generalized digital computer program listing is given that can simulate two-dimensional flow of ground water in heterogeneous aquifers under nonleaky and/or leaky artesian conditions. A special feature of the program is that a large simulation (up to an estimated 10,000 node problem) can be accomplished on a disk supported small computer with only 4,000 words (32 bits per word) of core storage. The computer program can handle time varying pumpage from wells, natural or artificial recharge rates, the relationships of water exchange between surface waters and the ground water reservoir, the process of ground water evapotranspiration, and the mechanism of flow from springs. A finite difference approach is used to formulate the equations of ground water flow. A modified alternating direction implicit method is used to solve the set of resulting difference equations. The discussion of the digital technique includes the necessary mathematical background, the documented program listing, sample computer input data, and explanations of job setup procedures. The program is written in Fortran and will operate with any consistent set of units.

Prickett, T.A., Naymik, T.G. and Lonnquist, C.G., "A 'Random-Walk' Solute Transport Model for Selected Groundwater Quality Evaluations", ISWS/BUL-65/81, 1980, Illinois State Water Survey Division, Champaign, Illinois.

A generalized computer code is given that can simulate a large class of solute transport problems in ground water. The effects of convection, dispersion, and chemical reactions are included. The solutions for ground water flow include a finite difference formulation. The solute transport portion of the code is based on a particle-in-a-cell technique for the convective mechanisms, and a random-walk technique for the dispersion effects. The code can simulate one- or two-dimensional nonsteady/steady flow problems in heterogeneous aquifers under water table and/or artesian or leaky artesian conditions. The program covers time-varying pumpage or injection by wells, natural or artificial recharge, and the flow relationships of water exchange between surface waters and ground water.

Schmidtke, K.D., McBean, E.A. and Sykes, J.F., "Stochastic Estimation of States in Unconfined Aquifers Subject to Artificial Recharge", Water Resources Research, Vol. 18, No. 5, Oct. 1982, pp. 1519-1530.

An extended Kalman filter model for characterizing minimum variance estimates of the piezometric heads and coefficients defining an unconfined aquifer subject to artificial recharge is developed. The system evolution model employs Hantush's characterization. Sensitivity analyses were used to test the estimation capability of the technique. The ability of the extended Kalman filter to use all available information from both the system model and measurements of the state to provide approximate minimum variance estimates of the state and confidence limits is shown. The extended Kalman filter version of Hantush's model is applied to a record data length of one year obtained at the study area located in Norwood, Ontario. The results of the study indicate that the meteorological conditions influence the water table elevations to a greater extent than the particular rate of recharge employed. Observation of the measured values indicates that greater recharge rates could be employed during the summer months, when rainfall is low and evapotranspiration is high, and lower recharge rates during the winter months to prevent the water table from inundating the weeping tiles.

Seward, L.B., Knowles, T.R. and Klemt, W.B., "Groundwater Models for Texas", Conference on Water for Texas, Sept. 19, 1974, Texas A&M University, College Station, Texas.

Sound ground water management decisions must be based on the geohydrology of the aquifer, water demands, and the aquifer response to many alternative plans of operation. The high-speed electronic digital computer is able to store voluminous complex hydrologic data and rapidly analyze many alternative management plans at a reasonable cost. The comparison by managers, hydrologists, and others of the aquifer's response in terms of water level change results in the selection of a management plan which is consistent with overall objectives at minimum cost. The construction of two digital ground water models, one on the Edwards aquifer in central Texas (Balcones Fault Zone) and one on the Carrizo aquifer of south Texas, is described. Several model applications, such as: (1) simulation of future water levels using projected recharge and pumpage rates, (2) determination of optimum pumpage in specific regions and those suited to artificial recharge, (3) prediction of future spring flow under projected conditions, and (4) studies of movement of water quality parameters are presented as illustrations of how such models can be used as tools in a water planning operation.

Shamir, U.Y. and Harleman, D.R.F., "Numerical and Analytical Solutions of Dispersion Problems in Homogeneous and Layered Aquifers", RR66-14, May 1966, Massachusetts Institute of Technology, Cambridge, Massachusetts.

From a review of previous work on dispersion in porous media it is concluded that at present the tools for solving most practical problems of mixing of fluids in artificial recharge, waste water disposal and ground water contamination are still missing. A numerical method is proposed for the solution of dispersion problems in steady three-dimensional potential flow fields in porous media. The numerical scheme uses the convective dispersion equation in curvilinear coordinates consisting of the flow streamlines and equipotentials. By means of this equation any problem of dispersion in a steady potential flow field, as well as a special type of problem with unsteady flow, can be solved. The numerical scheme is programmed for a computer and applied to a few simple cases, for which analytical solutions exist.

Signor, D.C., "Laboratory Facility for Studies Related to Artificial Recharge", Underground Waste Management and Artificial Recharge, Vol. 2, 1973, pp. 799-822.

Flow can be induced through repacked or field cores under constant flow or constant pressure. A flow-test data-acquisition and computation system provides punched-tape data storage and real-time computation and plotting of intrinsic permeability changes with column depth and time. Water quality data are recorded during the test, and characteristics of the porous media and suspended solids are determined. Accumulation of material in porous media interstices is observed visually with a scanning electron microscope. A U.S. Geological Survey laboratory at Lubbock, Texas, was designed and equipped to provide a means of testing flow through porous media columns. Artificial ground water recharge is subject to limitations caused by mechanisms which degrade the hydraulic conductivity of the porous media. Reduction of hydraulic conductivity may be caused by suspended solids, bacterial growth, chemical reaction of dissolved solids with the porous media or native water, and air entrainment or dissolution of gases in the interstices of the porous medium.

Sunada, D.K., Warner, J.W. and Molden, D.J., "Artificial Groundwater Recharge, San Luis Valley, Colorado", Completion Report No. 123, Mar. 1983, Colorado Water Resources Research Institute, Colorado State University, Fort Collins, Colorado.

Intense use of aquifers for irrigation waters has caused ground water storage depletion in many areas of the arid and semi-arid west, including the San Luis Valley in south central Colorado. Artificial recharge is a means of alleviating this problem. To show the practical benefits of artificial recharge to local water users, a demonstration recharge basin was operated in the San Luis Valley. Both numerical and analytical models were calibrated to the aquifer response to suggest operational policies. Analysis of the results of the demonstration project indicate that if recharge operations are conducted during the non-irrigation season when excess water is available, significant amounts of water can be added to storage and combat ground water depletion. One difficulty with the use of models is that the results obtained are hard to visualize by nontechnical persons. Recently a wide variety of microcomputers have become available which are relatively inexpensive and have the capability of readily evaluating solutions which describe ground water response to artificial recharge. Their portability and graphics features make them excellent demonstration tools. As part of this study, a computer program was written which uses Glover's (1960) solution for recharge from a rectangular basin to model artificial recharge. The program is totally interactive, extremely user friendly, and runs on an Apple II+ 48K microcomputer. The model describes ground water response to artificial recharge in an infinite, homogeneous aquifer and in a stream aquifer system and can also calculate discharge into a stream. The model is designed for use by both technical and nontechnical persons, and is an excellent means of transferring knowledge from ground water hydrologists to water users. The microcomputer model developed in this study could be used in other parts of the San Luis Valley to evaluate the benefits of artificial recharge in these other areas.

Swain, L.A., "Predicted Water-Level and Water-Quality Effects of Artificial Recharge in the Upper Coachella Valley, California, Using a Finite-Element Digital Model", Water Resources Investigations 77-29, Apr. 1978, U.S. Geological Survey, Menlo Park, California.

From 1936 to 1974, water levels declined more than 100 ft in the Palm Springs area and 60 ft in the Palm desert area of the upper Coachella Valley, California. Water from the Colorado River Aqueduct is presently being recharged to the basin. The dissolved solids concentration of native ground water in the recharge area is about 210 mg/l and that of recharge water ranges from 600 to 750 mg/l. A finite-element model indicates that without recharge the 1974 water levels in the Palm Springs area will decline 200 ft by the year 2000 because of pumpage. If the aquifer is recharged at a rate from about 7,500 ac-ft per year in 1973, increasing to 61,200 ac-ft per year in 1990 and thereafter, the water level in the Palm Springs area will decline about 20 ft below the 1974 level by 1991 and recover to the 1974 level by 2000. The solute-transport finite-element model of the recharge area indicates that the artificial recharge plume (bounded by the 300 mg/l line) will move about 1.1 miles downgradient of the recharge ponds by 1981 and about 4.5 miles from the ponds by 2000.

Tyagi, A.K., "Impact of Artificial Recharge on Groundwater Quality of Gypsum Aquifer in Harmon County, Oklahoma", 1983, Water Resources Research Institute, Oklahoma State University, Stillwater, Oklahoma.

The Strongly Implicit Procedure (SIP) is used to solve the ground water flow equation, replacing the Alternating Direction Implicit Procedure (ADIP) in the solute transport model developed by the U.S. Geological Survey. The SIP Model was then coupled with the dispersion equation solved by the Method of Characteristics (MOC). The comparisons of the SIP-MOC Model against the ADIP-MOC Model indicate that the convergence rate in the flow portion of the problem is significantly improved, saving computer time and the user's computer cost. The SIP Model handles problems which cannot be conveniently solved by the ADIP Model.

Vauclin, M., Khanji, D. and Vachaud, G., "Experimental and Numerical Study of a Transient, Two-Dimensional Unsaturated-Saturated Water Table Recharge Problem", Water Resources Research, Vol. 15, No. 5, Oct. 1979, pp. 1089-1101.

Transient, two-dimensional water flow was studied in relation to the recharge of a water table aquifer. The approach was based on the physics of water transfer in the complete domain defined by both the saturated and unsaturated zones of soil. Experimental data were first obtained in a slab of soil (3 m in length, 2 m in height, and 5 cm in thickness) in which the changes of water content and water pressure occurring in the flow domain were measured throughout an artificial recharge event. These were compared with the predictions of a numerical model based on the continuity of transfer between the unsaturated and saturated zones. The validity of the model was proved by the excellent agreement between simulated and experimental results. Furthermore, comparison with the results obtained by a viscous flow analog revealed the errors associated with the traditional free surface approach, which neglects transfer in the unsaturated zone.

Volker, R.E., "Digital Model as an Aid in Management of an Artificially Recharged Aquifer", Annual Engineering Conference, Townsville, May 10-14, 1976, Paper 523/2, 1976, Institute of Engineers, Sydney, Australia, pp. 228-232.

A description is given of a numerical model developed to simulate the behavior of a coastal aquifer. Results are presented which highlight the benefit of artificial recharge operations currently being carried out. The model also enables a comparison to be made of the magnitudes of recharge from various sources and allows a prediction of the effects of any changes in natural river flow conditions. Methods used to predict the extent of sea water intrusion are also presented.

Weaver, R.J. and Kuthy, R.A., "Field Evaluation of a Recharge Basin", Report NYSDOT-ERD-75-RR26, Feb. 1975, Engineering Research and Development Bureau, New York State Department of Transportation, Albany, New York.

The results are presented of a full-scale field infiltration test performed on a temporary 150- by 35-ft recharge basin, which was inundated by controlled discharge from a 4.5-million gallon storage tank. The data obtained from the test permitted the verification of a basin design method developed in a previous study. The theory developed in the original study was for a constant-head, suddenly applied case and used a value equal to half the maximum head in determining the infiltration rate. This theory and corresponding assumptions were found to lead to a desirable net safety factor of 1.40, as was originally anticipated. As a result of this work, the Soil Mechanics Bureau of the New York State Department of Transportation is preparing a manual covering both design and environmental considerations in using recharge basins.

Webb, S.N. and Watson, K.K., "Gravity Drainage of a Sand Profile Under a Time-Dependent Surface Flux", Australian Journal of Soil Research, Vol. 15, No. 1, Mar. 1977, pp. 9-16.

The replenishment of ground water supplies from recharge trenches often involves the use of water containing quantities of suspended solids which eventually become deposited on the sides and bottom of the trench with a consequent decrease in intake rate. This process is simulated numerically in one dimension by using a finite difference solution of the flow equation with a time-dependent surface flux boundary condition. Three algebraic functions are used to represent the flux change so that different physical characteristics of the system can be studied. The numerical results indicate that for a reasonably slow rate of flux reduction water content and pressure head profiles of a steady state configuration, appropriate to the flux at the time, are obtained. However, with a more rapid flux change unsteady state conditions remain established throughout the profile and predictions of water content and pressure head can only be made using an unsteady state analysis.

Webb, S.N. and Watson, K.K., "Diurnal Periodicity in Ground-Water Recharge Rates", Journal of the Irrigation and Drainage Division, American Society of Civil Engineers, Vol. 104, No. IR1, Proceedings Paper 13613, Mar. 1978.

In an investigation of the flow characteristics of a recharge pit used for artificial ground water recharge purposes, diurnal variations of a cyclic nature were detected in the intake rate. The output from soil water pressure sensors installed between the base of the pit and the water table indicated similar cyclic variations. The maximum intake rate occurred regularly each day at approximately 2 a.m. to 3 a.m., and the minimum approximately 12 hr later. It was concluded that these variations are due to the activity of algae on the base of the pit and in the top layer of sand. Diurnal peaks and troughs of the algae oxygen production result in changes in the hydraulic conductivity of the profile. This would appear to be related to changes in the volume of the occluded oxygen which exists as bubbles in the soil pores. These reduce the flow area available for recharge and thus cause a reduction in the intake rate.

Willis, R., "A Management Model for Determining Effluent Standards for the Artificial Recharge of Municipal and Industrial Wastewaters", Proceedings of the Symposium on Advances in Groundwater Hydrology, held at Chicago, Illinois, 1976, Sept. 1976, American Water Resources Association, St. Anthony Falls Hydraulic Laboratory, Minneapolis, Minnesota, pp. 287-308.

Two planning models were developed for the optimal land disposal of wastewater effluents in spray irrigation or basin spreading land disposal systems. The models consider the operation of the disposal systems over an individual (a static model) or a sequence of interdependent disposal cycles (a dynamic model). The objectives of the models were to estimate the required treatment of a given waste load to prevent the possible degradation of the soil or ground water system. The management models are predicated upon the partial differential equations characterizing mass transport in soils. The models incorporate time-dependent non-equilibrium adsorption and first-order chemical or biochemical reactions.

The Galerkin finite-element method transforms the system's transport and adsorption equations into systems of ordinary differential equations. The transformed equations, which are the response equations of the system, are imbedded within the constraint sets of the planning models. The planning models, structured as linear programming problems, determine the optimal effluent standards and the time and spatial variation of constituent concentrations within the soil system. The planning models were applied to a spray irrigation site disposing of food processing waste in western New York. Effluent standards, expressed as the maximum concentration of dissolved organics in the disposal system, were estimated for a range of ground water quality standards. The effluent standards may be used in the preliminary design and operation of conjunctive ground and surface wastewater treatment systems.

Willis, R., "Optimal Management of the Subsurface Environment: Parameter Identification", May 1976, Water Resources and Marine Sciences Center, Cornell University, Ithaca, New York.

Mathematical models are presented for assessing the potential impact of the artificial recharge of municipal wastewater effluents on the subsurface environment. Wastewaters are assumed to be recharged to a shallow unconfined ground water aquifer in surface spreading basins. A Galerkin finite element simulation model predicts the time and spatial variations of conservative and non-conservative constituent concentrations within the soil profile. The model incorporates time dependent non-equilibrium adsorption and chemical and/or biochemical reactions. The Galerkin procedure transforms the partial differential equations describing mass transport and adsorption into systems of first-order ordinary differential equations. The resulting equations are recursively solved by finite difference techniques. The simulation equations form the basis for the parameter identification model which is formulated as a quadratic programming problem using a generalized Newton-Raphson procedure. The model is used to estimate (1) the kinetic, dispersion, and adsorption parameters characterizing quality changes within the soil system, and (2) the initial distribution of a contaminant within the sorbed or solution phases at the inception of the spreading cycle. The simulation model and the parametric identification algorithm are validated using published analytical solutions of the adsorption and mass transport equations. The results indicate the computational efficiency of the identification algorithm for the large scale estimation of ground water quality parameters.

Wilson, L.G., Rasmussen, W.O. and O'Donnell, D.F., "Feasibility of Modelling the Influences of Pit Recharge on Groundwater Levels and Quality in Alluvial Basins", Completion Report, July 1976, Water Resources Research Center, University of Arizona, Tucson, Arizona.

The specific objectives of this study included: (1) examining the effect of pit recharge, using industrial blowdown effluent, on local ground water levels and quality in the Tucson basin; and (2) evaluating the applicability of two digital models for modelling the hydraulic response of the aquifer to pumping and recharge. Fourteen pit trials ranging from 1 day to 185 days duration were implemented during a two year period. The total volume of effluent recharged was 157 ac-ft. Pit recharge on a limited basis did not appear to offset the general decline in local ground water levels. However, recharge deteriorated the quality of ground water in shallow (150 ft) wells but not in a deeper (300 ft) well. A finite difference (FD) model and a finite element model (FEM) were used to simulate hydraulic conditions in a 2000 ft by 2000 ft aquifer region near the pit. An attempt was made to calibrate the models using water level data from a two-week aquifer test on a shallow well. Results were depicted using a three-dimensional graphics technique (SYMVU). Better results were obtained with the FD model than with the FEM, primarily because a finer mesh was used to discretize the region with the FD model than with the FEM. Results of the FEM were also biased by the rectangular configuration of the aquifer bounary. The finite element model was used to simulate aquifer response during a pit recharge test. Again, results were limited by the injudicious selection of elements and boundary configurations.

Witherspoon, P.A. and Narasimhan, T.N., "Seepage from Recharge and Landfill Ditches", Agricultural and Urban Considerations in Irrigation and Drainage; Selected Papers from the Irrigation and Drainage Division Specialty Conference, Fort Collins, Colorado, Apr. 22-24, 1973, 1974, American Society of Civil Engineers, New York, New York, pp. 39-66.

An investigation was made of steady-flow rectangular recharge pits in two dimensional flow regions using a numerical method of analysis. In the case of single layer systems, the effects of the dimensions of the flow region, pit size, hydraulic gradient and anisotropy on total flow rates from the pit were investigated. In addition, the possibility of reducing pit leakage by means of a low-permeability lining in the recharge pit was investigated. For the particular two-dimensional systems studied, it was found that the overall flow rate increases with increasing hydraulic gradient, but in a slightly non-linear manner. By varying the width and depth of the recharge pit it was possible to determine that optimum pit dimensions exist, beyond which only slight improvement in flow rates can be accomplished by increasing pit dimensions. For an isotropic, single layer system, the results of this work suggest that these optimum dimensions are attained when both the depth and the width of the pit are 0.1 to 0.2 times the thickness of the flow region. In the third dimension, the recharge pit is in the form of a long ditch or channel. The results imply that in artificial recharge operations, considerable costs could be avoided by not designing recharge pits with dimensions greater than the optimum values. On the other hand, in the case of landfill operations, the results suggest that increasing the depth and/or width of the pit will not necessarily cause proportional increases in the volume of leachates entering the ground water system.

Wood, W.W., "Toward a Rational Development of Artificial Recharge Research", Advances in Groundwater Hydrology: American Water Resources Association, Sept. 1976, pp. 243-247.

Artificial recharge can be an effective water management technique if the hydrogeologic conditions of the aquifer system are quantified and if rational design criteria are established. The successful application of artificial recharge depends upon the development of methods for evaluating the changes in aquifer characteristics in both the saturated and unsaturated zones as a function of chemical, biological, and physical changes associated with the introduction of foreign water into the subsurface environment. Also, additional methods need to be developed to predict resulting variations of water quality in time and space.

ECONOMIC FEASIBILITY

Broadbent, R.C. et al., "Artificial Recharge of Colorado River Water in Las Vegas Valley: Management Options and Potential Aquifer System Responses", Publication No. 41074, Jan. 1982, Desert Research Institute, University of Nevada, Reno, Nevada.

A study was undertaken to assess whether and how southern Nevada can take maximum advantage of excess Colorado River (CR) diversion capacity that is expected to exist over the next 20 years through the southern Nevada water supply system. Analysis of the water resource system defined a range of potential ground water (GW)-surface water management scenarios defined by the resource system model. Chemical equilibria calculations and sediment-water mixing experiments were also conducted. The study results led to conclusions that: the apparent least-cost management strategy is to continue GW pumping at the current rate without salvaging lost CR water, but failure to build a water reserve for future use and the problems inducible by such failure may make this strategy undesirable; artificial recharge of CR water is technically feasible but probably undesirable economically; reduction in current total GW pumpage to 40,000-45,000 ac-ft/yr would permit salvage of CR water by maximizing its use and allowing GW reservoir replenishment, however reduced GW pumpage would induce higher water costs.

Gisser, M. and Mercado, A., "Economic Aspects of Ground Water Resources and Replacement Flows in Semiarid Agricultural Areas", American Journal of Agricultural Economics, Vol. 55, No. 3, Aug. 1973, pp. 461-466.

In semiarid areas where farmers are faced with falling water tables, hydrology and economics must be combined in an integrated model. A single cell hydrologic model of an aquifer in a semiarid region is integrated with an agriculture (linear) negatively sloped demand function for irrigation water. Ground water and replacement flow are charged artificially into the ground. The aquifer benefits from a significant natural recharge and loses water via natural discharge, which is functionally related to water level. Concern is with the long-run steady-state solutions characterized by an eventual equilibrium in which, if necessary, with the help of artificial recharge, overdraft is reduced to zero. Essentially, the economic problem is to select the replacement flow project that will leave the farm with the maximum profit (net income) and an aquifer that is in steady state. Given pumping and replacement flow costs, only one project (as proposed to a variety) yields simultaneously maximum profit and steady-state solution. The model presented can aid policy makers in solving problems from overdraft leading to a falling water table and in examining the profitability of importing water from surplus areas and artificially recharging the aquifer.

Karlinger, M.R. and Hansen, A.J., Jr., "An Engineering Economic Analysis of Artificial Recharge in a Conjunctive Irrigation Plan in the Columbia Basin

Project, Washington", Open File Report 81-67, 1983, U.S. Geological Survey, Tacoma, Washington.

An appraisal-level engineering economic cost analysis was performed for two primary types of irrigation systems in a portion of the Columbia Basin Project--a surface water irrigation system in which water is supplied via canals and laterals, and a system in which surface water is brought to recharging wells and eventually to farms using the transmissive properties of the aquifer and pumping. At 1979 electric power rates, the artificial recharge irrigation scheme is a viable alternative to surface-distributed irrigation systems, but as electric rates increase, its viability decreases. At three times the 1979 rate, the recharge scheme is uneconomical.

Karlinger, M.R. and Hansen, A.J., Jr., "Engineering Economic Analyses of Artificial Recharge in the Columbia Basin Project, Washington", Water Resources Bulletin, Vol. 19, No. 6, Dec. 1983, pp. 967-975.

Appraisal-level, engineering economic analyses for two types of irrigation systems are compared: a surface-water irrigation system, and a ground water recharge irrigation system. The study area is in the southern part of the Columbia Basin Project, where development of irrigable land is in the planning stage. The proposed plan is to import water from Franklin D. Roosevelt Lake in northeastern Washington through an extension of the East Low Canal and to apportion the water to farm units using a lateral distribution system. An artificial recharge irrigation alternative would divert water from the lateral system and deliver it to recharge facilities, using the same supply facility and canal extension. The aquifer would then be used to convey water to individual agricultural units for irrigation pumping. An artificial recharge irrigation system is an economically viable alternative to surface-distributed irrigation in a conjunctive irrigation plan if electric power rates remain sufficiently low. As electric rates increase, this viability generally decreases until it becomes completely unattractive at rates of about $0.012/kwh.

Mawer, P.A. and O'Kane, J.P.J., "Economic Feasibility of Artificial Recharge", Conference on Artificial Groundwater Recharge, University of Reading, England, Sept. 21-24, 1970, Paper No. 2, 1970, The Water Research Association, Marlow, England.

The economics of streamflow regulation by surface storage and aquifer recharge are compared. The alternative use of artificial recharge as a water treatment process was not considered in detail. The comparison is based on the identification of the required recharge input rating and the required surface storage volume necessary to achieve any specified degree of regulation, and several unit cost parameters. A streamflow regulation model can be used to determine the economic relationships for surface and aquifer storage. A simple cost model of aquifer and surface storage compares the likely unit costs of the component works of the two alternatives.

Supalla, R.J. and Comer, D.A., "An Economic Evaluation of the Feasibility of Artificial Ground Water Recharge in Nebraska", OWRT-A-065-NEB(1), Jan.

1981, Office of Water Research and Technology, U.S. Department of the Interior, Washington, D.C.

The purpose of this project was to analyze the economic feasibility of recharge for irrigation use. This involved developing a procedure for estimating recharge benefits, application of the benefit estimating procedure to a proposed Nebraska case, and a review of the literature regarding recharge program costs. The benefit estimating procedure considers two types of benefits; those due to reduced pumping lifts and those due to extended aquifer life.

Supalla, R.J. and Comer, D.A., "The Economic Value of Ground Water Recharge for Irrigation Use", Water Resources Bulletin, Vol. 18, No. 4, Aug. 1982, pp. 679-686.

A methodology was sought for estimating ground water recharge benefits in irrigation use areas, and that methodology was used to estimate ground water recharge benefits for selected situations. These objectives were addressed using a Nebraska case. As presented, the methodology separates recharge benefits into two components: pumping costs savings and aquifer extension benefits. Simplified procedures designed for use by state and Federal water planning agencies are presented for each recharge benefits component. Experience indicates that the required data, time, and computer resources for the model were within the range which would permit use of the model for first round, reconnaissance level studies as well as for more comprehensive analyses. Results of recharge benefit calculations indicate that benefits in irrigation use areas probably range from less than $2.00 to over $6.00 an ac-ft. A comparison to published data on recharge costs concludes that single purpose recharge programs to augment water supplies for irrigation are certainly economically unfeasible. However, the analysis indicates that the benefits of recharge are probably substantial enough to merit their consideration as part of appropriate multipurpose water projects.

SITE SELECTION AND SYSTEM DESIGN

Ackermann, W.C., "The Future of Water Resources in Northeastern Illinois", Journal of the American Water Works Association, Vol. 67, No. 12, Dec. 1975, pp. 691-693.

Several possibilities were investigated for increasing water supplies before the year 2020 for six northeastern Illinois counties. A United States Supreme Court decree of 1966 limited Illinois to 3200 cu ft/sec of Lake Michigan water. Many towns in the area were expecting water deficits by the year 1990 unless additional water could be secured from Lake Michigan or developed from new or, at that time, unusable sources. The Illinois State Water Survey undertook a complete systems analysis of water resources and future water demands in northeastern Illinois. A second detailed analysis considered economic aspects of ground water development in the area. The cost of raw and treated ground water produced in quantities sufficient, in most cases, to meet projected demand to 2020 was estimated for 104 area townships. Water from shallow sand aquifers, shallow dolomite aquifers, and deep sandstone aquifers was considered. Unit cost of treated water varied from 22-53 cents per 1000 gallons. It was concluded that additional water could come from the Kankakee River, ground water recharge, reuse, and possibly from treated brackish water from deep aquifers. Eventually, it was concluded that a substantial quantity of Lake Michigan water will be needed, particularly in areas peripheral to those now using Lake Michigan water.

Aronson, D.A., Reilly, T.E. and Harbaugh, A.W., "Use of Storm-Water Basins for Artificial Recharge with Reclaimed Water, Nassau County, Long Island, New York--A Hydraulic Feasibility Study", Long Island Water Resources Bulletin 11, 1979, U.S. Geological Survey, Syosset, New York.

A survey of 205 storm-water basins in Nassau County, New York, shows that 14 of the 50 largest basins would be suitable for dual infiltration of reclaimed water and storm runoff. Each basin would be divided by an earthen partition so that half would be used for storm-water retention, the other half for reclaimed water. The 14 basins together could accommodate 38.6 mgd of reclaimed water at infiltration rates as high as 0.28 ft/hr, which is approximately 20 percent of the basins' theoretical maximum infiltration capacity. However, the basins cannot accommodate the amounts needed to offset the ground water deficit predicted for the 1990's in Nassau County. The tendency of certain basins to overflow during large-magnitude storms could be reduced by enlarging the basins to accommodate excess runoff, lowering the height of the dividing partition, diverting excess runoff to another basin, or temporarily halting inflow of reclaimed water. The regional water table rise that would result from artificial recharge at the 14 selected basins would partly offset the decline in the water table and in streamflow caused by sewering in parts of Nassau and Suffolk Counties.

Baffa, J.J., "Planning a Must for Groundwater Development", Water and Wastes Engineering, Vol. 12, No. 2, Feb. 1975, pp. 44-50.

Careful analysis is needed in ground water aquifer development. The quantity of ground water available is related to aquifer extent and properties, recharge characteristics, natural discharge, artificial recharge possibilities, drought occurrence, and withdrawals already being made. Quality is related to the naturally occurring relationships of solubility and ion exchange of soil rocks. Geophysical methods and bore holes can suggest the locations of aquifers. Resistivity methods can be used to locate ground water boundaries such as faults and dykes. Induced recharge can be accomplished by creating a reservoir in a natural streambed located in a recharge area. Rates of 0.1 to 0.4 ft/day have been reported for this type of recharge depending on local conditions. Artificial recharge of special recharge basins or wells is from diversion of stream flow from peak discharge periods or of storm water. Recharge rates of 30 ft per day and occasionally as high as 100 ft per day have been recorded. Injection wells operate in the range of 50 to 3000 gpm. In designing a ground water supply, account must be taken of the effects of urban and agricultural land use, sanitary landfills, dissolved solids present, biological activity and the possibility of purifying water before recharge. It is good practice to undertake pilot projects before undertaking large scale facilities; well performance, well levels, hydrodynamic dispersion of recharge waters, and water quality effects may be measured in order to estimate recharge potential.

Bianchi, W.C. and Nightingale, H.I., "Hammer Seismic Timing as a Tool for Artificial Recharge Site Location", Proceedings of the Soil Science Society of America, Vol. 39, No. 4, July-Aug. 1975, pp. 747-751.

Hand-carried hammer impact seismic timing equipment utilizes the physics of seismic wave refraction to explore stratigraphic changes deeper than those found on soils maps. In alluvial areas, artificial recharge and liquid waste disposal sites were selected from sandy, single-grained soil series noted for their high water-transmitting properties through the surface. The surface horizon seismic velocity exhibited by these soils in the San Joaquin Valley had a very good definition to the refraction off the second horizon. Accurate information on depth and horizontal continuity of this second horizon could be obtained in areal surveys. A third horizon may often be described, but its areal continuity is usually masked by intermediate discontinuous layers above it of equal or lesser refractivity. All these horizons are of hydraulic significance for ground water recharge by water spreading.

Bouwer, H., Lance, J.C. and Rice, R.C., "Renovating Sewage Effluent by Groundwater Recharge", Proceedings of the 40th Annual Meeting of the Western Snow Conference, Apr. 18-20, 1972, Phoenix, Arizona, 1972, Colorado State University, Fort Collins, Colorado, pp. 9-16.

In the Salt River Valley in central Arizona the performance of a system for renovating waste water by ground water recharge depends upon the local conditions of climate, soil and ground water. The hydrogeological conditions in Salt River Valley, central Arizona, are favorable for ground water recharge with infiltration basins in the river bed. It is estimated that tertiary treatment by this method would cost about $5 per ac-ft or less than one-tenth of equivalent in-plant treatment. This would be an additional source of water to meet future agricultural and other demands.

Cehrs, D., Soenke, S. and Bianchi, W.C., "A Geologic Approach to Artificial Recharge Site Selection in the Fresno-Clovis Area, California", Technical Bulletin No. 1604, Jan. 1980, Science and Education Administration, Water Management Research, Fresno, California.

In areas of alluvial sediments, the potential for the recharge of surface waters varies with the characteristics of the sediments and the total amount of area exposed to rechargeable water. The best available geologic sites with easy access to rechargeable water, and which are hydraulically above the area of need, are identified and located using surface and subsurface sedimentary geology. By using cursory survey techniques on geologic and soils maps, initial areas with good surface recharge potentials can be identified. Computer averaged water well logs can be used to delineate the subsurface. This subsurface data, used in conjunction with the surface material data, will demarcate the best potential areas for basins. Some areas with poor surface potential may have excellent subsurface recharge potential and be excellent sites for recharge wells. The final siting decision should be based on the hydraulic data, as the recharge efficiency and its economic criteria are controlled by the percolation rates of the surface and subsurface strata. The alluvial geology and related hydrology of the Fresno-Clovis, California, area were studied in relation to artificially recharged water and ground water dynamics.

Eakin, T.E., Price, D. and Harrill, J.R., "Summary Appraisals of the Nation's Ground-Water Resources--Great Basin Region", Professional Paper 813-G, 1976, U.S. Geological Survey, Carson City, Nevada.

Ground water withdrawals by wells in the Great Basin Region were about 1.1 million ac-ft in 1970. An appraisal of the regional ground water resource indicates the region could sustain an annual net pumpage of about 2.6 million ac-ft. Larger withdrawals could be sustained if only part of the pumped water was used consumptively, if conflicts with existing surface water rights are resolved, and if extensive treatment, artificial recharge, and reuse of water prove feasible. Ground water stored in the upper 100 ft of saturated deposits of the valley ground water reservoirs is estimated to be on the order of 300 million ac-ft. Total ground water storage probably exceeds several billion ac-ft. Only a few areas of the Great Basin Region have been studied in detail sufficient to enable adequate design of an areawide ground water development. These areas already have been developed. As of 1973, data for broadly outlining the ground water resources of the region had been obtained. However, if large-scale planned development is to become a reality, a program for obtaining adequate hydrologic and related data would be a prerequisite. Ideally, the data should be obtained in time to be available for the successively more intensive levels of planning required to implement developments.

Fidler, R.E., "Potential Development and Recharge of Groundwater in Mill Creek Valley, Butler and Hamilton Counties, Ohio, Based on Analog Model Analysis", Water Supply Paper 1893, 1970, U.S. Geological Survey, Washington, D.C.

Mill Creek Valley is part of the greater Cincinnati industrial area in southwestern Ohio. In 1964, nearly 30 percent of the water supply in the study area of about 27 square miles was obtained from wells in the glacial-outwash aquifer underlying the valley. Ground water demand has increased steadily since the late 1800's, and excessive pumpage during the years of World War II caused water levels to decline to critical levels. Natural recharge to the aquifer, from precipitation, is about 8.5 mgd. In 1964, the total water use was about 30 mgd, of which 8.1 mgd was obtained from wells in Mill Creek Valley, and the remainder was imported from outside the basin. By the year 2000 ground water pumpage is expected to exceed 25 mgd. Three hypothetical pumping plans were developed by projecting past pumpage data to the years 1980 and 2000. Various combinations of injection wells were tested on the model under different hypothetical conditions of pumpage. Based on analog model analysis, from 3 to 8 injection wells, with an approximate input of 2 mgd each, would reverse the trend in declining ground water levels and provide adequate water to meet anticipated demands.

Garza, S., Weeks, E.P. and White, D.E., "Appraisal of Potential for Injection-Well Recharge of the Hueco Bolson with Treated Sewage Effluent--Preliminary Study of the Northeast El Paso Area, Texas", Open File Report 80-1106, Sept. 1980, U.S. Geological Survey, Austin, Texas.

The U.S. Geological Survey, in cooperation with the Texas Department of Water Resources, made a preliminary study of specific factors related to recharging the Hueco Bolson in the northeast El Paso area with treated sewage effluent. The city is interested in the location and spacing of injection wells relative to: (1) maintaining the injected effluent in the aquifer for a predetermined amount of time (residence time) before it is pumped out, (2) recovery by pumping of as much of the injected effluent as possible, and (3) the long-term effects of injection on water level declines. A two-dimensional digital computer model was developed to project short-term hydraulic gradients under various conditions of pumping and injections. A corresponding range of interstitial velocities (294-773 ft per year) was estimated by assuming idealized piston-type flow. These velocities may be used to plan the location and spacing of production and injection wells under assumed time factors related to the required residence time for the injected water. The injection sites were selected near a proposed sewage treatment facility in an area that will allow flexibility in the locations of the production and injection wells. Maximum 20-year declines of about 35 ft were projected for areas several miles west and southwest of the facility under anticipated injection and pumping rates. The proposed injection water will require strict water quality controls, which may involve chlorination and the removal of suspended solids. Mixing of the proposed injection water with the native ground water probably will not clog the aquifer by mineral precipitation. The relatively large concentrations of sodium in the injection water may reduce the hydraulic conductivity of the clay layers in the aquifer, but the permeable sands should not be seriously affected. Plans for an artificial recharge program need to include an experimental installation to evaluate the system under field conditions.

Gillespie, J.B. and Slagle, S.E., "Natural and Artificial Ground-Water Recharge, Wet Walnut Creek, Central Kansas", Kansas Water Resources Board Bulletin No. 17, 1972, U.S. Geological Survey, Lawrence, Kansas.

The withdrawal of ground water for irrigation is accelerating rapidly in western and central Kansas. Natural recharge or artificial recharge to the aquifers is needed to ensure an adequate supply of water for the future. Wet Walnut Valley in central Kansas is a narrow alluvial valley in which an interrupted meandering stream dominates the hydrologic system. Wet Walnut Creek is generally a losing stream in most of the reach through the study area. The average annual streamflow leaving the area is 50,530 ac-ft, most of which occurs as storm runoff. Recharge to the aquifer from the losing reach of the creek during low-flow periods is only about 0.05 cu ft/sec/mi of channel because of a relatively impermeable layer of fine-grained material with an average thickness of about 2 ft that lies between the bottom of the channel and the sand and gravel of the lower alluvium. The maximum amount of water available for artificial recharge is streamflow, which averages about 50,000 ac-ft/yr. The average amount probably will be less than the potential for recharge after floodwater retarding structures and artificial recharge systems have been placed in service.

Hutchinson, C.B. and Wilson, W.E., "Evaluation of a Proposed Connector Well, Northeastern Desoto County, Florida", Water Resources Investigations 5-74, July 1974, U.S. Geological Survey, Tallahassee, Florida.

At a 24,000 ac citrus grove in northeast Desoto County, Florida, a connector well is proposed as a resource-management tool for capturing water normally lost through evapotranspiration and by excess runoff. Such a well would connect the surficial sand aquifer with the deep, highly transmissive limestone Floridan aquifer. Because of natural head differences, water would move by gravity flow from the sand into the Floridan aquifer, thus replenishing water withdrawn for irrigation from the Floridan aquifer. A 70-ac marsh was selected as the test site based on analyses of hydraulic conductivity, porosity, and water quality. The recharge rate through the connector well under steady-state conditions is estimated at 160 gpm. The proposed connector well is designed to have 10-in screens opposite zones in the 45-ft thick sand aquifer, be cased for 400 ft opposite confining beds and a secondary limestone aquifer, and be an open hole for about 250 ft in the Floridan aquifer. A graded-sand filter pack placed around the screened sections of the well will increase its efficiency.

Johnson, B., Mumford, L. and Romani, M., "Survey of Conjunctive Use and Artificial Recharge Activity in the United States", HEC-RR-21, Jan. 1984, Hydrologic Engineering Center, U.S. Army Corps of Engineers, Davis, California.

Artificial recharge plays a significant role in conjunctive use where surface water is replenished by ground water. Examples are presented of sites selected from across the United States, where conjunctive use systems exist and where artificial recharge either exists or is planned.

Mallory, M.J., Swain, L.A. and Tyley, S.J., "Potential for Using the Upper Coachella Valley Ground-Water Basin, California, for Storage of Artificially Recharged Water", Open File Report 80-599, Sept. 1980, U.S. Geological Survey, Menlo Park, California.

This report presents a preliminary evaluation of the geohydrologic factors affecting storage of water by artificial recharge in the upper Coachella Valley, California. The ground water basin of the upper Coachella Valley seems to be geologially suitable for large-scale artificial recharge. A minimum of 900,000 ac-ft of water could probably be stored in the basin without raising basinwide water levels above those that existed in 1945. Preliminary tests indicate that a long-term artificial recharge rate of 5 ft per day may be feasible for spreading grounds in the basin if such factors as sediment and bacterial clogging can be controlled. The California Department of Water Resources, through the Future Water Supply Program, is investigating the use of ground water basins for storage of State Water Project water in order to help meet maximum annual entitlements to water project contractors.

McCray, K., "The Ogallala--Half Full or Half Empty", Water Well Journal, Vol. 36, No. 7, July 1982, pp. 53-62.

The Ogallala Aquifer is an irregular unconfined aquifer within a 400 mile wide and 800 mile long area from South Dakota to Texas, including small sections of Colorado, Kansas, Nebraska, New Mexico, and Wyoming. Its 1 quadrillion gal of water are vital to agriculture and to the 2 million people living in the region. A survey indicates that 5 percent of the entire aquifer has been drained at a rate of 14 million ac-ft per year. The saturated thickness has declined more than 10 percent under a fourth of the total aquifer area. The greatest well level declines, 200 ft, have been in Floyd County, Texas. Average depletions are: Nebraska, 1%; Kansas, 8%; and New Mexico, 16%. Within 25 years Colorado's share of the Ogallala will be 40% depleted. There are differing opinions on the future of agriculture in the Ogallala region. Since less water will be available, it is expected that agricultural patterns will change. The beef cattle industry will shift from Texas and Oklahoma to Nebraska and Kansas; Texas and New Mexico will grow cotton and wheat. Consequences of a switch to dry farming may include great increases in erosion. Suggestions for improving water management are conjunctive water use, land leveling, skip row irrigation, night-time watering, improved sprinklers, demand watering, tax incentives, weather modification, artificial recharge, and proper well design and operation. An extensive system of 280 ft wide canals to tap rivers in Missouri and Arkansas has been proposed.

Mitten, H.T., "Preliminary Evaluation of the Potential for Artificial Groundwater Recharge in Eastern San Joaquin County, California", Open File Report 82-123, July 1982, U.S. Geological Survey, Sacramento, California.

In response to increasing demand on water supplies and declining water levels in San Joaquin County, the U.S. Geological Survey, in cooperation with the San Joaquin County Flood Control and Water Conservation District, is evaluating the potential for artificially recharging the aquifer system in eastern San Joaquin County, California. Through a well canvass and analyses of existing data on geology, soils, and drillers logs, this study (phase one of three phases) resulted in identification of 20 sites for exploratory test drilling in areas potentially favorable for artificial recharge. Ten of the sites are in areas adjacent to the Mokelumne River, six are in areas adjacent to the Calaveras River and Mormon Slough, and four are north of Littlejohns Creek.

Motts, W.S., "Optimum Utilization and Management of Water Bearing Formations Hydraulically-Connected to a Stretch of the Connecticut River", OWRR-A-021-MASS-(1), Dec. 1970, Water Resources Research Center, University of Massachusetts, Amherst, Massachusetts.

This investigation examined the ground water resources and ground water recharge potential of the Greater Springfield Complex--a high-density urban area surrounded by a rural peripheral area that will soon undergo urban development. Highly productive gravel and sand aquifers of glacial outwash underlie parts of the peripheral areas, especially west of East Mountain and in the towns of Granby, Hampden and Wilbraham, Massachusetts. Also, a buried channel of outwash yields copious ground water supplies which are used by the municipality of South Hadley. This buried channel overlain by poorly permeable silt and clay may contain large undeveloped supplies of ground water in the South Hadley and South Springfield areas. Potential artificial recharge sites occur along areas of glacial outwash in the Greater Springfield Complex. Programs could be developed to recharge surface water into the aquifers during winter and spring and withdraw the ground water during the heavy water demand periods of summer and fall. Recharge to wells by induced infiltration is possible along some of the tributaries to the Connecticut River such as the Chicopee River, Scantic River, and Great Brook.

Reeder, H.O. et al., "Artificial Recharge Through a Well in Fissured Carbonate Rock, West St. Paul, Minnesota", Water Supply Paper 2004, 1976, U.S. Geological Survey, St. Paul, Minnesota.

Tests were made near St. Paul, Minnesota, to determine the feasibility of artificially recharging a consolidated aquifer by well injection. The formations of primary concern at the recharge experiment site are the Prairie Du Chien group and the underlying Jordan sandstone, which is the major aquifer in the area. The Prairie Du Chien is made up primarily of well-fractured dolomite, which contains solution openings. The Jordan sandstone is more uniform and is considered to be homogeneous on a large scale, although small fractures and variations in material are recognized locally. The two layers are hydraulically interconnected and are artesian. A head difference of 2 ft is noted between water levels in a well open only in the dolomite and in a nearby well open only in the sandstone. For the tests, a production well was installed in the Prairie Du Chien group, and observation wells were installed nearby, both in the Prairie Du Chien and in the Jordan sandstone. Pumping tests were performed before and after injection, and the injection test itself involved recharge of 2,754,000 gal occurring at a rate of about 100 gpm for 20 days. Air entrainment during injection was avoided by use of a small-diameter pipe, in which pipe friction was sufficient to maintain positive pressure throughout the column. The injection test demonstrated that it is hydrologically feasible to recharge the Prairie Du Chien group and the Jordan sandstone artificially through wells completed in the Prairie Du Chien group.

Schaefer, D.H. and Warner, J.W., "Artificial Recharge in the Upper Santa Ana River Area, San Bernardino County, California", USGS/WRD-75-054, July 1975, U.S. Geological Survey, Menlo Park, California.

By 1990 the San Bernardino Valley Municipal Water District will be receiving 102,000 ac-ft per year of imported northern California water to be used for artificial recharge to the local ground water system. The upper Santa Ana River area is suited for artificial recharge because it is largely underlain by permeable river-channel deposits. Some sandy clay, silt, and cemented sand and gravel layers may locally retard downward percolation of recharge water. However, test drilling indicates none of these is extensive enough to impede recharge in the spreading grounds. Analyses of ground water movement during current spreading operations, test drilling data, and application of Baumann's (1965) equation for calculating the theoretical recharge mound indicate that: (1) barriers to ground water movement are not evident in the river channel deposits; (2) depth to a basement complex is less in the area of the eastern spreading basins than in the western basins; (3) the spreading grounds are capable of accepting as much as 80,000 ac-ft per year; and (4) the water being recharged should move through the aquifer toward the areas of pumpage.

Schaefer, D.H., "Ground-Water Conditions and Potential for Artificial Recharge in Lucerne Valley, San Bernardino County, California", Water Resources Investigations 78-118, Jan. 1979, U.S. Geological Survey, Menlo Park, California.

The water level in two areas of Lucerne Valley in California has declined more than 100 ft since 1917, including 60 ft from 1954 to 1976.

These declines are the result of pumping for the irrigation of alfalfa. The lowering of water levels has caused many shallow domestic wells to go dry. Well yields in the valley generally are between 10 and 1,000 gpm. About 240,000 ac-ft of ground water was withdrawn between 1950 and 1976. About 1,750,000 ac-ft remains in storage. Water of poor quality underlies the valley around Lucerne Lake. There was no definable movement of this water from 1954 to 1976, but the possibility exists for future movement toward centers of pumping. Lucerne Valley may be hydrologically suitable for artificial recharge operations.

Sinclair, W.C., "Experimental Study of Artificial Recharge Alternatives in Northwest Hillsborough County, Florida", USGS/WRD/WRI-77/043, May 1977, U.S. Geological Survey, Tallahassee, Florida.

Extensive water withdrawal from the Floridan aquifer in the urban Tampa Bay area has induced leakage from the overlying surficial aquifer adversely affecting the water table and lake levels. Artificial recharge could reduce the impact of these adverse effects. Four experiments were conducted to investigate possible recharge alternatives; sinkhole recharge, water spreading, connector wells, and subsurface tile drainage to a deep well. Experiments indicate that all four methods can be effective. However, the sinkhole recharge experiment moved the greatest volume of water into the Floridan aquifer.

Smith, H.F., "Artificial Recharge and Its Potential in Illinois", Artificial Recharge and Management of Aquifers, Symposium of Haifa, Mar. 19-26, 1967, Publication No. 72, 1967, International Association of Scientific Hydrology, pp. 136-142.

A definite relationship is shown between the grain size of filter blankets and the amounts of sediment that penetrate the aquifer through recharge pits. High rates of recharge have been achieved under certain conditions where the filter grain size was optimum, the pit the correct shape, and the permeability high. Criteria established for evaluating the recharge potential of aquifers include thickness of aquifer, areal extent of aquifer, and minimum economical withdrawal rate, as well as limits of turbidity, temperature, and mineral and bacteriological quality of recharge water. These recharge criteria were applied to areas of Illinois where aquifers were known to exist. Estimates were made of the potential artificial recharge into surface sands and gravels of the state, particularly in northeastern Illinois where population is dense and where withdrawal exceeds natural and induced recharge.

Sunada, D.K., Warner, J.W. and Molden, D.J., "Artificial Groundwater Recharge, San Luis Valley, Colorado", OWRT-A-050-COLO(1), Mar. 1983, Office of Water Research and Technology, U.S. Department of the Interior, Washington, D.C.

Intense use of aquifers for irrigation waters has caused ground water storage depletion in many areas of the arid and semiarid west, including the San Luis Valley in south central Colorado. Artificial recharge is a means of alleviating this problem. To show the practical benefits of artificial recharge to local water users, a demonstration recharge basin was operated in the San Luis Valley. Both numerical and analytical models were calibrated to the aquifer response to suggest operational policies. Analysis of the results of the demonstration project indicate that if recharge operations are conducted during the non-irrigation season when excess water is available, significant amounts of water can be added to storage and combat ground water depletion. The microcomputer model developed in this study could be used in other parts of the San Luis Valley to evaluate the benefits of artificial recharge in these other areas.

Taylor, O.J., "Artificial-Recharge Experiments in the Alluvial Aquifer South of Fountain, El Paso County, Colorado", Water Resources Circular No. 31, 1975, Colorado Water Conservation Board, Denver, Colorado.

Artificial recharge experiments were conducted in the alluvium along Fountain Creek and south of Fountain, Colorado. The saturated thickness of the sand and gravel aquifer ranges from 0 to 34 ft. The transmissivity of the aquifer ranges from 0 to 40,000 sq ft per day. The water table slopes downstream indicating a southerly direction of ground water movement. Five artificial recharge pits were excavated in the unsaturated zone above the alluvial aquifer. Each pit site was instrumented to measure inflow, stage fluctuations, and water table fluctuations. The 15 experiments, each of approximately 12 days' duration, indicated that artificial recharge varied with pond stage. The average adjusted artificial recharge rate for the five pits ranged from 0.1 to 4.7 ft per day. The annual amount of recharge from operation of the five pits was estimated as 16,000,000 cu ft. Observed transmission losses in the canal system indicate that the alluvial aquifer is also recharged through the present irrigation practices. Other promising locations for artificial recharge pits are in coarse material in the relatively thick unsaturated zone above the aquifer. Another advantage of artificial recharge in this area is that it would delay the loss of recharged water to underflow or flow from the aquifer to Fountain Creek. Large-capacity wells in this aquifer yield as much as 2.78 cfs and could be used to withdraw the water stored in the alluvial aquifer.

Taylor, O.J., "Summary Appraisals of the Nation's Ground-Water Resources--Missouri Basin Region", Professional Paper 813-Q, 1978, U.S. Geological Survey, Lakewood, Colorado.

The Missouri Basin Region, about one-sixth of the contiguous United States, utilizes large water supplies for irrigation, industrial, public supply, and rural use. Ground water resources occur in sand and gravel alluvium, glacial deposits, dune sand, basinfill deposits of sand and gravel, sandstone, siltstone, fractured sandy clay, limestone, and dolomite. Ground water is undeveloped in many areas. Unconsolidated and semiconsolidated aquifers have potential for conjunctive use with surface water, reuse of available supplies, artificial recharge, and salvage of evapotranspiration. Sandstone aquifers have potential for artificial recharge, induced interaquifer leakage, conjunctive use with surface water, and temporary mining of ground water. Limestone and dolomite aquifers have potential for development of a large water supply, induced increase in recharge, and induced interaquifer leakage. Saline ground water occurs throughout the region, but is most abundant in deep aquifers in Wyoming, North Dakota, and South Dakota. Improved water use in the Missouri Basin Region would require a periodic inventory of current supplies in precipitation, streamflow, surface-water storage, and ground water; an awareness of the many possible alternatives to meet demands; and implementation of sound water management plans.

Tyley, S.J., "Artificial Recharge in the Whitewater River Area, Palm Springs, California", Open File Report, Jan. 1973, U.S. Geological Survey, Menlo Park, California.

Water levels in the Palm Springs, California, area declined as much as 75 ft from 1945 to 1970. To relieve this overdraft the two primary agencies responsible for water supply to the area, the Coachella Valley County Water District and the Desert Water Agency, contracted with the State of California to begin in 1973 to purchase water imported through the California aqueduct. A study in the area revealed that the percolation basin method seems to be the most feasible way to recharge the proposed 61,000 ac-ft per year of imported Colorado River water in the Whitewater River area north of Palm Springs. The Windy Point area provides the most logical artificial recharge sites. No significant clay or silt layers exist in the alluvial fill to impede the downward movement of artificially recharged water to the water table. Infiltration rates in the Windy Point area ranged from 4 to 24 ft per day. A realistic long term average infiltration rate would be about 5 ft per day. Mixing in the aquifer system of the Colorado River water with native surface water and ground water should have no detrimental effects on infiltration rates. The quality of the ground water as affected by artificial recharge will depend on the ratios of quality and quantity of imported to local recharge water and on the degree to which the artificially recharged water mixes with the native water.

University City Science Institute, "A Study to Evaluate Potential Solutions to the Problem of Insuring an Adequate Water Supply for New Castle County, Delaware", Mar. 1971, Philadelphia, Pennsylvania.

This study concludes that current water demand exceeds current safe yield in New Castle County, Delaware, that the deficit may increase by 2 to 3 million gallons per day in the near future, and that, from the economic and engineering viewpoints, the proposed White Clay Creek dam and reservoir has the greatest prospect of any proposed project for increasing water supply. Previous engineering, economic, and feasibility studies of the water resources and demand projections in the county are reviewed. Aspects discussed include: projected water demand and county needs, an economic analysis of the White Clay Creek dam and reservoir proposal, other currently feasible alternatives, ground water development, a study of the ecological and aesthetic aspects of the reservoir site, and a projection of economic conditions and financing alternatives. Other alternatives discussed include an interbasin transfer from the Schuylkill and Susquehanna Rivers to the White Clay Creek, proposed by the Mason-Dixon Task Force, and a large-scale artificial ground water recharge project.

Warner, J.W. and Moreland, J.A., "Artificial Recharge in the Waterman Canyon-East Twin Creek Area, San Bernardino County, California", Open File Report, Nov. 1972, U.S. Geological Survey, Menlo Park, California.

The feasibility of recharging 30,000 ac-ft a year of imported water from northern California in the Waterman Canyon-East Twin Creek area in San Bernardino County, California, will depend on the effectiveness of Fault K as a barrier to ground water movement near the land surface. Fault K, about three-fourths of a mile southwest of the south branch of the San Andreas Fault, is nearly parallel to the San Andreas Fault. This fault is not evident from surface expression but is postulated from geophysical and water-level data. Results of test drilling and infiltration tests reveal that the subsurface material at the spreading grounds is permeable enough to allow recharged water to percolate to the water table. Data indicate that Fault K extends into the Waterman Canyon-East Twin Creek area and may impede the lateral movement of recharged water. Fault K has no known surface expression and therefore probably does not affect the highly permeable younger alluvium. If that is so, Fault K will be less effective as a barrier to ground water movement as the recharge mound rises.

West, S.W. and Broadhurst, W.L., "Summary Appraisals of the Nation's Ground-Water Resources--Rio Grande Region", Professional Paper 813-D, 1975, U.S. Geological Survey, Reston, Virginia.

The Rio Grande is an interstate and international stream which begins in the high mountains of Colorado, flows across New Mexico, and forms the boundary between Texas and Mexico. Annual precipitation on the region is about 86 million ac-ft; however, all but 4 million ac-ft is returned to the atmosphere by evapotranspiration. The ground water reservoirs contain an aggregate of 5,800 million ac-ft of fresh and slightly saline water in storage, which could be withdrawn through wells. In contrast, the surface reservoirs have a combined storage capacity of only 18 million ac-ft. Withdrawal of ground water in 1970 was 2.7 million ac-ft, of which 88 percent was used for irrigation. The region appears to offer several possibilities for utilizing underground space for purposes other than the withdrawal of water, such as waste disposal, artificial recharge, water quality control, and development of geothermal energy.

Weston, L.K. and Swain, R.E., "Artificial Recharge in South Platte River Basin", <u>Journal of the Irrigation and Drainage Division, American Society of Civil Engineers</u>, Vol. 105, No. IR2, Proceedings Paper 14606, June 1979, pp. 117-127.

A reconnaissance study of the potential for artificially recharging the alluvial aquifer along 110 miles of the South Platte River from Fort Morgan to Julesburg, Colorado, was made based on published hydrologic and geologic information. Analyses were made determine the availability of water for recharge; anticipated infiltration rates; potential storage capacity of the ground water reservoir; the degree of ground water mounding beneath potential recharge sites; the magnitude of annual drain-out loss to the river; possible water quality problems; and the potential effects on the South Platte River flows. Artificially recharging the alluvial aquifer appears technically feasible. Water-logging conditions are likely to develop in areas that lie within 1 or 2 miles of the river. Due to the high dissolved solids concentration of the recharge water, the quality of the ground water in some areas may deteriorate and preclude its use for domestic and municipal water supplies.

Wilson, L.G. et al., "The Design of an Experimental Artificial Recharge Well at Tucson, Arizona", <u>Annals of Arid Zone</u>, Vol. 8, No. 2, Sept. 1969, pp. 267-274.

Tucson, Arizona, obtains its entire water supply by pumping ground water. Recharge occurs during influent seepage in ephemeral rivers, but it is lagging behind withdrawal and water levels are dropping at a rate of about 1 meter per year. Artificial well recharge appears to be the best regional water conservation mechanism. In designing an artificial recharge well, it is important to investigate subsurface materials at the proposed site. Driller's logs were examined, a water supply well and three observation wells were drilled, geophysical exploration techniques were utilized and subsurface flow characteristics determined. Since the materials in the vadose zone were found to be highly permeable, a dual recharge well was constructed for recharge of the unsaturated region as well as the phreatic zone. The design and operation of the well are described in detail. Injected water was industrial effluent from a nearby power plant. Recharge tests showed injection rates ranging over 23-102 cubic meters per hour. Such wells may be effective tools for water conservation in the Tucson Basin.

Wood, W.W., "Development of Technical Site Selection Criteria for Artificial Recharge", Proceedings of the Symposium on Wastewater Reuse for Groundwater Recharge, held Sept. 6-7, 1979 at California State Polytechnic University at Pomona, May 1980, California State Water Resources Control Board, Office of Water Recycling, Sacramento, California, pp. 73-79.

Artificial recharge has often been neglected as a water management technique for lack of suitable site selection criteria. Six general factors must be quantified to develop these criteria: rate of recharge, water quality, aquifer storage capacity, rate of recovery from aquifer, quality of extracted water, and aquifer plugging. More work is needed in the first five areas; however, some predictions can be made in each of them and they are also being evaluated in other areas in the field of hydrology. Of particular importance is predicting the onset of aquifer plugging by chemical, physical and biological mechanisms, which results in reductions in hydraulic conductivity. Plugging of artificially recharged aquifers by biological growth is a serious problem that occurs due to in-situ growth of bacteria at or near the water-aquifer interface. Chemically induced plugging involving clay dispersion and flocculation also results in significant reduction of the aquifer's hydraulic conductivity. Plugging by mineral precipitation has not been a serious problem. Because of the lack of predictive capabilities in the areas of plugging, two empirical approaches have been developed. The first is to remove all particulate material and organic substances and chemically adjust the water to match that in the aquifer. The second is to let the aquifer provide the water quality improvement. A combination of the two is often used.

CASE STUDIES

Allen, W.B., Miller, J.B. and Wood, W.W., "Availability of Water in Kalamazoo County, Southwestern Michigan", Water Supply Paper 1973, 1972, U.S. Geological Survey, Washington, D.C.

Water facts are appraised that aid in planning the orderly development of the water resources of Kalamazoo County, Michigan, and serve as a guide for management. The report defines the relationship between surface and subsurface water units and map variations in water-yielding characteristics of principal ground water units. The availability of water in streams, the location of surface storage sites, and the use of storage to augment streamflow are described. Average annual recharge was determined by means of water budgets utilizing precipitation, temperature, streamflow and water level data. The physical and chemical quality of ground and surface water are related to water use. Present and potential development of the principal subsurface reservoirs and existing pumping centers are described using mathematical models. Areas favorable for development of additional centers of pumping and their practical limits of development are defined. Problems may arise with excessive subsurface reservoir development. These problems include changes in streamflow, the need for artificial recharge ponds, and deteriorating water quality.

Anderson, G.S., "Artificial Recharge Experiments on the Ship Creek Alluvial Fan, Anchorage, Alaska", USGS/WRI-77-38, June 1977, U.S. Geological Survey, Anchorage, Alaska.

During the summers of 1973 and 1974, water from Ship Creek was diverted at an average rate of approximately 6 cubic feet per second to an 11-acre recharge basin. Maximum sustained unit recharge for the basin was approximately 1.4 ft per day. During 1975 a second basin of 8 acres was also used for recharge, and the total diversion rate was increased to as much as 30 cu ft per sec. During 1973 and 1974, when only one recharge basin was in operation, a maximum rise of 18 ft was observed in the ground water table near the basin. In 1975, when both basins were being used, the maximum rise was 30 ft in the same area. During 1973 and 1974, the water level rise was 12 and 8 ft in the unconfined and confined systems. Water level data and changes in saturation conditions interpreted from borehole geophysical logs indicate that most of the recharged water remained in the unconfined aquifer. Thus the benefits of recharge would not persist into late winter, the critical period of water availability in Anchorage, unless diversion to the basins could be continued until January or February.

Aronson, D.A., "The Long Island Experience with Water Retention and Recharge Management", Proceedings of a Workshop on Managing the Demand for Recreation and Tourism held in Ocean City, Maryland, Sept. 27-29, 1978, 1978, Co-sponsored by the Cooperative Extension Services of Delaware, Maryland, New Jersey and Virginia, pp. 48-51.

Ground water is the sole source of fresh water for the residents of Nassau and Suffolk Counties on Long Island, New York. The ground water reservoir is recharged only by local precipitation seeping from the land surface to the water table. Rapid urbanization of Nassau County caused a reduction of natural recharge as storm runoff from paved surfaces was carried to coastal waters through storm sewers. In 1935, Nassau County began to construct large basins to impound storm water and to return it to the ground water reservoir by infiltration through underlying deposits. The use of storm water basins not only helps to conserve storm runoff and to augment the fresh water reservoir, but it also has eliminated the need for long, costly trunk storm sewers to carry storm runoff to tidewater. Storm water basins on Long Island are open pits excavated in permeable sand and gravel deposits of glacial origin. Basins dispose mostly of storm runoff from residential, industrial, and commercial areas and from highways. As of 1978, more than 2,500 storm water basins were in operation in Nassau and Suffolk Counties.

Bianchi, W.C. and Lang, G.J., "The City of Fresno's Leaky Acres Ground Water Recharge Project--Construction and Performance", Journal of the American Water Works Association, Vol. 66, No. 3, Mar. 1974, pp. 176-180.

The urban area hydrology and geologic setting of Fresno, California, produces ground water depressions that require stabilization. Artificial ground water recharge by water spreading on selected agricultural soils can be operationally and economically successful for urban water supply. Construction, structures, and water control operations are closely related to agricultural irrigation practice. Artificial recharge can be accomplished into the Fresno depression at a total cost of $25.00-$25.50/ac-ft (at the mains) by water spreading in the urbanized area. This is about half of what filtration and treatment would cost. The actual cost of water spreading amounts to only $2.50-$3.00/ac-ft, with the rest going to water purchase and pumping costs to put the water into the city's mains.

Bloyd, R.M., Jr., "Underground Storage of Imported Water in the San Gorgonio Pass Area, California", Open File Report, July 1969, U.S. Geological Survey, Menlo Park, California.

Estimated long-term average annual precipitation in the San Gorgonio Pass Water Agency drainage area is 332,000 ac-ft, and estimated average annual recoverable water is 24,000 ac-ft, or less than 10 percent of the total precipitation. Estimated average annual surface outflow is 1,700 ac-ft, and estimated average annual ground water recharge is 22,000 ac-ft. Estimated subsurface outflow from the area in 1967 was 6,000 ac-ft into the Redlands area, 220 ac-ft into Potrero Canyon, and 800 ac-ft into the fault systems south of the Banning storage unit. An anticipated 17,000 ac-ft of water per year will be imported by 1980. Information developed in this study indicates it is technically feasible to store imported water in the eastern part of the Beaumont storage unit without causing adverse conditions and without losing any significant quantity of stored water.

Brown, R.F., Signor, D.C. and Wood, W.W., "Artificial Ground Water Recharge as a Water-Management Technique on the Southern High Plains of Texas and New Mexico", Report 220, Aug. 1978, Texas Department of Water Resources, Austin, Texas.

Artificial ground water recharge of the Ogallala Formation of late Tertiary age is a water management technique that may be of significant value on the southern High Plains of Texas and New Mexico. Under specific conditions artificial recharge by use of either water spreading basins or injection wells is a proven method of storing water that is available from the playa lakes on the southern High Plains of Texas and New Mexico. Spreading basins are the most economical method of recharge in most areas of the southern High Plains; however, in many areas where the surficial materials have low vertical permeability, and where the spreading basin method cannot be used successfully, water can be recharged through injection wells. Artificial recharge of playa lake water is more likely to be successful if the water is collected from near the surface of the playa lakes, treated and clarified in a settling basin, and recharged through a system that minimizes the clogging effect of suspended sediment. Reduction of the infiltration rate is minimized if the recharged water contains little or no suspended sediment. Case histories of recent recharge experiments on the Southern High Plains, the results of laboratory studies of sediment flocculation of playa lake water, and a cost analysis of recharge systems are presented.

Bruington, A.E., "Wastewater Use and Groundwater Recharge in Los Angeles County", Proceedings of the Seminar on Advanced Wastewater Treatment and Disposal, June 10, 1971 at Hauppauge, New York, 1972, Los Angeles County Flood Control District, California, pp. 105-127.

Injection well ground water recharge of the West Coast Basin Barrier Project, in operation for over ten years, has successfully halted saltwater intrusion into the water table in Los Angeles County. Paralleling the coast at an average distance of 30 miles is a mountain range which receives 40 inches of rain per year while the coast averages 9 inches. Runoff from the mountains enters the very large ground water basin of the Coastal Plain composed of accumulated marine sediments. In spite of water imported from the Colorado River, the county continues to depend on ground water for about one-third of its supply. Artificial spreading basins formed on flat land and in river beds to increase ground water recharge tend to become clogged with silt and recharge at very slow rates. The nine mile long row of 97 wells at the coastline has been more successful. Injection of 35,000 ac-ft of Colorado River industrial and commercial water per year has raised the ground water surface, forming a ridge which salt water will not penetrate. Experiments with thoroughly treated wastewater injection show no transfer of virus to test wells only 20 ft away. Biological oxygen demand (BOD) is not a problem if the water is injected with oxygen. Costs are competitive with Colorado River water. Reclaimed water has not been used on a large scale due to potential hazards from exotic new chemical wastes, capital outlay costs, and public opinion.

Buros, O.K. et al., "Florida Style Water Supply and Treatment", Water Engineering and Management, Vol. 129, No. 4, Apr. 1982, pp. 30-33.

Florida's abundant precipitation (163 million ac-ft per year) recharges the very productive Floridan aquifer underlying most of the state and the smaller Biscayne aquifer under the southeast coast. Ground water provides almost 90 percent of the state's public supply. Florida's most prevalent problems with its water supply are hardness (high Ca, low Mg), high hydrogen sulfide levels, salt water intrusion, and trihalomethane (THM) formation. As a result, treatment methods have evolved. Water softening with lime or lime-soda ash is widespread. Many softening plants use potato starch as a flocculant aid to improve the sludge handling characteristics. Surface water quality is variable; it is often high in organic color. Treatment usually required is coagulation, settling, filtration, and activated carbon. Aeration has been a common process for removing hydrogen sulfide from waters. However, Orlando is using a new process based on oxidation of sulfides with chlorine prior to passage through a granular activated carbon bed. Methods of handling the salt water intrusion problem include artificial recharge of an aquifer with river water, control of large withdrawals in sensitive areas, raising canal control elevations, and relocation of wells further inland. In areas of abundant brackish water supplies desalination by electrodialysis or reverse osmosis is in use. The combination of high dissolved organic matter and bromide ions in Florida waters favors THM formation. Three possible treatment strategies are available: removal of THM precursors, removal of THM after chlorination, and alternate disinfectants.

Coe, J.J., "Ground-Water Storage for California Water Project", Journal of the Irrigation and Drainage Division, American Society of Civil Engineers, Vol. 105, No. IR3, Proceedings Paper No. 14832, Sept. 1979, pp. 305-315.

Environmental opposition and escalated construction costs associated with construction of dams and reservoirs in Northern California have caused the California Department of Water Resources to examine the feasibility of using ground water basins in the San Joaquin Valley and Southern California to store State Water Project water during periods of above normal runoff in Southern California for later withdrawal and delivery in subsequent dry years. Conjunctive use of surface and ground water is not a new concept. However, for many years, there has been a reluctance to initiate conjunctive use because of concerns over legal and institutional aspects such as: (1) what are the priorities between overlying users and public agencies to available storage space and what are the priorities between public agencies; (2) who owns the water after it is spread or injected and placed in storage; (3) how should state and regional conjunctive use programs be made compatible with local water management operations. In California, two recent State Supreme Court decisions have provided some guidance.

Cohen, P. and Durfor, C.N., "Artificial-Recharge Experiments Utilizing Renovated Sewage-Plant Effluent--A Feasibility Study at Bay Park, New York, U.S.A.", Artificial Recharge and Management of Aquifers, Symposium of Haifa (Mar. 19-26, 1967), Publication No. 72, 1967, International Association of Scientific Hydrology, pp. 193-199.

The U.S. Geological Survey, in cooperation with the Nassau County Department of Public Works, is conducting a series of artificial recharge experiments on Long Island, New York to obtain scientific and economic data needed to evaluate the feasibility of injecting highly treated sewage plant effluent into a proposed network of "barrier" injection wells that are intended to prevent or retard the landward movement of salty water from the Atlantic Ocean into major aquifers. Tertiary sewage treatment was developed to produce an effluent that meets requirements commonly accepted for potability. An experimental injection well and injection plant have been completed. Remote sensing downhole geochemical probes permit the measurement of water quality and hydraulic head changes at several points within the injection well and the filter pack.

Cooley, R.L. et al., "Influence of Surface and Near-Surface Caliche Distribution on Infiltration Characteristics and Flooding, Las Vegas Area, Nevada", Project Report No. 21, Feb. 1973, Center for Water Resources Research, University of Nevada, Reno, Nevada.

This study was conducted on the Red Rock Canyon fan which is a large alluvial fan bordering the Las Vegas basin on the east flank of the Spring Mountains. The objectives were: (1) to determine the types, degrees of development, and distribution of calcium carbonate deposition on the Red Rock Canyon fan as these affect infiltration and runoff; (2) to investigate the engineering projects designed for flood control in the urban area in relation to the expected runoff; (3) to obtain geomorphic and hydrologic information essential to the selection of alternative sites for artificial recharge and flood impoundments; and (4) to coordinate the study with local agencies currently planning flood control projects. Because adequate precipitation and runoff data were not available and could not be collected during the short term of the project, the study was concentrated upon the potential for the factors studied to influence runoff.

Dennis, S.A., Dvoracek, M.J. and Peterson, S.H., "Artificial Recharge of Hydrologically Isolated Aquifers", Winter Meeting of the American Society of Agricultural Engineers, Dec. 8-11, 1970, Chicago, Illinois, Paper No. 70-718, 1970, Department of Agricultural Engineering, Texas Tech University, Lubbock, Texas.

Hydrologically isolated aquifers have unique problems related to their recharge. This type of aquifer may have limited natural recharge. One of the largest unconfined, hydrologically isolated, aquifers is the Ogallala, which underlies the southern High Plains, of Texas, one of the largest agricultural areas in the United States. Over five million acres are irrigated using Ogallala ground water alone. The major crops of the area are cotton, grain sorghum, wheat, and vegetables. Pumpage annually ranges from 4 to 8 million ac-ft. If a portion of the annual runoff water could be recharged into the Ogallala, the rate at which the ground water is being depleted could be decreased. A major problem affecting all recharge methods is the quality of water. Since the most common source is runoff water, physical as well as chemical/biological pollution or contamination is possible. To date primary emphasis has been devoted to physical contamination (sediment). Pressure injection recharge wells do not seem to be affected by the use of highly turbid water. With periodic maintenance, pits, trenches, and spreading basins are best utilized in areas when a highly permeable layer lies at or very near the surface.

Dewiest, R.J.M., "Artificial Recharge Through Augmented Bank Storage", Symposium on Artificial Recharge and Management of Aquifers, Mar. 1967, Haifa, Israel.

Nonsteady flow is examined for a ground water recharge project in which rainfall runoff is returned to a water table aquifer in the vicinity of Princeton, New Jersey. The recharge is accomplished by damming a river and its tributary and by reversing the hydraulic gradient between the river and its connected ground water basin. Most of the natural recharge to the ground water table in the area occurs during the rainy season from November to late May. With the resulting build-up of ground water level, effluent discharge to the rivers increases. This increase ultimately is a net loss of fresh water to the ocean. This loss would be reduced significantly by construction of dams. Increased water levels in rivers would automatically decrease the hydraulic gradient from ground water reservoir to the rivers and hence limit the effluent discharge. Simultaneously, the ground water reservoir would be pumped and this would make available more storage space for retention of a higher percentage of winter-spring rainfall or for quick recharge from flash summer runoff. A geohydrologic description of the watershed is given and various physical parameters entering the study are discussed.

Directo, L.S. and Lindahl, M.E., "River Water Quality for Artificial Recharge", Journal of the American Water Works Association, Vol. 61, No. 4, Apr. 1969, pp. 175-180.

The Platte River was sampled in 1966-1967 at three locations near Ashland, Nebraska to investigate seasonal variations in its physical, chemical, and biological qualities in a study of the feasibility of using the river water for recharge of the aquifer supplying the wells for the city of Lincoln. All the samples were taken using the equal transit rate method of depth integration. Streamflow data were obtained from USGS gaging station records. The mean specific conductance of 439 micromhos/cm was exceeded 60 percent of the time. The mean solids concentration was 1090 mg/l. This is expected to cause clogging problems in use of the water for recharge unless solids are removed. The sodium absorption ratio averaged 1.05 mg/l and iron concentration 0.41 mg/l, neither of which should cause any problem. The phosphate concentration averaged 1.07 mg/l with an algae count of 910, which could result in heavy algal growth on sand recharge beds. Coliform density was 6,610/100 ml at Ashland. All the collected data, including temperature, pH, alkalinity, and hardness are shown graphically. River water is similar in chemical quality to the ground water in the well field, and should cause no compatibility problems. The high solids, algae, and coliform concentrations may cause clogging.

Dvoracek, M. and Peterson, S.H., "Recharging the Ogallala Formation Using Shallow Holes", Ogallala Aquifer Symposium, Contribution No. 78, Apr.-May 1970, Texas Tech University, Lubbock, Texas.

Natural ground water recharge cannot sustain the Southern High Plains portion of the Ogallala formation. Nearly all the water used in this region is derived from the Ogallala formation. The southern part of the Ogallala is hydrologically isolated from all outside areas; natural recharge to this portion of the aquifer originates only from precipitation falling on the aquifer. The region has only limited and erratic rainfall, poorly established drainage, and very few lakes or reservoirs; therefore, it offers little opportunity for natural recharge. The volume of water withdrawn, as shown by the water level decline, is many times greater than the estimated natural recharge of about 1 inch per year. In tests, more than 15 ac-ft of runoff water was recharged in shallow holes completed above the water table. Increases in rates of recharge were observed during the tests. A condition of soil piping is probably responsible for the increasing rates and loss of material from cavities near the recharge wells.

Emmons, P.J., "Artificial-Recharge Tests in Upper Black Squirrel Creek Basin, Jimmy Camp Valley, and Fountain Valley, El Paso County, Colorado", Water Resources Investigations 77-11, July 1977, U.S. Geological Survey, Lakewood, Colorado.

Artificial recharge tests were conducted in the allvium in upper Black Squirrel Creek basin, the alluvium in Jimmy Camp Valley, and in the alluvium overlying the Widefield aquifer which is located in an ancestral channel in Fountain Valley, Colorado. Nine artificial recharge pits with areas of approximately 9,200 sq ft each were excavated in the unsaturated zones above the three aquifers. Each artificial recharge site was instrumented to measure inflow, stage fluctuations, and water table fluctuations. Artificial recharge tests of approximately 10 days' duration were conducted at each of the nine artificial recharge sites and one extended test of approximately 30 days' duration was conducted in each of the three study areas. Periphyton growth, present in most of the artificial recharge ponds, was insufficient to cause a noticeable decline in the rate of infiltration.

Florida Department of Natural Resources, "Permitting of Artificial Recharge Systems", Florida Administrative Code Chapter 16 H5, Secs. 5.01-5.31, 1978, Suwannee River Water Management District, White Springs, Florida.

The rules and regulations described are designed to regulate artificial recharge systems within the Suwannee River Water Management District. Artificial recharge, as used here, is defined as the intentional introduction of water into any underground formation. Included in the definition are fluids discharged from land application facilities, and disposal, storage, and connector well facilities. Excluded are wastewaters from residential septic tank systems, effluent from certain sewage treatment plants, and facilities for agricultural irrigation. In most instances, work may not commence on an artificial recharge system without a permit. Certain wells, however, are exempt from the permit requirement. Rules detailing how to file a permit application and under what conditions construction may be exempted are stated. The District may require public hearings before a permit is issued; it is also entitled to give orders relating to drilling techniques designed to prevent waste and pollution. In addition to the permit requirement, a specific casing and sealing program shall be submitted to the District for approval. Details in regard to this program are set out. Periodic well testing is required, as are written records of all construction work performed.

Ford, R.S. and Hills, E.E., "Evaluation of Ground Water Resources: Livermore and Sunol Valleys", Bulletin No. 118-2, June 1974, California State Department of Water Resources, Sacramento, California.

Livermore and Sunol Valleys are part of the rapidly urbanizing metropolitan region surrounding San Francisco Bay. These two valleys contain three basic resources: land, gravel, and water. In the late 1940s and during the 1950s, water demand exceeded supply and ground water levels declined. This trend has been stopped by the availability of a new water supply to the area from the Del Valle Reservoir. The present studies of the area will develop a better understanding of the ground water resources leading to development and testing of alternative plans for conjunctive use of surface, ground, and waste waters. The geology and hydrology of the area were reported in sufficient detail to permit planning for use of the ground water.

Foxworthy, B.L., "Hydrologic Conditions and Artificial Recharge Through a Well in the Salem Heights Area of Salem, Oregon", Water Supply Paper 1594-F, 1970, U.S. Geological Survey, Tacoma, Washington.

In the Salem Heights area of Salem, Oregon, pumping from wells that tap a permeable zone of limited extent in basalt of the Columbia

River group caused serious year-to-year declines of ground water levels. To determine the feasibility of reducing these declines by artificially augmenting the natural recharge and to develop techniques applicable to a municipal program of artificial recharge, 24.5 million gallons of surplus water was injected under pressure into one of the municipal wells during three periods, ranging in duration from 1 to 15 days, at an average injection rate of about 830 gpm. The recharge water contained abundant dissolved air and, at times, excessive sediment; in other respects it was of excellent quality and was compatible with the native ground water. The artificial recharge had no apparent deleterious effects on the quality of the ground water. Pressure rise from the injected water spread rapidly through the permeable aquifer, but the residual buildup of ground water levels was soon masked by a seasonal rising trend of levels. The geologic and hydrologic conditions, however, preclude the escape of substantial volumes of the recharge water. The conditions appear to be favorable for further artificial recharge. Changes in the specific capacity of the recharge well provide valuable guidance for subsurface injection operations.

French, J.J., "Ground-Water Storage in the Johnson Valley Area, San Bernardino County, California", Water Resources Investigations 77-130, Apr. 1978, U.S. Geological Survey, Menlo Park, California.

The Mojave Water Agency includes several desert basins where ground water in storage is many times as great as the average annual recharge. The Johnson Valley area in San Bernardino County, California, was evaluated for: (1) the quantity of ground water in storage; (2) the chemical quality of the ground water; and (3) the potential for storage of recharge water in the unsaturated zone. Johnson Valley contains about 250,000 ac-ft of water in storage, of which about half can be considered recoverable. About 250,000 ac-ft of void space in the unsaturated alluvium is available for storage of imported water, but not all of the recharged water could be recovered. The quality of the water in storage is satisfactory for public consumption, although water from some areas has high fluoride concentrations and should be mixed with water of low fluoride concentration.

Garrett, A.A. and Londquist, C.J., "Feasibility of Artificially Recharging Basalt Aquifers in Eastern Washington", Open File Report, 1972, U.S. Geological Survey, Tacoma, Washington.

Artificial recharge methods are evaluated as a means of augmenting natural recharge to offset the effects of intensive pumping of ground water in parts of eastern Washington. Ground water in alluvial fill in the coulees is considered the most suitable source of water for artificial recharge. This shallow ground water, replenished periodically by surface flow, is likely to contain little sediment and dissolved air, which have been clogging agents in previous injection operations. Of 22 township-size areas where overdraft is evident, only three were considered technically suited to an artificial recharge program. For use of water from Carb Creek Valley, the estimated cost of artificial recharge ranges from about $6 per ac-ft for recharging 2,000 ac-ft per year to about $4 per ac-ft (1970 values) for recharging 6,000 ac-ft per year if the source and injection sites were both on the valley floor (the most favorable situation).

Gillespie, J.B., Green, D.W. and Slagle, S.E., "Artificial Groundwater Recharge, Wet Walnut Creek, Central Kansas", Progress Report No. 2, Jan. 1970, Kansas Water Resources Board, Topeka, Kansas.

The capability to artificially recharge the alluvial aquifer underlying Wet Walnut Creek Valley, central Kansas, was tested at two sites--one an alluvial slope near the edge of the valley and the other a shallow tributary channel with ephemeral flow. Most of the site-development work is finished except the dam for pond 1 at the tributary-channel site. At the alluvial-slope site, after water pumped from wells has been ponded for several days, the infiltration rate is about 0.5 ft per day. The infiltrated water is perched above a zone of silty clay from where it percolates to the aquifer. At the tributary-channel site, the infiltration rates vary with different channel-bottom conditions.

Gillespie, J.B., Hargadine, G.D. and Stough, M.J., "Artificial-Recharge Experiments Near Lakin, Western Kansas", Bulletin No. 20, 1977, Kansas Water Resources Board, Topeka, Kansas.

The irrigation economy of the High Plains of Kansas probably will decine if the depletion of ground water from storage in the major aquifers continues to accelerate. Experiments have been made in western Kansas to investigate artificial recharge as a possible means of alleviating the problem. Possible sources of water for artificial recharge are storm runoff, irrigation tail water, and imported surface water. The artificial recharge experimental site near Lakin, Kansas, is underlain by about 1.5 ft of loessial soil, 12 ft of loess, and 170 ft of undifferentiated Pliocene (Ogallala Formation) and Pleistocene deposits overlying shale of the Dakota Formation. Depth to the water table is about 65 ft. The unconsolidated material in the unsaturated zone consists mainly of silt and clayey silt with some layers of sand and gravel. Unconsolidated material in the upper part of the saturated zone consists mostly of fine sand. Artificial recharge experiments were conducted in three small ponds using ground water and one small pond using surface water from the Arkansas River. As the infiltrated water percolated to the water table, changes in chemical quality were monitored by sampling through porous cups installed at various depths in the unsaturated zone. After infiltration of 120 ft of water, leaching had nearly stabilized and the chemical quality of water reaching the water table was similar to that of the input water. The data collected indicate that artificial recharge of alluvial aquifers by water spreading could be accomplished on most of the High Plains of Kansas, provided that water for recharge was available.

Gleason, U.E., "Water Projects Go Underground", Ecology Law Quarterly, Vol. 5, No. 4, 1976, pp. 625-668.

Recently, California courts have affirmed the public's right to store, protect and retrieve imported water in natural underground basins. This will alleviate the need for additional dams by providing a ready means to capture, during the rainy season, the excess water which usually escapes unused. Underground water storage is accomplished through the process of artificial recharge. California is now preparing to use underground water storage facilities in its multi-billion dollar state water system. One legal problem still requiring clarification is the exchange storage situation which results when a ground water right owner forgoes the extraction of a certain amount of ground water by substituting imported surface water supplies. An additional problem is establishing a priority list for underground storage in areas where the storage space is limited. Obviously the overlying communities should get first priority. Perhaps one of the most significant consequences of the action in California is the current concern that the state's existing resources, both natural and man-made, should be used to their fullest extent.

Green, M.G., "Artificial Recharge to the Edwards Limestone Aquifer in South Texas", Proceedings of Dubrovnik Symposium (Oct. 1965), Hydrology of Fractured Rocks, Volume 2, Publication 74, 1967, International Association of the Science of Hydrology, pp. 465-481.

Because of critical water conditions in South Texas, the Fort Worth District of the U.S. Army Corps of Engineers, in cooperation with state agencies, investigated the feasibility of providing recharge to the Edwards Limestone aquifer as a flood control and water conservation measure. A proposed plan is presented. The study consisted of: (1) a survey of available water for recharge and the problem of open-water evaporation; (2) the determination of maximum infiltration rates along the streams; (3) a further development of the hydrology and geology of the Edwards Limestone; (4) utilization of radioactive tracers; (5) fluctuation of the "bad-water" line; (6) methods to control springflows; (7) feasibility of reducing leakage from an existing reservoir located in the Balcones Fault zone; and (8) geologic investigation of ten damsites. The study shows that the construction of five dams (three dams to be considered primarily for recharge, flood control, and recreation) should be considered as highly desirable. Two other water-conservation dams would provide an additional 89,000 ac-ft annually. It is estimated that the proposed plan of improvement will meet all the water demands in the area to about the year 2000, beyond which better use of return flows and development of additional water supplies outside the Edwards Reservoir area will be needed.

Griffis, C.L., "Artificial Recharge in the Grand Prairie, Arkansas", Bulletin 810, Aug. 1976, Agricultural Experiment Station, University of Arkansas, Fayetteville, Arkansas.

The Grand Prairie is located in east-central Arkansas between the White River and Bayou Meto. The area is primarily agricultural, with rice being one of the major crops. This rice requires a considerable amount (22 inches) of water each year, and most of the water for this purpose is pumped from underground wells. As a result, the supply of water in the water-bearing strata underground is being depleted faster than it is being replaced. For this reason, some means of artificially replenishing the water in the aquifer is needed. Bench-scale tests showed that a sand-filled, gravity-flow recharge well was not suitable for recharging untreated runoff water in the Grand Prairie. A series of studies then was made to search for a low-cost water treatment system. The water treatment techniques studied included the use of alum, cationic and anionic polyelectrolytes, lime, and magnesium hydroxide. In addition, some studies included combinations of two or more of the chemicals, and in one series of tests the sludge formed in one batch of treated water was used as seed material for the next batch. No treatment method was able to clarify the water for a cost less than the target cost of $5 per ac-ft.

Other sources for recharge water were examined and the most likely candidates were found to be the drainage water from flooded rice fields, and rainwater collected on the fields during the off-season for rice production. Use of the first water source presents problems. The latter possibility is now under study.

Guyman, G.L., "Application of Artificial Recharge Technology for Managing the Water Resources--Anchorage, Alaska", IWR-24, June 1972, Institute of Water Resources, University of Alaska, Anchorage, Alaska.

In spite of an abundant water resource in Alaska, water supply shortages exist in many communities primarily because of quality considerations. Water temperatures are particularly important because of distribution costs and problems associated with low ambient winter temperatures. As a result there are management needs for water resources in Anchorage, Alaska. Past and current knowledge related to artificial recharge is examined in relation to increasing water table elevation and temperatures of the ground waters underlying Anchorage, Alaska. Long term infiltration processes are considered from the standpoint of the design and operation of off-stream artificial recharge basins. Significantly, the geometry of recharge ponds is related to the temporal infiltration relationship. Because of a noticeable lack of knowledge, it is difficult to determine aquifer response resulting from artificial recharge. The damped character of ground water basins seems to preclude warming of ground waters by selective recharge of warm summer stream flows.

Hargis, D.R. and Peterson, F.L., "Artificial Recharge Practices in Hawaii", OWRR-A-028-HI(1), Oct. 1970, Water Resources Research Center, University of Hawaii, Honolulu, Hawaii.

Artificial recharge practices in Hawaii include stream water recharge through wells and shafts, storm drainage disposal through wells and pits, induced leakage from ditches and reservoirs and deliberate spreading of excess irrigation water, incidental ditch and reservoir leakage and other non-deliberate recharge of irrigation water, and cesspool seepage. The most significant source of artificial recharge to ground water bodies in Hawaii is probably recharge incidental to irrigation. Deliberate artificial recharge to ground water bodies consists of wells recharging stream water at Hanapepe, Kauai and Puukapu, Hawaii, wells and pits which dispose of storm drainage at Wailuku and Kahului, Maui and Hilo, Hawaii, and excess irrigation water by induced leakage from ditches and reservoirs and water spreading over fields and in gulches at the Hawaiian Commercial and Sugar Company on the East Maui isthmus. In addition, the Department of Land and Natural Resources of the State of Hawaii will begin disposal of treated sewage through three deep wells at Waimanalo, Oahu as soon as the sewerage system is completed.

Hargis, D.R. and Peterson, F.L., "Effects of Well Injection on a Basaltic Ghyben-Herzberg Aquifer", Ground Water, Vol. 12, No. 1, Jan.-Feb. 1974, pp. 4-9.

The practice of artificially recharging wastewater such as storm runoff, sewage effluent, and various industrial wastes into the subsurface is of growing importance in Hawaii. In 1970 construction of a collection basin and four deep injection wells for the disposal of storm runoff from a residential development began in Kahului, Maui. This presented an opportunity to evaluate the response of a basaltic Ghyben-Herzberg aquifer to well injection, and to measure the rate of movement of injected water in the aquifer. The results of pumping and injection tests of one completed well and one test hole indicate that the finished injection wells should be able to transmit sediment-free water at rates in excess of 5,500 gpm per well. Observation well measurements indicate that injected water moves at a gross velocity of one ft per minute in the basalt aquifer in the vicinity of the injection wells.

Harmeson, R.H., Thomas, R.L. and Evans, R.L., "Coarse Media Filtration for Artificial Recharge", Journal of the American Water Works Association, Vol. 60, No. 12, Dec. 1968, pp. 1396-1403.

When ground water conditions in Peoria were approaching a critical state, a study was made of various schemes for artificial recharge used in this country and abroad. Preliminary tests of the effects of artificially recharging the glacial drift aquifer were conducted in 1941 in an abandoned gravel pit near the banks of the Illinois River. Ground water levels in the aquifer responded both rapidly and substantially. Because bacterial degradation of ground water quality occurred in one of four observation wells during these preliminary tests, it was assumed that a sand filtration layer, based on water treatment practice, would be required in recharge pits to protect the ground water from bacterial contamination. Multiple correlations and regression analysis of data provide an equation relating the variables studied. This equation can be used to indicate the filtration efficiency of coarse media with relation to the parameters of filter layer depth, coarse media diameter, and rate of recharge.

Hickey, J.J. and Barr, G.L., "Hydrogeologic Data for the Bear Creek Subsurface-Injection Test Site, St. Petersburg, Florida", Open File Report 78-853, 1979, U.S. Geological Survey, Tallahassee, Florida.

Lithologic, hydraulic, geophysical, and water quality data were collected at the Bear Creek subsurface-injection test site at St. Petersburg, Florida. The data were collected to determine the feasibility of subsurface injection of storm runoff. An exploratory hole and five observation wells were constructed between October 1974 and April 1976. The lithology of the upper 185 ft at the test site is predominantly sand and marl. From 185 ft to 3,504 ft, limestone and dolomite predominate. Also, gypsum is present below 1,290 ft. Vertical intrinsic permeability, porosity, and compressibility of cores are reported. A 73-hr withdrawal test discharging 3,450 gpm was run in the test injection well. At the site, the chloride concentration in water from 192 to 340 ft, ranged from 150 to 680 mg/l, and from 500 to 1,267 ft ranged from 16,000 to 20,000 mg/l. Eleven additional wells near the test site were sampled for water quality.

Hirashima, G.T., "Availability of Streamflow for Recharge of the Basal Aquifer in the Pearl Harbor Area, Hawaii", Water Supply Paper 1999-B, 1971, U.S. Geological Survey, Washington, D.C.

The Pearl Harbor area is underlain by an extensive basal aquifer that contains large supplies of fresh water. Because of the presence of a cap rock composed of sedimentary material that is less permeable than the basaltic lava of the basal aquifer, seaward movement of ground water is retarded. The cap rock causes the basal water to stand at a high level; thus, the lens of fresh water that floats on sea water is thick. Discharge, which includes pumpage from the wells and shafts, averaged 250 million gpd during 1931-65. Because the water level did not decline, recharge must have been approximately equal to discharge. Increases in pumping may cause increases in chloride content unless the increased discharge is balanced by increased recharge. Average annual direct runoff from the 90 sq mi Pearl Harbor area is 47.27 million gpd or 11.1 in; this is 13.3 percent of the average annual rainfall. Average annual direct runoff in streams at the 800- and 400-ft altitudes is 29 and 38 million gpd, respectively. Because streams are flashy, only 60 percent of the runoff can be economically diverted to recharge areas.

International Association of Scientific Hydrology, "International Survey on Existing Water Recharge Facilities", Publication No. 87, 1970, Gentbrugge, Belgium.

The International Association of Scientific Hydrology, with the assistance of the Burgear Company (France), established an international inventory of installations for the artificial recharge of underground water-levels. The aims pursued are solutions to the problems of: (1) dry season pumping at higher rates than natural aquifer recharge capacities; (2) water storage allowing pumping rates greatly superior to natural ground water resources; (3) recharge of overpumped ground water reservoirs; (4) improvement of water quality by natural filtration; (5) fresh water barriers against saline intrusion or infiltrations from polluted rivers; and (6) flood control. Each country included in the recharge installation inventory conducts the process in various geologic lithologies ranging from diluvium gravels to basalts. This wide variety of lithologies effectively represents worldwide geohydrologic conditions. The origin of injected water, techniques of recharge, structures used for artificial recharge and pumping systems are outlined for each particular installation listed. The inventoried countries include: Germany (Western), Australia, England, France, Hungary, Iran, Israel, Jamaica, Japan, Morocco, Mexico, Holland, South Africa, Switzerland, Czechoslovakia and the United States of America.

International Center for Arid and Semi-Arid Land Studies, "Ogallala Aquifer Symposium", Special Report No. 39, 1970, Texas Tech University, Lubbock, Texas.

This publication is a collection of 20 papers presented at a 1970 symposium on the Ogallala aquifer. The formation furnishes agricultural, domestic and municipal water to a one-quarter million square-mile area, and is one of the nation's major water-bearing units. The aquifer extends from the Northern High Plains to central west Texas, where livestock, grain and fiber are produced by the management of the Ogallala. The purpose of this symposium was to provide a unified picture of the work accomplished and to define areas of additional research need on the geohydrological problems of the formation. Papers include significance, review, geology, ground water resources, waste disposal, well logging, natural recharge, linearity, drainage analysis, digital simulation, numerical modeling, dynamic modeling, specific-yield determination, estimation of coefficient of permeability, water transfer, assessments, herbicide movement, and artificial ground water recharge of the Ogallala aquifer.

Jenkins, C.T. and Hofstra, W.E., "Availability of Water for Artificial Recharge, Plains Groundwater Management District, Colorado", Groundwater Series Circular No. 13, 1970, Colorado Water Conservation Board, Denver, Colorado.

The Plains Groundwater Management District (Colorado) and the U.S. Geological Survey began a cooperative study in March 1969 to investigate the availability and distribution of surface water for artificial recharge to the Ogallala aquifer. The mean annual surface runoff out of the Plains District is estimated to be about 12,000 ac-ft per year, compared to ground water withdrawals of about 142,000 ac-ft (1968). The surface runoff amounts to about 0.13 in per year of a total of 16 in of precipitation. If impounding structures were located in or upgradient from the ground water withdrawal area, the amount of water available would be less than that available at the boundaries of the District. Extensive recharge operations therefore may not be feasible. Water in some of the stream channels contributes substantial recharge to the aquifer under present conditions. Runoff from about 20 percent of the area of the district is caught in playas and does not contribute to surface runoff out of the district. Some of this water recharges the aquifer but most of it is lost by evaporation. A significant amount of water probably could be salvaged by pumping water from playas or tailwater ponds into irrigation systems.

Jenkins, C.T. and Hofstra, W.E., "Problems of Artificially Recharging the Ogallala Formation in Colorado", The Ogallala Aquifer--A Symposium, International Center for Arid and Semi-arid Land Studies Special Report No. 39, 1970, Texas Tech University, Lubbock, Texas, pp. 154-164.

Two studies relating to artificial recharge were made by the U.S. Geological Survey in the Northern High Plains of Colorado in 1969. The studies consisted of an evaluation of water available in Kit Carson County, Colorado, and test drilling near Akron, Colorado. Less than 1 percent of the average annual precipitation, which is about 16 inches, leaves the area as surface runoff. This limited amount of water available for increasing recharge probably precludes the construction of large recharge projects. However, the study indicated that significant natural recharge occurs in some ephemeral stream channels. The test-drilling study showed that many poorly permeable lenses within the Ogallala formation inhibit recharge. Some water might be salvaged from playas either by pumping the water for reuse onto cropped land or by recharging it to the aquifer through wells.

Jones, O.R. and Schneider, A.D., "Groundwater Management on the Texas High Plains", Water Resources Bulletin, Vol. 8, No. 3, June 1972, pp. 516-522.

The effects of major water management practices on the pumping requirement from the Ogallala aquifer are discussed. Demand on the aquifer may be reduced as much as 15 percent by recycling irrigation runoff, 25 percent by recycling irrigation runoff and irrigating with water from playas, and 29 percent by recycling irrigation water in combination with irrigation from playas and artificial recharge of playa water to the aquifer. Other practices that can result in more efficient use of precipitation and ground water are limited irrigation, land farming, soil profile modification, and improved irrigation systems, thereby reducing the pumping demand on the Ogallala. Additional water supplies can be obtained by water harvesting, weather modification, and water importation. The overdraft on the aquifer can be reduced by the application of sound water management practices on an area-wide basis.

Koehler, J.H., "Artificial Recharge in the Northern Part of Chino Ground Water Basin, Upper Santa Ana Valley, California", Water Resources Investigations Report 82-4122, June 1983, U.S. Geological Survey, Menlo Park, California.

This study was made to help management design and implement a recharge-recapture system for State Water Project water in the Chino Basin. Nine test holes were drilled in the study area. Analyses of data from these test holes and drillers' logs of water wells indicate the presence of clay deposits. The clay deposits cannot be correlated between holes which indicates that they are in the form of discontinuous beds or lenses. The existence and location of two ground water barriers (Barrier "J" and Red Hill Barrier) have been postulated in previous reports. Water-level data indicate that Barrier "J" is probably not effectively stopping the movement of ground water. Data are insufficient to determine the effectiveness of the Red Hill barrier. Five existing recharge facilities in the study area were previously constructed to control floodflow. Infiltration tests were conducted at three of the facilities, and results of these tests indicate infiltration rates of 2.6 ft per day at Day Creek, 2.0 ft per day at East Etiwanda, and 1.3 ft per day at San Sevaine. A total of about 9,000 ac-ft of State Water Project water was recharged between June 1980 and July 1981. Rising water levels in wells indicate that recharge water is percolating down to the water table.

Lichtler, W.F., Stannard, D.I. and Kouma, E., "Artificial-Recharge Investigation Near Aurora, Nebraska, 2-Year Progress Report", Open File Report 79-1492, 1979, U.S. Geological Survey, Lincoln, Nebraska.

This report presents the results of the first 2 years of a 4-year investigation of potential for artificial recharge and recharge methods that might be used to mitigate excessive aquifer depletion in Nebraska. A quaternary sand and gravel aquifer near Aurora, Nebraska, was recharged by injecting water through a well at a rate of approximately 730 gpm for nearly 6 months. Total recharge was 530 ac-ft. Recharge was intermittent during the first 2 months, but was virtually continuous during the last 4 months. Buildup of the water level in the recharge well was 17 ft. The rate of buildup indicates that the well could have accepted water by gravity flow at more than 3,000 gpm for at least one year. The cause of a continuing slow rise in water levels in the recharge well in contrast to nearly stable water levels in observation wells as close as 10 ft from the recharge well is as yet uncertain. The recharge water and the native ground water appeared to be chemically compatible. Infiltration rates from 24-ft diameter surface impoundments ranged from 0.04 to 0.66 ft per day. The higher rates may have resulted in part from leakage down incompletely sealed holes that were drilled to install monitoring equipment.

Lichtler, W.F., Stannard, D.I. and Kouma, E., "Investigation of Artificial Recharge of Aquifers in Nebraska", Water Resources Investigations 80-93, 1980, U.S. Geological Survey, Lincoln, Nebraska.

Progressive declines of ground water levels in some areas of Nebraska prompted this investigation into the technical feasibility of recharging aquifers through wells, impoundments, pits, and canals. Information gained from a literature search and from preliminary tests was used to design several artificial recharge experiments in Nebraska from 1977 to 1979. In well experiments, 0.46 billion gallons of water from an aquifer recharged by the Platte River was transported by pipeline and injected through a well into a sand and gravel aquifer near Aurora. Recharge was at about 730 gpm during tests of 6- and 8-months duration. No evidence of clogging of the aquifer due to chemical reactions, air entrainment, or bacteria was detected in either test. In the 6-month test, evidence of clogging due to fine sediment in the recharge water was detected; however, analysis of this test indicated that recharge could have continued for several years before rehabilitation would have become necessary. Results of the 8-month test confirmed results of the earlier test until casing failure in the supply well and subsequent sediment deposition in the recharge well caused a rapid water-level rise in the recharge well. In surface spreading experiments, maximum infiltration rates from 24-ft diameter ring infiltrometers near Aurora and Tryon were 0.4 and 11 ft per day, respectively. Results indicate that large-scale surface spreading is feasible only where low permeability layers are absent in the subsurface. Infiltration rates from reuse pits ranged from 0.01 to 1.6 ft per day, indicating highly variable subsurface permeability. Flow measurements in an irrigation canal near Farwell indicated an infiltration rate of 0.37 ft per day.

Manbeck, D.M., "Making Water Available for Irrigation", Proceedings of the Irrigation Short Course, Jan. 24-25, 1977, 1977, Nebraska Center for Continuing Education, University of Nebraska, Lincoln, Nebraska, pp. 117-124.

Several areas in Nebraska are experiencing a decline in water tables caused by the increasing withdrawal of water for irrigation purposes. This drop in the levels of underground water is the primary reason for considering artificial recharge. Three methods of artificial recharge are considered: (1) surface application methods which involve the placement of water on the ground in order to increase the amount of water which can infiltrate into the ground and move to the water table; (2) placement of water into pits or shafts which penetrate permeable layers at shallow depth; and (3) recharge wells that allow water to move directly from the surface to the underground water bearing formations. Most recharge systems lose their capacity to recharge with time. The decreases in recharge rates are dependent on the type of method used; and the amount of suspended material, bacteria, entrained air and chemical quality of the water being recharged. Three field research projects on artificial recharge of Nebraska's ground water are currently in progress. One project involves studying the amount of water which moves downward beneath the reservoir of a flood-retarding structure. An associated project is measuring the amount of water infiltrating through the bottom of a stream. The third project is using recharge wells.

Marlette, R.R., "Artificial Recharge Through Injection Wells in a Sandstone Aquifer", Groundwater, Proceedings of General Assembly of Bern (Sept.-Oct. 1967), Publication No. 77, 1968, International Association of Scientific Hydrology, pp. 300-308.

Pump tests were made in an aquifer which is located within the city of Lincoln, Nebraska. Water level changes caused by withdrawal and recharge of single wells and of an entire wellfield were computed by using formulas for ground water flow to wells. Computed and measured water levels of pumped wells were compared. Well designs for the recharge system, operating procedures, and the use of tracers and piezometric levels to define the recharge dynamics are described. Prevention of saline water intrusion into the wellfield by using injected water is also discussed.

Meyer, W. and Patrick, L., "Effects of Artificial-Recharge Experiments at Ship Creek Alluvial Fan on Water Levels at Spring Acres Subdivision, Anchorage, Alaska", Open File Report 80-1284, 1980, U.S. Geological Survey, Tacoma, Washington.

The effect of the artificial recharge experiments on water levels at Spring Acres subdivision, Anchorage, Alaska, was evaluated using two digital models constructed to simulate ground water movement and water-level rises induced by the artificial recharge. The models predicted that the artificial recharge would have caused water levels in the aquifer immediately underlying Spring Acres subdivision to rise 0.2 ft from May 20 to August 7, 1975. The models also predicted a total rise in ground water levels of 1.1 ft at this location from July 16, 1973 to August 7, 1975, as a result of the artificial recharge experiments. Water level data collected from auger holes on March 1975 by a consulting firm for the contractor indicated a depth to water at 6-7 ft below the land surface at Spring Acres subvidision at this time. Water levels measured in and near Spring Acres subdivision several years before and after the 1973-75 artificial recharge experiments showed seasonal rises of 2 to 12.4 ft. A depth to water below the land surface of 2.6 ft was measured 600 ft from the subdivision in 1971, and in the subdivision in 1977. Average measured depth to water in the area was 7.0 ft from early 1976 to September 1979.

Mocek, M.J., Claborn, B.J. and Wells, D.M., "Conjunctive Use of Water in West Texas--Benefits to Noncooperators", Report WRC 71-6, Sept. 1971, Water Resources Center, Texas Tech University, Lubbock, Texas.

Utilization of the partially depleted Ogallala formation in the High Plains as a terminal underground storage reservoir for irrigation water in that area is described. The primary advantage would be reduction of the requirement for surface storage of water in west Texas. Water could be delivered to west Texas by the Trans-Texas Canal on an 11-month basis. From a small regulating reservoir, the water could then be delivered to west Texas irrigators on a 12-month uniform flow basis with only short interruptions for maintenance of the distribution system. When the farmers were not irrigating, they could recharge this delivered water into the underground formation for storage through their irrigation wells. During irrigation seasons the farmers could combine pumpage from their wells with the imported water concurrently being delivered to them for application to their crops. Utilization of the underground formation as a storage reservoir in this manner would eliminate the necessity for holding delivered irrigation water in large surface reservoirs until needed at the farms; consequently, evaporation losses would be greatly reduced.

Moreland, J.A., "Artificial Recharge in the Upper Santa Ana Valley, Southern California", Open File Report, Apr. 1972, U.S. Geological Survey, Menlo Park, California.

Artificial recharge has long been an integral part of water supply management in the upper Santa Ana Valley, California. About 7,500 acres of unlined channels and 2,590 acres in 60 off-channel recharge facilities are available for water spreading. To estimate potential infiltration rates for the recharge facilities, 63 single-ring infiltrometer tests were conducted at 22 recharge facilities. Measured infiltration rates ranged from about 0.1 ft per day at Patton Basin to 106 ft per day at Lynwood Basin. Grain-size distribution was determined on 25 soil samples collected at infiltrometer test sites. Regression analysis showed a direct relation between the logarithm of the infiltration rate and the logarithm of the 20th percentile particle diameter, corresponding to 20 percent finer on the grain-size graph. Regression analysis also showed an inverse relation between the logarithm of the infiltration rate and the logarithm of the sorting coefficient. Silt clogging greatly affects infiltration rates. Some clogging was observed in nearly all facilities, and significant reduction of infiltration rates was measured in seven basins. Estimated recharge rates range from less than 1 ft per day in areas underlain by older alluvium to 4 ft per day in areas underlain by coarse-grained younger alluvium.

Moreland, J.A., "Evaluation of Recharge Potential Near Indio, California", Water Resources Investigations 33-74, Feb. 1975, U.S. Geological Survey, Menlo Park, California.

The feasibility of utilizing sewage effluent to recharge the ground water in a 25-square mile area northwest of Indio, California, was evaluated. The depth to water in the area ranged from about 50 ft to more than 200 ft. Dissolved solids concentrations were greater than 500 mg/l in the shallow aquifers in the eastern and southern parts of the area. The permeability of the shallow sediments ranges from 15 to 50 ft per day. The eastern part of the area is underlain by fine-grained sediments that effectively separate the deep and shallow aquifers and would not be conducive to artificial recharge. In the western part of the area, sediments in the upper 100 ft are primarily sand. Considering all but economic factors, the most hydrologically favorable area for recharge is west of Washington Street and north of the Whitewater River. Using three spreading pits on a rotating basis and assuming long-term infiltration rates of 2 ft per day, an area of 33 acres would be required to infiltrate 8,000 ac-ft per year. At the recharge site, a water-level rise of about 36 ft due to recharge is expected.

Motts, W.S. et al., "Feasibility of Increasing Water Supplies and Preventing Environmental Damage by Artificial Recharge in Massachusetts", Publication No. 132, Jan. 1983, Water Resources Research Center, University of Massachusetts, Amherst, Massachusetts.

Successively more detailed first, second and third-level

approximation methods show that artificial recharge (AR) is a viable, cost effective and environmentally advantageous method for many localities in Massachusetts. First-level approximation (overview) methods indicate highly or potentially favorable AR conditions adjacent to 52.8 percent of the 650 municipal wells which occur in 126 (or 38.3 percent) of 329 state towns. Furthermore, 41.6 percent of the towns now without municipal wells could have future AR potential if ground water supplies are developed. The second-level approximation method, a basin-wide study of the Ipswich Basin, showed favorable AR conditions in the towns of Danvers, Middleton, and Ipswich. Detailed third-level approximation studies showed favorable AR conditions in Amherst, Holden, Pittsfield, and Fairhaven. All water development plans including AR should be designed to obtain a water supply with lowest economic and environmental cost. The following plan for a typical Massachusetts community proceeds through successively more costly alternatives: (1) conservation and/or analysis of distribution and storage systems, (2) ground water, (3) artificial recharge and water salvage, (4) development and treatment of ferromanganese water, and (5) as a last resort, more expensive alternatives including surface water reservoirs, water diversion, and tertiary treatment of contaminated sources.

Muir, K.S., "Seawater Intrusion, Ground-Water Pumpage, Ground-Water Yield, and Artificial Recharge of the Pajaro Valley Area, Santa Cruz and Monterey Counties, California", Water Resources Investigations 9-74, Oct. 1974, U.S. Geological Survey, Menlo Park, California.

Seawater intrusion is occurring in Pajaro Valley, California from several miles north to several miles south of the mouth of the Pajaro River and in a small area about 4 miles north of the river. The intrusion extends inland about 1 mile. Two water-bearing zones are being intruded--the depth intervals 100-200 ft and 300-600 ft. Ground water pumpage averaged 49,100 ac-ft per year for the 9-year period 1963-71. The long-term ground water yield of the Pajaro Valley is about 44,000 ac-ft per year. Artificial recharge can be effected through modified streambeds at filtration rates as high as 3 ft per day. Injection wells may have recharge capabilities of as much as 500 gpm.

Norvitch, R.F., Thomas, C.A. and Madison, R.J., "Artificial Recharge to the Snake Plain Aquifer in Idaho; An Evaluation of Potential and Effect", Water Information Bulletin No. 12, Aug. 1969, U.S. Geological Survey, Washington, D.C.

The major factors involved in using surplus water for artificial recharge of the Snake Plain aquifer in southern and southeastern Idaho are the availability of water, the effects of mixing on ground water quality and aquifer properties, and the effects of artificial recharge on ground water levels and discharge. About 5,100,000 ac-ft of water is recharged to the aquifer annually by irrigation. An average flow of about 1.3 million ac-ft or more occurs at a rate of once every 2 years. There probably will be no chemical-quality problems involved in recharge. A transient state, electric-analog model of the Snake Plain aquifer was constructed. Model predictions show that by recharging a total of 3.7 million ac-ft of water over a 10-yr period at 4 different places on the plain, water-level rises of less than 1 to more than 5 ft, measured 21 months after recharge stopped, would occur. If water were added at 62,000 ac-ft per month for 3 continuous months at each place, once every 2 yr, over a 10-yr period, 3.3 million ac-ft would go into storage in the aquifer and 0.4 million ac-ft would flow out of the springs.

Peterson, F.L. and Hargis, D.R., "Effect of Storm Runoff Disposal and Other Artificial Recharge to Hawaiian Ghyben-Herzberg Aquifers", OWRR-A-028-HI(1), Nov. 1971, Water Resources Research Center, University of Hawaii, Honolulu, Hawaii.

In 1970 the Kahului Development Company began construction of a collecting basin and four deep injection wells for the disposal of storm runoff from a residential development in Kahului, Maui. This presented a unique opportunity to evaluate the suitability of the site for artificial recharge and to study the possible effects recharge of storm runoff might have on the local ground water body, both from a water quality and a hydraulic standpoint. This report discusses such tests. The most serious potential water quality problem may be a reduction in injection efficiency owing to possible well clogging by heavy sediment loads. The general water quality effects of injecting storm runoff into the ground water body will be to decrease the dissolved solids concentration of the ground water in the vicinity of the wells.

Sasman, R.T., "The Future of Groundwater Resources in Dupage County", Ground Water, Vol. 12, No. 5, Sept.-Oct. 1974, pp. 277-282.

Dupage County, one of the major counties of the Chicago metropolitan area, obtains all of its water supply from ground water resources. In 1972, pumpage was 56.6 mgd, 18 percent more than the combined potential yield of shallow aquifers and the practical sustained yield of deep aquifers. Total pumpage from shallow aquifers almost equals the potential yield. In some areas, yields of shallow wells have declined drastically as a result of excessively heavy pumpage. Pumpage from deep aquifers is more than double the practical sustained yield. Demands for water are projected to increase more than 250 percent by the year 2020. Alternative methods of developing supplies to meet the anticipated demands include artificial recharge of shallow and/or deep aquifers, mining of deep aquifers, increased diversion of water from Lake Michigan, demineralizing Mt. Simon sandstone water, and importation of underdeveloped surface or ground water resources in north central and northwestern Illinois.

Sherwani, J.K., "Public Policy for the Management of Groundwater in the Coastal Plain of North Carolina", Report No. 158, July 1980, Water Resources Research Institute, University of North Carolina, Raleigh, North Carolina.

A capacity use area was established in the Coastal Plain of North Carolina for the purpose of management of ground water resources. However, no effective water management measures have yet been attempted. The need for management is emphasized by the fact that in some parts of the region, the aquifer system is reaching its production potential to permit open-pit mining for phosphate ore. The areas of large concentrated withdrawals and extensive lowering of the potentiometric surface have created conditions for possible water quality deterioration. The projections of water needs for municipal, industrial, and agricultural uses to the year 2000 are developed. As phosphate mining constitutes the largest use, a detailed analysis of alternative arrangements of dewatering wells is conducted. Supplemental irrigation requirements are considered in some detail. Ground water quality is of major importance in the optimal management of the aquifer system on a regional basis. The most important problems that are likely to result are: (1) upconing of brackish water, (2) lateral flow from lenses of poor quality water within the aquifer, and (3) downward movement of estuarine water. These are examined. The role of artificial recharge in aquifer management is explored. The value of water in alternate uses is estimated. The interdependence of water users and external costs imposed on others is considered. The need for the creation of a water district for the management of the aquifer system on a regional basis and the role of permits and user charges are outlined.

Sinclair, W.C., "Hydrogeologic Characteristics of the Surficial Aquifer in Northwest Hillsborough County, Florida", Information Circular No. 86, 1974, Florida Bureau of Geology, Tallahassee, Florida.

To study the feasibility of artificial recharge of the shallow aquifer in Hillsborough County, Florida, 59 holes were augered to the top of the limestone Floridan aquifer. Lithologic and gamma logs of the holes were used in conjunction with laboratory analyses of samples to define the hydrogeology of the unconsolidated deposits of the surficial aquifer overlying the limestone. The surficial aquifer comprises an upper fine sand unit which averages about 15 ft thick and a lower sequence of sandy clay and clayey sand layers which average about 25 ft thick. Median grain size, specific yield, and vertical permeability of the surficial aquifer decrease downward. The coefficient of vertical permeability of the sand is about 100 gpd per sq ft, but the coefficient of vertical permeability of the lower sand and clayey sand is much lower, ranging from 0.01 to 0.1 gpd per sq ft. A confining layer of dense clay underlies the surficial aquifer, separating it from the Floridan aquifer. The potentiometric surface in the surficial aquifer stands an average of 10 ft above that in the Floridan. Leakage of about 140,000 gpd per sq mi from the surficial aquifer to the Floridan occurs through the confining layer as well as through perforations in the confining layer.

Singer, J.A., "Geohydrology and Artificial-Recharge Potential of the Irvine Area, Orange County, California", Open File Report, Jan. 1973, U.S. Geological Survey, Menlo Park, California.

The Irvine area is in hydraulic continuity with the rest of the coastal plain in Orange County, California. Rapid facies change and the large percentage of silt and clay in the section locally result in confining conditions. The aquifer, most of which is included in the Fernando formation, is as much as 1,300 ft thick beneath parts of the plain. The alluvium overlying the Fernando formation averages about 200 ft in thickness and also contains significant amounts of silt and clay. Transmissivities range from 25,000 to 100,000 gpd per ft in the Irvine area, values which are much lower than those in the rest of the coastal plain in Orange County. Water levels have recovered as much as 60 ft from the low levels of the early 1950's. In the winter nonpumping season water tends to move toward upper Newport Bay and the rest of the coastal plain. During the summer pumping season a cone of depression develops, reversing the winter gradient. The average dissolved-solids content of the ground water is about 800 mg/l. The most prevalent cations are sodium and calcium; the most prevalent anions are bicarbonate and sulfate, no long-term degradation of water quality has occurred, with the exception of a slight increase in dissolved solids. No areas in the Irvine area are suitable for the large-scale spreading of water for artificial recharge. Clay and silt predominate in the section beneath the Tustin Plain, and in the foothill areas either bedrock is close to the surface or the alluvium is fine grained.

Sloan, R.F., "Ground Water Resource Management in Kansas", Kansas Water News, Vol. 22, No. 1 and 2, Winter 1979, pp. 2-6, 23.

The Kansas Ground Water Management District Act of 1972 encouraged the formation of local ground water management districts each with its own unique hydrogeologic characteristics. Local decision making would be supported by state guidance, technical expertise and financial assistance to obtain the optimum product of most benefits/least costs. A brief synopsis of five Kansas ground water management districts is given. Ground water quantity and quality problems have given rise to associated management strategies suitable for each district. This has resulted in stabilization of ground water levels or decelerating drawdown to acceptable rates, monitoring of salt water intrusion, irrigation scheduling for optimum ground water use, and augmenting available ground water supplies through weather modification and artificial recharge.

Smith, A., "Nebraska Studies Ways to Recharge Aquifer Storage", Johnson Drillers Journal, Vol. 49, No. 5, Sept.-Oct. 1977, pp. 1-3.

The importance of recharge studies to the national ground water situation is being demonstrated by current research in Nebraska. The problem of declining water levels in that state is so severe that the future of continued ground water use for irrigation in some regions depends largely on the results of injection well studies now being made in Hamilton County between Aurora and Grand Island. This project is being sponsored by the Old West Regional Commission and conducted by the Nebraska Water Resources Center and the U.S. Geological Survey. The experimental system consists of a withdrawal well located three miles west of an injection well; observation wells continuously monitor changes in aquifer levels as injection proceeds. In both the injection and the withdrawal wells, electric logs were valuable in accurately locating aquifer zones of best producing and recharge potential. The Nebraska study is only a part of a wide-ranging Old West Regional Commission artificial recharge project which includes Montana, North and South Dakota, and Wyoming as well.

Sophocleous, M., "The Declining Ground Water Resources of Alluvial Valleys: A Case Study", Ground Water, Vol. 19, No. 2, Mar.-Apr. 1981, pp. 214-226.

Present and future ground water withdrawals in the area of the Pawnee Valley, Kansas, were evaluated for their impact on water resources. The depletion of the water resource in the area is documented, and a preliminary hydrologic budget for the area is calculated. Also, a mathematical model is adapted and calibrated to simulate the operation of the hydrogeologic system in order to evaluate schemes for managing the ground water resource. The average natural recharge in the area was 0.5 in/yr, and the amount of appropriated ground water in the area exceeded the natural recharge figure by about 11 times. The simulations indicated that without any additional development or without very wet periods, water level declines will continue indefinitely. To reduce water depletion it will be necessary to reduce water waste, to impose a not more than 40 percent saturated-thickness depletion allowance for the next several years, to impose a freeze on the number of irrigation wells, and to engage in artificial recharge.

Speiker, A.M., "Future Development of the Ground-Water Resource in the Lower Great Miami River Valley, Ohio--Problems and Alternative Solutions", Professional Paper 605-D, 1968, U.S. Geological Survey, Washington, D.C.

The valley-train sand and gravel aquifers of the great Miami River Valley below Dayton, Ohio, can be expected to yield an initial withdrawal of 300 mgd--nearly three times their present rate of pumpage--provided that future developments are sufficiently distant from the present pumping centers. Most of the used ground water returned to the river as sewage could be recycled as induced recharge to the aquifers. This recycling would increase the total potential yield of the aquifers to much more than the initial withdrawal yield. The only limit on this recycling would be imposed by deterioration of quality of the used water or by the cost of treatment necessary to maintain adequate quality. Probably the greatest future problems will be declining ground water levels and local overdraft, both resulting from increased water use, ground water contamination, and the application of existing law to situations involving water rights. Overdraft can best be avoided by careful spacing of future ground water developments and location of these supplies in favorable hydrogeologic environments. Reuse of water by industry is an alternative to developing new supplies, for it can materially reduce water consumption. Ground water contamination, most of which is derived from water induced from streams by pumping, can best be controlled by maintaining water of good quality in the streams. Plans for management of the ground water resource should include careful consideration of the close interrelation of surface water and ground water.

Sutherland, J.C. and Bruce, R.M., "Surface Water-Esker Recharge Study East Lansing-Meridian Township Michigan", Water Resources Bulletin, Vol. 13, No. 6, Dec. 1977, pp. 1215-1225.

The East Lansing-Meridian Water and Sewer Authority studied a sand-gravel esker near the existing water treatment plant to determine its potential as an independent surface water supply. The nearby Red Cedar River also was investigated as a possible source of water for immediate treatment or for recharge of the esker. Although the bedrock aquifer (Grand River and Saginaw Formations) yields water adequate for the next 20 years, potential savings in treatment (hardness, iron) and pumping costs, estimated at $30,000 per year for a present demand of 5 mgd, are attractive incentives for a surface water-esker recharge program. Operation savings also would be realized by constructing additional bedrock wells in new areas. The river-esker-recharge and new wellfield alternatives were compared for cost effectiveness. Land costs make the recharge alternative more expensive. The land is undeveloped suburban property with potential for recreational use in conjunction with water supply. More places of outdoor retreat and aesthetics are needed in the Lansing Metropolitan area. A portion of the land costs would have to be borne by these or other interests for the river recharge scheme to be economically feasible.

Swain, L.A., "Predicted Water-Level and Water-Quality Effects of Artificial Recharge in the Upper Coachella Valley, California, Using a Finite-Element Digital Model", USGS/WRI-77-29, Apr. 1978, U.S. Geological Survey, Menlo Park, California.

From 1936 to 1974, water levels declined more than 100 ft in the Palm Springs area and 60 ft in the Palm Desert area of the upper Coachella Valley, California. Water from the Colorado River aqueduct is presently being recharged in the Whitewater River subbasin to help retard this trend of water level decline. Water quality in the basin varies from a sodium sulfate type in the northern three subbasins (Desert Hot Springs, Mission Creek, and Garnet Hill) to a calcium bicarbonate type in the southernmost Whitewater River subbasin. The purpose of the study was to: (1) analyze the present water quality conditions of the ground water basin; (2) determine the water level changes that will result from recharging the aquifer; and (3) determine the water quality changes that will result from recharging the aquifer. The study included: (1) collecting, evaluating, tabulating, and computer processing all available ground water quality data for the study area; (2) describing the areal distribution of water quality throughout the ground water basin; (3) updating the pumpage and other hydrologic data compiled for the analog model of the study area; (4) developing a digital hydraulic (flow) model utilizing the finite-element method; and (5) developing a digital solute-transport (water quality) model utilizing the finite-element method. This report presents the results of the 3-year study and the predictions of the possible effects of ground water recharge on the water resources of the upper Coachella Valley.

Trapp, H., Jr., Pascale, C.A. and Foster, J.B., "Water Resources of Okaloosa County and Adjacent Areas, Florida", Water Resources Investigations 77-9, Aug. 1977, U.S. Geological Survey, Tallahassee, Florida.

Okaloosa County, in the northwest Florida panhandle, uses the Floridan aquifer for water supply, although it also has abundant surface water and ground water in the surficial sand-and-gravel aquifer. Water levels have declined locally more than 90 ft in the upper limestone of the Floridan aquifer. The Floridan aquifer is overlain by the Pensacola clay confining bed, and the Bucatunna Clay subdivides it into two limestone units. Water in the upper limestone is generally of good quality. The lower limestone probably contains saline water. Average daily stream discharge is about 2,500 million gallons. Stream discharge does not diminish excessively during droughts, owing to high base runoff. Water levels in the Floridan aquifer will decline as long as pumping increases in the present areas of withdrawal. The decline could be alleviated by redistribution of pumping, artificial recharge, and the use of the sand-and-gravel aquifer or streams.

Vanlier, K.E., Wood, W.W. and Brunett, J.O., "Water-Supply Development and Management Alternatives for Clinton, Eaton, and Ingham Counties, Michigan", Water Supply Paper 1969, 1973, U.S. Geological Survey, Washington, D.C.

The water resources of the Tri-County Region, consisting of Clinton, Eaton, and Ingham Counties, an area of 1,697 square miles in Michigan's lower peninsula were investigated. The population is nearing 400,000 and by 1990 will be near 600,000. Average daily water use is slightly more than 30 million gallons in 1973; by 1980 it will be about 50 million gallons, and by 1990 it will probably be about 70 million gallons. The Tri-County Region is drained by seven river systems. Low flows in cubic feet per second are as follows: Grand River, 180; Maple River, 34; Looking Glass River, 28; Red Cedar River, 30; Portage Creek, 15; Battle Creek, 20; and Thornapple River, 24--a total of 331 cubic feet per second or about 220 million gpd. The principal source of ground water is a complex aquifer system composed of the Saginaw and Grand River formations and some of the overlying glacial sediments. This aquifer yields between 300 and 700 gpm to individual wells. In some parts of the region, the glacial sediments are favorable for development of moderate to large supplies of water. Some of the chief factors and methods that must be considered when planning water systems are combined use of ground and surface water, artificial recharge, treatment of wastes, use of storage reservoirs, and importation of water from the Great Lakes.

White, C.R., "Meeting Water Demands in the Chino-Riverside Area", Bulletin No. 104-3, May 1971, California State Department of Water Resources, Sacramento, California.

The California Chino-Riverside area occupies 641 sq mi throughout portions of Riverside, San Bernardino and Los Angeles Counties. The 1965 water demand for the area was approximately 450,000 ac-ft; it is estimated to reach 850,000 by 2015. To meet this projected demand an investigation of alternative supply sources was conducted. Nearly 12 million ac-ft of water is stored in ground water basins in the area, which is replenished at an annual average of 394,000 ac-ft. Current ground water extraction accounts for 70 percent of the demand with imported water from the Bunker Hill-San Timoteo Basin and the Colorado River covering 18 percent and 6 percent of the total. Four plans to meet future demand are considered. First, a maximum amount of imported water for artificial recharge while using ground water for demand. Plan 2 also maximizes imported water for direct delivery while reducing ground water pumping. Both plans would increase ground water storage by 2015. Plan 3 uses a maximum of ground water without depleting the last 50 ft. Plan 4 would stabilize existing water levels, supplementing pumping with imported water. Water sources, water quality, and costs were estimated for each plan. The least expensive alternatives are 1, 3, 4 and 2, respectively.

Wood, W.W., "Use of Laboratory Data to Predict Sulfate Sorption During Artificial Ground-Water Recharge", Ground Water, Vol. 16, No. 1, Jan.-Feb. 1978, pp. 22-31.

Sulfate sorption data determined from laboratory investigations were used to predict movement of sulfate during a field experiment of artificial ground water recharge in a basin near Lubbock, Texas. Laboratory studies confirmed that sulfate sorption is controlled in the system by surface area, content of iron oxides and hydroxides, and pH. Predicted sulfate distribution in the unsaturated zone was made by assuming constant one-dimensional flow. Where these assumptions were met, predictions were generally good at shallow depths and for short times. At greater depths and longer times, these assumptions combined with other factors, such as changes in effective porosity, variable mineralogy, changing hydrodynamic dispersion coefficients, and variable infiltration rates contributed to the failure of the predicted values to match the observed data. It appears that sulfate distribution in the unsaturated zone during artificial recharge can be predicted by using laboratory data if the flow conditions in the field can be described adequately.

MISCELLANEOUS

Agrawal, B.K., "Effect of Viscosity Ratio on the Recovery of Fresh Water Stored in Saline Aquifers", M.S. Thesis, Dec. 1975, Department of Petroleum Engineering, Louisiana State University, Baton Rouge, Louisiana.

Previous studies have shown the technical feasibility of storing fresh water in saline aquifers where the viscosities of injected and native fluids are essentially identical. This study was on the effect of viscosity difference on the extent of mixing between injected and native fluids in two artificially consolidated sandstones of differing thicknesses. When a lower viscosity fluid miscibly displaces a higher velocity fluid, it tends to finger into the higher viscosity fluid region. This causes an irregular boundary and larger interface between the two, with increased mixing. In a cyclic operation of injection/storage/production of a less viscous fluid in a medium containing a more viscous fluid, fingers during injection were found to retrace themselves during production. During a cyclic process the amount of mixing which occurs may be described in terms of an effective dispersivity which is a function of the viscosity ratio. This coefficient may be used in existing computer programs to predict the portion of uncontaminated injected fluid to be later recovered. This recovery efficiency is approximately the same no matter which fluid is the more viscous. The effective dispersivity may be used together with the arithmetic average viscosity to predict the performance of a cyclic process in which the fluids differ in density and viscosity. The results of this study are particularly applicable to the underground storage of heat in the form of hot water.

Banks, H.O., "National Water Commission Views Groundwater Protection", <u>Water and Sewage Works</u>, Vol. 124, No. 10, Oct. 1977, pp. 73-75.

Management of ground water resources and protection of ground water quality has received considerable attention from the National Water Commission. Studies conducted by the Commission on state water and ground water laws, management, and administration indicate that state ground water law is often inadequate to allocate the resource, and is not responsive to the problem of excessive use. The commission does not feel that Federal regulation is desirable, but that a new legal system should be instituted in some states to govern ground water. Laws authorizing the establishment of agencies to manage both surface and ground water should be enacted to coordinate withdrawals, control the rate of aquifer depletion, replenish aquifers through artificial recharge techniques, and protect them from pollution. Ground water pollution may be more significant than pollution of surface water, but often the same quality control expertise must be applied to both. Agencies responsible for surface water pollution controls should also be responsible for ground water abatement. Additionally, interstate aquifers should be managed and regulated in the same manner as intrastate aquifers. Finally, more technical data and other information is needed to make sound decisions regarding ground water regulation and management.

Bianchi, W.C. and Cehrs, D., "Ground Water Reservoir Management Through Artificial Recharge", Water Well Journal, Vol. 38, No. 2, Feb. 1984, pp. 43-46.

If ground water management through artificial recharge is to become a fact in the foreseeable future, the public, its elected officials and the water managers who are responsible to them, must come to a consensus on whether to establish ground water as a renewable resource or continue to exploit it through overdraft and/or pollution. Application of existing recharge technology is more easily justified when all the hidden costs are included in the value of the water placed in ground water storage.

Bourgeois, P.O., "Notes on Developing Ground Waters", Water and Pollution Control, Vol. 110, No. 10, Oct. 1972, pp. 24-28.

Advantages in developing ground water resources over surface waters for water supply and some methods for doing so are briefly discussed. Principles of artificial recharge, induced infiltration, and ground water recirculation are given. It is emphasized that well construction should be executed properly--and that when constructed by competent conscientious experts, a well should be a work of art and engineering. Analysis of geological conditions forms the first part of a ground water development program. Test drilling and pumping, to establish actual transmissivities, individual well potential yields, and complete well field yields, is then carried out. Horizontal wells and infiltration galleries are mentioned, to which the same general hydrologic and geophysical laws apply as to vertical wells. Ground water pollution is discussed, with detergent effluents from septic tanks singled out as a major pollutant. The value of record-keeping by the individual drilling or owning a well or well field is emphasized.

Brown, R.F. and Signor, D.C., "Artificial Recharge--State of the Art", Ground Water, Vol. 12, No. 3, May-June 1974, pp. 152-160.

The largest potential underground reservoir for the storage of potable water is in the unsaturated zone. Artificial recharge has many similarities to liquid-waste disposal through deep wells. In both, the problem is to place liquid in a permeable lithologic unit at an economic rate, to predict movement and the chemical reactions and physical changes that take place while the liquid is in the reservoir. In both artificial recharge and liquid-waste storage, the nature of the storage must be known, particularly that of the unsaturated zone. Water commonly is recharged by surface spreading through basins or by induced recharge from adjacent streams and lakes or through injection wells. Research in recharge through basins has been dominated by mathematical models based on idealized conditions and empirical relations, derived by experimental sequencing of recharge operations, and operational controls in the pretreatment of recharge water. Recharge by injection wells has been undertaken in a variety of hydrologic environments. In Israel, efforts have been directed toward the analyses of diffusion and dispersion of the injected water. Much research in the United States has been directed toward the movement of bacteria and organic matter through an aquifer and toward the chemical modeling of changes in recharged water as it moves.

Brown, R.F. and Signor, D.C., "Groundwater Recharge", Water Resources Bulletin, Vol. 8, No. 1, Feb. 1972, pp. 132-149.

An evaluation of current knowledge of processes and problems of artificial recharge indicates that a great deal of additional research is necessary before recharge feasibility can be evaluated in most situations. Experience in using recharge wells on the southern High Plains of New Mexico and Texas yields predictions of success when using good quality water, and failure when recharge water contains high concentrations of particulate matter. Surface spreading is a more suitable method when water has a high sediment content, but may not be feasible in some hydrogeologic situations. Theoretically, well construction should be important to the success of injection recharge operations, but little experimental work exists to support this view. Results of recharge experiments on the southern High Plains are reviewed.

Gelhar, L.W., "The Aqueous Underground", Technology Review, Vol. 74, No. 5, Mar.-Apr. 1972, pp. 45-53.

Problems associated with the development of ground water resources are reviewed. As demands increase on underground sources, the need for fundamental knowledge as well as understanding of many local geologic details becomes apparent. With the gradual development of water-reclamation schemes based on the recharge of aquifers, the water quality aspects of artificial recharge are becoming important. Although ground water is generally protected from gross pollution, low-level contamination which may present a hazard in drinking water is now common. The contamination increases gradually, reflecting the slow movement in the aquifer, and is thus less dramatic than surface water pollution. Thus it is more insidious--and potentially just as serious. Future developments and management of the system will require, as a basis, information on the long-term response of the aquifer system to possible hydrologic changes. An effective program to control ground water pollution might comprise measures ranging from public information programs to total bans on certain hazardous materials and might include strict controls on subsurface disposal practices and land use.

Gillette, F.L., Jr., "Proposed Legislation for Artificial Groundwater Recharge", Environmental Protection Agency Report, July 1972, College of Law, University of Florida, Gainesville, Florida.

Much research has been conducted in the area of artificial ground water recharge. The present scientific emphasis is placed on the total systems approach of recycling and reclaiming as much waste production as possible, especially water resources. Suggested legislation is presented, which is general in nature with details to be worked out by regulation. This article is divided into four parts: technological considerations, legal considerations, present legislation, and proposed legislation. Part one evaluates the technological complexities involved and lays the foundation on which the proposed legislation is predicated. Part two elaborates the legal problems involved and stresses the various possible liabilities incurred by anyone who attempts to use the process of artificial ground water recharge. Parts three and four review current and proposed legislation.

Pettyjohn, W.A., "Artificial Recharge--A Potential Solution for Many Water Problems", Water Well Journal, Vol. 38, No. 9, Sept. 1974, pp. 48-51.

Artificial recharge is a means of augmenting the natural infiltration of surface water into a ground water reservoir by wells, water spreading, or altering natural conditions. Water commonly used for artificial recharge includes natural stream flow, high flows diverted from a nearby water course, cooling waters, diverted storm drainage water, or sludge effluent (treated or untreated). Increases in production of ground water supplies, and water quality control (through prevention of saline water encroachment or through addition of higher quality water to contaminated water) are uses of artificial recharge. Several examples of artificial recharge programs, including the program instituted in Minot, North Dakota in 1964 and the recharge system of Dayton, Ohio, illustrate the advantages of artificial recharge.

Smith, R.E. and Parlange, J.Y., "Percolation Rates with Surface Layers of Gravel or Clay Flocs on Soil Fed Clear and Turbid Waters", Water Resources Research, Vol. 14, No. 3, June 1978, pp. 539-541.

Suspended solids in water can plug the surfaces of aquifer recharge basins, and they are often chemically flocculated and removed in costly separate floc-settling basins. This work evaluated floc and gravel filters on recharge surfaces as substitutes for settling basins. Disturbed soil in permeameters were treated with layers of small gravel and constant feedwater heads of either clear tapwater, untreated turbid runoff water, or turbid runoff water treated to form and retain flocs. Untreated turbid water reduced percolation rates more than the other feedwaters, and gravel layers markedly reduced percolation of all feedwaters. The percolation rate of water with flocs was only slightly less than that of tapwater. These results suggested that untreated turbid water or gravel layers are unacceptable for recharge surfaces. However, combinations of soil, gravel, and clay may have considerable potential for reducing percolation in water retention basins. Flocculation and settling seldom remove all fine particles from turbid water; thus a filter layer of flocs may extend the effective life of a recharge surface and eliminate costly separate floc-settling basins.

van Hylckama, T.E.A., "Changes in Vegetation Diversity Caused by Artificial Recharge", Vegetatio, Vol. 39, No. 1, 1979, pp. 53-57.

Efforts to increase the rate of artificial recharge through basins often necessitates scraping and ditching before and during operations. Such operations can result in more or less drastic changes in vegetation (depending on what was there before), characterized by diminished numbers of species and lowered diversity. Two examples, one from the Netherlands and one from Texas, are presented showing how similar treatments cause similar changes in two completely different plant societies.

Walenta, T.R., "Some Legal Problems Concerning Ground Water Management", Symposium of the American Water Resources Association, Proceedings Series No. 4, Nov. 1967, San Francisco, California, pp. 39-48.

The many problems connected with water and related land resources have been generated by an expanding population, with its consequent needs for water far beyond any before contemplated. Some of these problems include: (1) the need to classify all ground waters as public waters subject to regulation and control by the state; (2) determination of critical ground water areas; (3) protection of the prior appropriator in extracting ground water from the basin or aquifer; (4) artificial recharge of ground water areas and their use for storage purposes; (5) control and abatement of ground water pollution and the need for water quality programs; and (6) cooperation between state and Federal agencies in developing all water resources, a new form of creative Federalism.

Weatherford, G., Malcolm, K. and Andrews, B., "California Groundwater Management: The Sacred and the Profane", Natural Resources Journal, Vol. 22, No. 4, Oct. 1982, pp. 1031-1043.

California pumps more ground water than any other state in the union, yet it lacks a coordinated statewide management program for the resource. California relies on an annual developed water supply of 42 million acre feet (maf), of which about 15 maf is pumped ground water. Misuse of ground water has caused problems, including threatened depletion in some areas, seawater intrusion, contamination by toxic compounds, and land subsidence. There is no coherent or comprehensive ground water law or policy. Ground water rights are defined by court-made law, and the regulation of ground water occurs selectively in some water districts. Ground water rights law involves correlative rights, appropriative rights, mutual prescription, and notions of equitable apportionment. Recharging of aquifers with imported water has been one way California has dealt with water supply problems.

INTERNATIONAL EXAMPLES

Agarwal, A., "The Rising Cost of Making Deserts Bloom", New Scientist, Vol. 76, No. 1073, Oct. 1977, pp. 96-97.

Israel has adopted extensive programs to utilize marginal water resources such as storm runoff, brackish water, and treated waste water for irrigation and for artificial ground water recharge. The Dan Region Sewage Reclamation Project, designed to treat 130 mcm of sewage from the greater Tel Aviv area, is scheduled for completion in 1985. Treated sewage has also been used to prevent seawater intrusion into fresh ground water. One of the major goals of the Israeli Water Program is to keep water consumption constant without limiting agricultural growth. Water saving practices in use include drip and sprinkler irrigation, electronically controlled irrigation systems, and irrigation with brackish water. Since Israel's water system now operates as a closed cycle, salt and other dissolved materials such as pesticides, fertilizer nitrates, detergents, and minerals are not regularly flushed from the aquifer. Desalination plants are being considered for aquifer purification.

Birtles, A.B. and Nutbrown, D.A., "Some Water Resource System Studies in the UK", Application of Mathematical Models in Hydrology and Water Resources Systems, Proceedings of the Bratislava Symposium, Czechoslovakia, Sept. 1975, IAHS-AISH Publication No. 115, 1975, International Association of Hydrological Sciences, pp. 28-35.

Various case studies which utilize systems analysis for solving water resource system problems are presented. Aspects of these studies include problems associated with artificial recharge, the combined use of ground and surface waters, and saline intrusion. Particular emphasis is placed on the interaction of the ground and surface water components. Aquifers typified by both fissure and intergranular flow are involved in the studies. Described is the use of modeling techniques in the optimization of the design and operation of such systems. Considered is a digital model constructed to describe the transient effects of natural infiltration and abstraction on the movement of ground water, to be used as a tool to assist with the long-term management of the chalk aquifer of the South Downs, United Kingdom, the primary aim being to estimate the maximum yield of the aquifer. Described next is a digital ground water model used in the design of a river regulation scheme in which ground water abstracted from the Triassic sandstones aquifer in the Vale of York, England, is used in combination with the Yorkshire House to provide a safe regulated field.

Bourguet, L., "International Survey on Management of Artificial Recharge--Analysis and Synthesis of Responses (Inventaire International Des Amenagements D'Alimentation Artificielle - Depouillement Et Synthese Des Responses)", International Association of Scientific Hydrology Bulletin, Vol. 16, No. 3, Sept. 1971, pp. 51-102.

An analysis is presented of the 121 answers to an international survey of ground water recharge. Aims, benefits, types of recharge, aquifers, and size of installations are discussed. Correlations and regressions were found relating permeability, size of equipment, and injection rate, and relating recharge to capital investment.

Chaturvedi, M.C. and Srivastava, V.K., "Induced Groundwater Recharge in the Ganges Basin", Water Resources Research, Vol. 15, No. 5, Oct. 1979, pp. 1156-1166.

The Ganges Basin in India has a serious problem of water availability. The basin area, which is only 1/12 that of the United States, has a population greater than the total U.S. population, and it is increasing at a rate of 2.5 percent per annum. About 77 percent of the population is engaged in agriculture, which is totally dependent on irrigation. Almost 85 percent of rainfall takes place in 2-3 monsoon months. The surface storage possibility is extremely limited, but ground water recharge appears feasible, since sedimentary alluvial formations extend to depths of thousands of meters. Three alternative schemes of ground water recharge were proposed. One involves pumping heavily along perennial rivers prior to monsoons so as to lower ground waters to promote induced ground water recharge. The second proposes a similar approach along nonperennial rivers. The third involves irrigation during monsoons with ground water lowered adequately in the nonmonsoon period so that enough induced ground water recharge takes place to provide adequate supplies for nonmonsoon months. A simulation-optimization model was developed to study the surface flow-ground water interaction and was applied to study comparative cost effectiveness of the three alternate approaches. Sensitivity analysis was also carried out. It was shown that the third scheme is the most attractive.

Chernov, A.S., "Groundwater Regime in Zones of Operating Artificial Recharge Systems", Water Resources (English Translation), Vol. 8, No. 6, Nov.-Dec. 1981, pp. 637-642.

Artificial ground water recharge has been carried out by means of infiltration basins in the caucasus since 1958. Regime observations were organized in 1977 to study the natural ground water regime ground water and its transformation in the area of the intake works under the effect of artificial recharge and long operation of the intake, to analyze the regime of surface waters, to refine the adopted technological scheme of artificial ground water recharge, to refine the optimal operating regime of the infiltration works, and to analyze the experience of artificial formation of ground waters under conditions of mountain-type river valleys. The study identified two types of ground water regimes--natural and disturbed. The main factor forming the natural regime of near-channel ground waters is the river hydrology. The disturbed regime is mostly affected by operation of the intake and artificial infiltration from the basins. Meteorological factors of precipitation and evaporation play only a secondary role except during basin cleaning. An important characteristic of the regime of ground water level in the zone studied is its highly dynamic character and the considerable range of variations on comparatively small areas near the basins and river. It is advisable to regenerate the infiltration basins twice a year in the fall and spring, when the necessary capacity of the intakes can be provided due to seepage from the river without artificial recharge. The mineral content of the ground waters was slightly higher than that of surface waters and was mostly in the form of calcium carbonate. Deterioration of the physical-chemical and bacteriological properties of the ground waters was not observed.

Dilyunas, I.P., "Experience in Artificial Ground Water Recharge for Water Supply in Lithuania", Soviet Hydrology: Selected Papers, Vol. 15, No. 4, 1976, pp. 340-343.

Lithuania has large ground water resources because of its location in the temperate climatic belt with ample precipitation (500-850 mm/yr), its dense drainage network (averaging 0.4 km/sq km), and the widespread occurrence of sandy-gravelly mantle deposits in the quaternary glacial stratum, which is 100-150 m or more thick. Systematic investigations of ground water recharge methods for water supply have been conducted since 1965. An alluvial aquifer 9-20 m thick in sandy-gravelly deposits is exploited in the area of the Eygulyays water intake. The ground waters are recharged artificially from two infiltration basins with infiltration areas of 4,250 and 3,790 sq m at a filling depth of 3.2-3.4 m. Their beds consist of coarse-grained sand underlain by gravelly-pebbly deposits. Infiltration occurs without backwater: the depth from the bottom of the basins to the ground water remains at 6-6.5 m throughout the infiltration cycle. The percolation rate during the infiltration cycle varies from 2.8 to 0.05 m/day, and averages 0.35 m/day. Artificial ground water recharge is justified in all the cases examined. This is a fairly efficient water supply method under the conditions prevailing in Lithuania, and it should be used to replenish the exploitable ground water reserves.

Edworthy, K.J. and Downing, R.A., "Artificial Groundwater Recharge and Its Relevance in Britain", Journal of the Institution of Water Engineers and Scientists, Vol. 33, No. 2, Mar. 1979, pp. 151-172.

The design criteria for artificial recharge works were reviewed, and the cost of incorporating recharge in water resource schemes in Britain was discussed. The conclusions were: (1) artificial recharge experiments carried out in Britain over the last 10 years have confirmed that: (a) the technique is applicable in the main aquifers, (b) average recharge rates of between 0.3 and 0.5 m/d can be anticipated where lagoons are used to recharge aquifers such as the Permo-triassic sandstones and the lower Greensand--in the case of wells, recharge rates similar to abstraction rates are likely, and (c) the quality of a water is improved by recharge through an unfissured aquifer; (2) studies with mathematical models have suggested that recharge of the chalk where it crops out and is drained by perennial streams, should be feasible in many circumstances; (3) it has been shown theoretically that even a limited amount of recharge can make a significant difference to minimum storage levels in coastal aquifers; (4) development of artificial recharge on a regional scale would in many cases cost between 8 p and 10 p/cu m; (5) maximum use of aquifer storage requires artificial recharge; and (6) artificial recharge is often introduced into water resource systems for reasons of water quality. The quality of the supply to consumers can be manipulated to maintain acceptable standards, even where the source of supply is periodically susceptible to contamination. The natural purification afforded by aquifers, the long residence time, and the mixing of different waters in aquifers all act as safeguards for supply.

Edworthy, K.J., Stott, D.A. and Wilkinson, W.B., "Research into the Effects of Artificial Groundwater Recharge, Lea Valley, London, England", Water Resources Bulletin, Vol. 14, No. 3, June 1978, pp. 554-575.

Large-scale ground water abstraction from the Cretaceous Chalk/Lower Tertiary basal sands aquifer system of the London Basin in the last 150 years has developed storage of more than 1,000,000 cu m. Limited operational recharge was undertaken in the 1950's encouraging further detailed study of the wider possibilities. Following a comprehensive hydrogeological reappraisal, an economic and engineering study, and pilot-scale experiments, the Lea Valley was shown to have the greatest potential for recharge. Artificial recharge into the Chalk at a rate of 9,000 cu m/d appears possible, through acidized 900 mm diameter boreholes. An understanding of the degree of interconnection between the Chalk and Basal sands and of the causes of changes in quality of recharged water during storage was shown to be of particular importance to the successful operation of any recharge scheme using this system. A two-layer numerical ground water model of the 800 sq km area and a surface/ground water simulation model was used to assist with the design of the 84,000 cu m/d prototype scheme, and it also will help in assessing its efficiency. The models ultimately will be used to manage the operation of the first stage development now coming into use.

Ghassemi, F., "Mathematical Model Application in Ground Water Studies of Iran", Ground Water, Vol. 17, No. 4, July-Aug. 1979, pp. 359-365.

Mathematical models of ground water aquifers with complex geologic and hydrologic factors have been used effectively since 1960. In these models, first the past hydrogeological conditions are simulated, then it is possible to impose on the model the future conditions by increasing redistributed pumping, artificial recharging, or many other factors which may change the existing hydrogeological conditions. Different numerical methods, such as finite difference, finite element, and dynamic programming, are used in mathematical simulation of the aquifers. The study of different aquifers in Iran by classical methods has been started systematically since 1962, and during this period more than 200 aquifers were studied in reconnaissance phases and about 80 aquifers in semi-detailed phases. In 1967, the mathematical model technique was introduced to ground water researchers of Iran. Actually, 39 aquifers were studied by 2- or 3-dimensional flow models covering nearly 40,000 sq km of Iranian aquifers. In these studies finite difference methods were used, and only a few aquifers were studied by dynamic programming, but the finite element method has never been used in Iran.

Gupta, S.K., "Wells Recharged by Gravity in Ahmedabad", Johnson Drillers Journal, Vol. 49, No. 6, Nov.-Dec. 1977, pp. 4-5.

An unusual system for artificial recharge to utilize water available from the Sabarmati River during periods of excess flow has been put into practice for the city of Ahmedabad, India. In effect, a siphon is used to take water from the water table aquifer on the bank of the river and deliver the water into an injection well penetrating nearby artesian strata which furnish 90 percent of the city's water supply. The recharge water is obtained from a relatively shallow well about 21 meters deep. The siphon arrangement conducts water to the 238 m recharge well by gravity flow without the use of a pump. The water level in the supply well is of such height to permit the siphon to work only when the river carries excess water during the monsoon rains which occur from June to October. The system is metered and utilizes a control valve. Recharge was begun on May 20, 1977, and has continued uninterrupted without the aid of any external energy source.

Hunt, B., "An Analysis of the Groundwater Resources of Tongatapu Island, Kingdom of Tonga", Journal of Hydrology, Vol. 40, No. 1/2, Jan. 1979, pp. 185-196.

An analysis was made of the ground water resources of Tongatapu Island, the Ghyben-Herzberg approximation was used to estimate thicknesses of a fresh water lens floating on seawater. Finite-difference calculations were used to estimate rainfall recharge rates, and calculations were made to investigate the dispersion of chloride ions across the freshwater-salt-water interface. These calculations suggest that artificial recharge might be a useful device to control chloride concentrations in the freshwater aquifer.

Izatt, D., Fox, G.B. and Tague, M., "Lagoon Recharge of the Folkestone Beds at Hardham, Sussex--1972-75", Journal of the Institution of Water Engineers and Scientists, Vol. 33, No. 3, May 1979, pp. 217-236.

> This paper summarizes experiments carried out at an experimental lagoon site at Hardham Church Farm, 1 km south of Pulborough, Sussex during the period August 1972 to March 1975. The experiments involved artificial recharge of the Folkestone Beds aquifer with raw River Arun water. Recharge rates ranged from 2.1 m/day to 0.8 m/day per unit bed area, reflecting the complexity of lagoon recharge behavior and characteristics experienced. The cycles lasted up to 110 days, depending on the clogging of the lagoon beds. The main clogging agent was suspended solids present in the river water (coarse to medium silt), although algal growth in the lagoons made a significant contribution. Chemical clogging was insignificant, severe clogging was limited to the top few centimeters of the beds, although some sediment penetrated to at least 15 to 20 cm. Other factors influencing lagoon recharge behavior were temperature of the recharge water, growth of the recharge mound, mutual interference between the lagoons, and bed-cleaning activities. Few changes in the hydrochemistry of the recharged water occurred, namely, a slight reduction in pH, total hardness, alkalinity and calcium, and a marked reduction in dissolved iron and manganese. The lagoon beds were cleaned by manual scraping, drag-line, wheeled shovel loader, and tracked shovel loader.

James, S.E. and Henry, J.L., "Effects of Siltation on Rates of Recharge", Technical Paper No. 23, 1977, Australian Water Resources Council, Queensland Irrigation and Water Supply Commission, Brisbane, Australia.

> The eastern part of the Burdekin-Haughton Delta on the coast of Queensland, Australia, is serviced by an artificial ground water recharge scheme involving the diversion of flow from the Burdekin River through a system of channels, trenches and pits. The major operating problem of the scheme is a reduction in infiltration rates, thought to be associated with clogging of sediments by material suspended in the recharge water. A description of the region, its climate, geology, hydrology, water usage and recharge works is provided, together with an account of a data collection study of the composition of the recharge water and associated effects on infiltration rates.

Joseph, J.B., "Artificial Groundwater Recharge", Water Services, Vol. 85, No. 1024, June 1981, pp. 303-304, 307-308.

> Artificial ground water recharge is used in Britain for the disposal and reclamation of wastewater. Surface spreading or injection through wells and boreholes can be used. A typical surface spreading scheme capable of recharging 45 M liters/day will require 1.5 to 2 ha of land. While boreholes require less land, they require large scale drilling equipment and complex technology and operation. Chemical as well as physical problems can occur during recharge. Reaction can lead to a deterioration in water quality and clogging or collapse of the aquifer. One advantage of artificial recharge is that a system for resource augmentation can be constructed in phases whose financing and size can be adjusted to meet changes in the demand growth pattern. For example, phase I of the Lea Valley scheme has a yield of 60 M liters/day and was

designed and constructed in three years. Water is available for recharge at a marginal cost, typically 25 British Pounds Sterling/M liter. The standard cost for treated water is about 130 British Pounds Sterling. Artificial recharge can also be used for the prevention or reversal of saline intrusion.

Karuk, B.P. and Aliferov, A.N., "Use of Natural Underground Reservoirs for Control of Surface Runoff", Water Resources (English Translation), Vol. 9, No. 2, Mar.-Apr. 1982, pp. 169-171.

Various studies have been conducted in the USSR to examine the possibility of artificial replenishment of ground waters and creation of underground reservoirs in a number of regions in the Ukraine. The geological and hydrogeological conditions of the Crimean Plains permit creation of large underground reservoirs with a total volume of about 1.5-2.0 cubic km. All water surpluses of the North Crimean Canal can be fed into the underground horizons composed of porous, fractured limestones. Depending on the water quality in the source and requirements imposed by various consumers on water from underground reservoirs, two approaches to the problem of conditioning water being fed into the underground horizons are possible. First, all water is purified to the required degree of quality. Secondly, water is fed directly without pretreatment or with medium treatment to remove suspended material in settling ponds or in the rock stratum. This is possible under the following conditions: the water from the underground reservoirs is used mainly for irrigation or industrial water supply; there is no threat of polluting adjacent aquifers which are the source of domestic potable water supply; and there is no threat of clogging and additional pollution of the enclosing rocks. With consideration of imminent water management problems in a number of regions of the country it is expedient to accelerate a study of the feasibility and creation of underground reservoirs.

Perlina, A.M. and Berdanov, V.M., "Modification of Water Quality During Artificial Groundwater Recharge", Soviet Hydrology: Selected Papers, No. 4, 1971, pp. 391-396.

Artificial recharge of ground water by surface water from rivers, lakes, and reservoirs is being widely practiced in a number of countries to help meet domestic water demands. In the USSR, artificially recharged ground water is being used in water supply systems of Riga, Kaunas, Novokuznetsk, Ivano-Frankovsk and other cities. The quality of water supplied to filtration plants should meet certain requirements. The suspended-particle concentration should not exceed 10-20 mg/l, depending on particle size of the soil filter. Chromaticity, depending on its components, should not exceed 60 deg. The dissolved oxygen content of water, which governs permissible permanganate oxidation, should be within limits of 15 mg/l.

Piet, G.J. and Zoeteman, B.C.J., "Organic Water Quality Changes During Sand Bank and Dune Filtration of Surface Waters in the Netherlands", Journal of the American Water Works Association, Vol. 72, No. 7, July 1980, pp. 400-404.

Artificial recharge of polluted water began years ago in the Netherlands before any data concerning the behavior of organic chemicals in the soil were available. The current application of advanced analytical techniques is shedding new light on the potential of infiltration processes to remove chemical compounds from polluted raw waters and on the possible dangers of soil contamination in the recharge area. With these new insights, the reactions of organic chemicals--organic halogen compounds, organic nitrogen compounds, and odor-producing compounds--during the passage of water through the soil have gained not only a scientific but also a political impact.

Sychev, K.I., Khordikainen, M.A. and Usenko, V.S., "The Main Directions of Hydrogeological Investigations for the Substantiation of Artificial Recharge", Water Management and Development, Proceedings of the United Nations Water Conference; Mar Del Plata, Argentina, Mar. 1977, Pergamon Press, New York, New York.

Artificial recharge of ground water and storage of surface runoff in natural subsurface reservoirs are the most effective methods of preventing ground water depletion and of insuring their availability to the water supply. Such measures are used in the Soviet Union for: (1) increasing operating water intake yields, (2) regulating water resources, (3) improving water quality, (4) cooling water heated in air-conditioning systems, (5) forming fresh water lenses among saline ones, and (6) protecting aquifers from saline water intrusions and effluent contamination. Increasing ground water quantity has helped purify polluted surface water. Artificial recharge is usually more economical than other water supply methods, and is 20-30 percent cheaper than surface water purification. The Soviet Union emphasizes: bed regulation of surface runoff; improving permeability of bottom deposits; induced seepage; and to a limited extent construction of special systems such as recharge basins, canals, infiltration galleries, ponds, temporary basins, and clearing of river bottom deposits. A program of research and development was carried out in 1971-75. Regional investigations, including mapping, identified the most suitable areas for recharge projects, and a substantiating methodology was devised which assesses: (1) water quantity and quality; (2) hydrogeological, sanitary, and technological conditions; (3) method selection; (4) recharge structure operation; (5) effects of recharge on ground water quality; (6) water intake efficiency; and (7) economic effectiveness.

Webb, S.N. and Watson, K.K., "Analysis of the Movement of Water from Recharge Channels and Pits", Technical Paper No. 24, 1977, Australian Water Resources Council, New South Wales University, Kensington, Australia.

As part of a series of studies of the operation and problems of the artificial ground water recharge scheme on the Burdekin-Haughton Delta of Queensland, Australia, an analysis was undertaken to elucidate the patterns of movement of water from the recharge channels and pits. The results are presented in two parts: numerical analyses, with theoretical background and outlines of the computer programs; and field studies, with details of data collection and analysis. The numerical analyses included two dimensional unsteady flow and steady state studies, and a one dimensional simulation of surface clogging due to suspended material in the recharge water. Field studies investigated diurnal recharge variations, and pressure, water content, and temperature profile development patterns, for one of the pits.

Williams, D.E., "The Dashte-Naz Ground Water Barrier and Recharge Project", Ground Water, Vol. 15, No. 1, Jan.-Feb. 1977, pp. 23-31.

The danger from salination poses a significant threat to development of ground water resources for domestic or irrigation use in near-coastal environments throughout the world. The Dashte-Naz farm near the Caspian Sea in northern Iran is no exception. Large-scale ground water development in this region has been limited to wells south of the Caspian Sea near the base of the Alborz Mountains where the sediments are coarse and natural recharge high. In the northern regions, however, exploitation of the ground water resources for agriculture and livestock purposes has proven fruitless, with most of the high-yielding wells turning saline after only several irrigation pumping seasons. The source of the salination comes not from encroachment of Caspian Sea water, but rather from sediments saturated with connate or fossil waters. A wide zone of dispersion separates the salinated aquifers from the freshwater aquifers nearer the mountains. The Dashte-Naz ground water barrier and recharge project was conceived for purposes of overcoming the problem of exploitation of fresh ground water reserved in the vicinity of saline water fronts. The project presently under construction is a pilot research project on the practicality of using an injection well barrier for storing water underground in areas immediately adjacent to saline ground water reservoirs. The results from this project will be used to design other water supply projects in Iran not only along the Caspian Sea coast, but also along the Persian Gulf.

Zoeteman, B.C.J., Hrubec, I.J. and Brinkmann, F.J.J., "The Veluwe Artificial Recharge Plan Water Quality Aspects", Journal of the Institution of Water Engineers and Scientists, Vol. 30, No. 3, May 1976, pp. 132-137.

Water quality aspects were discussed considering two types of artificial recharge systems: (1) underground tunnels; and (2) open canals or ponds. Both systems were studied in a pilot plant located at Leiduin, the dune area near Amsterdam, The Netherlands. Effects of different grades of water treatment on performance of both types of recharge systems were determined. A pretreatment system was proposed that consists of at least coagulation, sedimentation, ozonation, secondary coagulation, biological filtration, breakpoint chlorination, and activated carbon filtration. In this way it is expected that high recharge rates of approximately 10 m/day and long recharge runs of several years in case of exclusion of light can be realized. As the quality of the water to be recharged is comparable to that of drinking water, contamination of the soil and pollution of local water resources by eventual losses of recharged water will be very limited. Those substances that are accumulated in the soil are concentrated mostly in the top-layer of the recharge medium, and practically all will be removed during cleaning of mechanically clogged sand layers. Underground storage of water gives different advantages over open storage in reservoirs; examples are reduction of temperature fluctuations, exclusion of light, limited claim on space, and better protection of the stored water, e.g., against pollution from the atmosphere. The advantages make artificial recharge a preferable water storage technique to open reservoirs in case a free choice between the two techniques can be made.

APPENDIX B

LIST OF COUNTIES BY STATE AND AGGREGATED SUB-AREAS

STATE	COUNTY	AGGREGATED SUB-AREA (ASA) NUMBER
Alabama	Autauga	307
	Baldwin	308
	Barbour	306
	Bibb	307
	Blount	308
	Bullock	307
	Butler	307
	Calhoun	307
	Chambers	306
	Cherokee	307
	Chilton	307
	Choctaw	308
	Clarke	308
	Clay	307
	Cleburne	307
	Coffee	307
	Colbert	602
	Conecuh	307
	Coosa	307
	Covington	307
	Crenshaw	307
	Cullman	308
	Dale	307
	Dallas	307
	De Kalb	602
	Elmore	307
	Escambia	307
	Etowah	307
	Fayette	308
	Franklin	602
	Geneva	307
	Greene	308
	Hale	308
	Henry	306
	Houston	306
	Jackson	602
	Jefferson	308
	Lamar	308
	Lauderdale	602
	Lawrence	602
	Lee	306
	Limestone	602
	Lowndes	307
	Macon	307
	Madison	602
	Marengo	308
	Marion	308

STATE	COUNTY	AGGREGATED SUB-AREA (ASA) NUMBER
Alabama (continued)	Marshall	602
	Mobile	308
	Monroe	307
	Montgomery	307
	Perry	307
	Pickens	308
	Pike	307
	Randolph	307
	Russell	306
	St. Clair	307
	Shelby	308
	Sumter	308
	Talladega	307
	Tallapoosa	307
	Tuscaloosa	308
	Walker	308
	Washington	308
	Wilcox	307
	Winston	308
Arizona	Apache	1501
	Cochise	1503
	Coconino	1502
	Gila	1503
	Graham	1503
	Greenlee	1503
	Maricopa	1503
	Mohave	1502
	Navajo	1501
	Pima	1503
	Pinal	1503
	Santa Cruz	1503
	Yavapai	1503
	Yuma	1502
Arkansas	Arkansas	801
	Ashley	802
	Baxter	1101
	Benton	1104
	Boone	1101
	Bradley	802
	Calhoun	802
	Carroll	1101
	Chicot	802
	Clark	802
	Cleburne	1101
	Cleveland	802

STATE	COUNTY	AGGREGATED SUB-AREA (ASA) NUMBER
Arkansas (continued)	Columbia	1107
	Conway	1104
	Clay	801
	Craighead	801
	Crawford	1104
	Crittenden	801
	Cross	801
	Dallas	802
	Desha	802
	Drew	802
	Faulkner	1104
	Franklin	1104
	Fulton	1101
	Garland	802
	Grant	802
	Greene	801
	Hempstead	802
	Hot Spring	802
	Howard	1107
	Independence	1101
	Izard	1101
	Jackson	801
	Jefferson	802
	Johnson	1104
	Lafayette	1107
	Lawrence	1101
	Lee	801
	Lincoln	802
	Little River	1107
	Logan	1104
	Lonoke	801
	Madison	1101
	Marion	1101
	Miller	1107
	Mississippi	801
	Monroe	801
	Montgomery	802
	Nevada	802
	Newton	1101
	Ouachita	301
	Perry	1104
	Phillips	801
	Pike	802
	Poinsett	801
	Polk	1107
	Pope	1104
	Prairie	801

STATE	COUNTY	AGGREGATED SUB-AREA (ASA) NUMBER
Arkansas (continued)	Pulaski	1104
	Randolph	1101
	St. Francis	801
	Saline	1104
	Scott	1104
	Searcy	1101
	Sebastian	1104
	Sevier	1107
	Sharp	1101
	Stone	1101
	Union	802
	Van Buren	1101
	Washington	1104
	White	801
	Woodruff	801
	Yell	1104
California	Alameda	1804
	Alpine	1807
	Amador	1803
	Butte	1802
	Calaveras	1803
	Colusa	1802
	Contra Costa	1804
	Del Norte	1801
	El Dorado	1802
	Fresno	1803
	Glenn	1802
	Humboldt	1801
	Imperial	1806
	Inyo	1807
	Kern	1803
	Kings	1803
	Lake	1802
	Lassen	1802
	Los Angeles	1806
	Madera	1803
	Marin	1804
	Mariposa	1803
	Mendocino	1801
	Merced	1803
	Modoc	1802
	Mono	1807
	Monterey	1805
	Napa	1804
	Nevada	1802
	Orange	1806

STATE	COUNTY	AGGREGATED SUB-AREA (ASA) NUMBER
California (continued)	Placer	1802
	Plumas	1802
	Riverside	1806
	Sacramento	1802
	San Benito	1805
	San Bernardino	1806
	San Diego	1806
	San Francisco	1804
	San Joaquin	1803
	San Luis Obispo	1805
	San Mateo	1804
	Santa Barbara	1805
	Santa Clara	1804
	Santa Cruz	1805
	Shasta	1802
	Sierra	1802
	Siskiyou	1801
	Solano	1804
	Sonoma	1804
	Stanislaus	1803
	Sutter	1802
	Tehama	1802
	Trinity	1801
	Tulare	1803
	Tuolumne	1803
	Ventura	1806
	Yolo	1802
	Yuba	1802
Colorado	Adams	1007
	Alamosa	1301
	Arapahoe	1007
	Archuleta	1403
	Baca	1103
	Bent	1102
	Boulder	1007
	Chaffee	1102
	Cheyenne	1010
	Clear Creek	1007
	Conejos	1301
	Costilla	1301
	Crowley	1102
	Custer	1102
	Delta	1402
	Denver	1007
	Dolores	1402
	Douglas	1007

STATE	COUNTY	AGGREGATED SUB-AREA (ASA) NUMBER
Colorado (continued)	Eagle	1402
	Elbert	1007
	El Paso	1102
	Fremont	1102
	Garfield	1402
	Gilpin	1007
	Grand	1402
	Gunnison	1402
	Hinsdale	1402
	Huertano	1102
	Jackson	1007
	Jefferson	1007
	Kiowa	1102
	Kit Carson	1010
	Lake	1102
	La Plata	1403
	Larimer	1007
	Las Animas	1102
	Lincoln	1102
	Logan	1007
	Mesa	1402
	Mineral	1301
	Moffat	1401
	Montezuma	1403
	Montrose	1402
	Morgan	1007
	Otero	1102
	Ouray	1402
	Park	1007
	Phillips	1010
	Pitkin	1402
	Prowers	1102
	Pueblo	1102
	Rio Blanco	1401
	Rio Grande	1301
	Routt	1401
	Saguache	1301
	San Juan	1403
	San Miguel	1402
	Sedgwick	1007
	Summit	1402
	Teller	1102
	Washington	1007
	Weld	1007
	Yuma	1010
Connecticut	Fairfield	104

STATE	COUNTY	AGGREGATED SUB-AREA (ASA) NUMBER
Connecticut (continued)	Hartford	105
	Litchfield	104
	Middlesex	105
	New Haven	104
	New London	104
	Tolland	104
	Windham	104
Delaware	Kent	203
	New Castle	203
	Sussex	205
District of Columbia	District of Columbia	206
Florida	Alachua	304
	Baker	303
	Bay	307
	Bradford	304
	Brevard	304
	Broward	305
	Calhoun	306
	Charlotte	304
	Citrus	304
	Clay	304
	Collier	305
	Columbia	304
	Dade	305
	De Soto	304
	Dixie	304
	Duval	304
	Escambia	307
	Flagler	304
	Franklin	306
	Gadsden	306
	Gilchrist	304
	Glades	305
	Gulf	306
	Hamilton	304
	Hardee	304
	Hendry	305
	Hernando	304
	Highlands	305
	Hillsborough	304
	Holmes	307
	Indian River	305
	Jackson	306
	Jefferson	306

STATE	COUNTY	AGGREGATED SUB-AREA (ASA) NUMBER
Florida (continued)	Lafayette	304
	Lake	304
	Lee	305
	Leon	306
	Levy	304
	Liberty	306
	Madison	304
	Manatee	304
	Marion	304
	Martin	305
	Monroe	305
	Nassau	303
	Okaloosa	307
	Okeechobee	305
	Orange	304
	Osceola	305
	Palm Beach	305
	Pasco	304
	Pinellas	304
	Polk	305
	Putnam	304
	St. Johns	304
	St. Lucie	305
	Santa Rosa	307
	Sarasota	304
	Seminole	304
	Sumter	304
	Suwannee	304
	Taylor	306
	Union	304
	Volusia	304
	Wakulla	306
	Walton	307
	Washington	307
Georgia	Appling	303
	Atkinson	303
	Bacon	303
	Baker	306
	Baldwin	303
	Banks	303
	Barrow	303
	Bartow	307
	Ben Hill	303
	Berrien	304
	Bibb	303

STATE	COUNTY	AGGREGATED SUB-AREA (ASA) NUMBER
Georgia (continued)	Bleckley	303
	Brantley	303
	Brooks	304
	Bryan	303
	Bulloch	303
	Burke	303
	Butts	303
	Calhoun	306
	Camden	303
	Candler	303
	Carroll	306
	Catoosa	601
	Charlton	303
	Chatham	303
	Chattahoochee	306
	Chattooga	307
	Cherokee	307
	Clarke	303
	Clay	306
	Clayton	306
	Clinch	304
	Cobb	306
	Coffee	303
	Colquitt	304
	Columbia	303
	Cook	304
	Coweta	306
	Crawford	306
	Crisp	306
	Dade	601
	Dawson	307
	Decatur	306
	De Kalb	306
	Dodge	303
	Dooly	306
	Dougherty	306
	Douglas	306
	Early	306
	Echols	304
	Effingham	303
	Elbert	303
	Emanuel	303
	Evans	303
	Fannin	601
	Fayette	306
	Floyd	307
	Forsyth	306

STATE	COUNTY	AGGREGATED SUB-AREA (ASA) NUMBER
Georgia (continued)	Franklin	303
	Fulton	306
	Gilmer	307
	Glascock	303
	Glynn	303
	Gordon	307
	Grady	306
	Greene	303
	Gwinnett	306
	Habersham	306
	Hall	306
	Hancock	303
	Haralson	307
	Harris	306
	Hart	303
	Heard	306
	Henry	303
	Houston	303
	Irwin	304
	Jackson	303
	Jasper	303
	Jeff Davis	303
	Jefferson	303
	Jenkins	303
	Johnson	303
	Jones	303
	Lamar	303
	Lanier	304
	Laurens	303
	Lee	306
	Liberty	303
	Lincoln	303
	Long	303
	Lowndes	304
	Lumpkin	306
	McDuffie	303
	McIntosh	303
	Macon	306
	Madison	303
	Marion	306
	Meriwether	306
	Miller	306
	Mitchell	306
	Monroe	303
	Montgomery	303
	Morgan	303
	Murray	307

STATE	COUNTY	AGGREGATED SUB-AREA (ASA) NUMBER
Georgia (continued)	Muscogee	306
	Newton	303
	Oconee	303
	Oglethorpe	303
	Paulding	307
	Peach	303
	Pickens	307
	Pierce	303
	Pike	306
	Polk	307
	Pulaski	303
	Putnam	303
	Quitman	306
	Rabun	303
	Randolph	306
	Richmond	303
	Rockdale	303
	Schley	306
	Screven	303
	Seminole	306
	Spalding	306
	Stephens	303
	Stewart	306
	Sumter	306
	Talbot	306
	Taliaferro	303
	Tattnall	303
	Taylor	306
	Telfair	303
	Terrell	306
	Thomas	306
	Tift	304
	Toombs	303
	Towns	601
	Treutlen	303
	Troup	306
	Turner	304
	Twiggs	303
	Union	601
	Upson	306
	Walker	601
	Walton	303
	Ware	303
	Warren	303
	Washington	303
	Wayne	303
	Webster	306

STATE	COUNTY	AGGREGATED SUB-AREA (ASA) NUMBER
Georgia (continued)	Wheeler	303
	White	306
	Whitfield	307
	Wilcox	303
	Wilkes	303
	Wilkinson	303
	Worth	306
Idaho	Ada	1703
	Adams	1703
	Bannock	1703
	Bear Lake	1601
	Benewah	1701
	Bingham	1703
	Blaine	1703
	Boise	1703
	Bonner	1701
	Bonneville	1703
	Boundary	1701
	Butte	1703
	Camas	1703
	Canyon	1703
	Caribou	1703
	Cassia	1703
	Clark	1703
	Clearwater	1704
	Custer	1704
	Elmore	1703
	Franklin	1601
	Fremont	1703
	Gem	1703
	Gooding	1703
	Idaho	1704
	Jefferson	1703
	Jerome	1703
	Kootenai	1701
	Latah	1704
	Lemhi	1704
	Lewis	1704
	Lincoln	1703
	Madison	1703
	Minidoka	1703
	Nez Perce	1704
	Oneida	1601
	Owyhee	1703
	Payette	1703
	Power	1703

STATE	COUNTY	AGGREGATED SUB-AREA (ASA) NUMBER
Idaho (continued)	Shoshone	1701
	Teton	1703
	Twin Falls	1703
	Valley	1703
	Washington	1703
Illinois	Adams	704
	Alexander	705
	Bond	705
	Boone	703
	Brown	704
	Bureau	704
	Calhoun	704
	Carroll	703
	Cass	704
	Champaign	506
	Christian	704
	Clark	506
	Clay	506
	Clinton	705
	Coles	506
	Cook	403
	Crawford	506
	Cumberland	506
	De Kalb	703
	De Witt	704
	Douglas	506
	Du Page	403
	Edgar	506
	Edwards	506
	Effingham	506
	Fayette	705
	Ford	704
	Franklin	705
	Fulton	704
	Gallatin	505
	Greene	704
	Grundy	704
	Hamilton	506
	Hancock	704
	Hardin	505
	Henderson	703
	Henry	703
	Iroquois	704
	Jackson	705
	Jasper	506
	Jefferson	705

STATE	COUNTY	AGGREGATED SUB-AREA (ASA) NUMBER
Illinois (continued)	Jersey	704
	Jo Daviess	703
	Johnson	505
	Kane	403
	Kankakee	704
	Kendall	704
	Knox	704
	Lake	403
	La Salle	704
	Lawrence	506
	Lee	703
	Livingston	704
	Logan	704
	McDonough	704
	McHenry	403
	McLean	704
	Macon	704
	Macoupin	704
	Madison	705
	Marion	705
	Marshall	704
	Mason	704
	Massac	505
	Menard	704
	Mercer	703
	Monroe	705
	Montgomery	705
	Morgan	704
	Moultrie	705
	Ogle	703
	Peoria	704
	Perry	705
	Piatt	704
	Pike	704
	Pope	505
	Pulaski	505
	Putnam	704
	Randolph	705
	Richland	506
	Rock Island	703
	St. Clair	705
	Saline	505
	Sangamon	704
	Schuyler	704
	Scott	704
	Shelby	705
	Stark	704

STATE	COUNTY	AGGREGATED SUB-AREA (ASA) NUMBER
Illinois (continued)	Stephenson	703
	Tazewell	704
	Union	705
	Vermilion	506
	Wabash	506
	Warren	703
	Washington	705
	Wayne	506
	White	506
	Whiteside	703
	Will	403
	Williamson	705
	Winnebago	703
	Woodford	704
Indiana	Adams	406
	Allen	406
	Bartholomew	506
	Benton	506
	Blackford	506
	Boone	506
	Brown	506
	Carroll	506
	Cass	506
	Clark	505
	Clay	506
	Clinton	506
	Crawford	505
	Daviess	506
	Dearborn	502
	Decatur	506
	De Kalb	406
	Delaware	506
	Dubois	506
	Elkhart	404
	Fayette	503
	Floyd	505
	Fountain	506
	Franklin	503
	Fulton	506
	Gibson	506
	Grant	506
	Greene	506
	Hamilton	506
	Hancock	506
	Harrison	505
	Hendricks	506

STATE	COUNTY	AGGREGATED SUB-AREA (ASA) NUMBER
Indiana (continued)	Henry	506
	Howard	506
	Huntington	506
	Jackson	506
	Jasper	704
	Jay	506
	Jefferson	505
	Jennings	506
	Johnson	506
	Knox	506
	Kosciusko	506
	Lagrange	404
	Lake	403
	La Porte	403
	Lawrence	506
	Madison	506
	Marion	506
	Marshall	404
	Martin	506
	Miami	506
	Monroe	506
	Montgomery	506
	Morgan	506
	Newton	704
	Noble	404
	Ohio	502
	Orange	506
	Owen	506
	Parke	506
	Perry	505
	Pike	506
	Porter	403
	Posey	506
	Pulaski	506
	Putnam	506
	Randolph	506
	Ripley	506
	Rush	506
	St. Joseph	404
	Scott	506
	Shelby	506
	Spencer	505
	Starke	403
	Steuben	404
	Sullivan	506
	Switzerland	502
	Tippecanoe	506

STATE	COUNTY	AGGREGATED SUB-AREA (ASA) NUMBER
Indiana (continued)	Tipton	506
	Union	503
	Vanderburgh	505
	Vermillion	506
	Vigo	506
	Wabash	506
	Warren	506
	Warrick	505
	Washington	506
	Wayne	503
	Wells	506
	White	506
	Whitley	506
Iowa	Adair	1009
	Adams	1009
	Allamakee	703
	Appanoose	1011
	Audubon	1009
	Benton	703
	Black Hawk	703
	Boone	703
	Bremer	703
	Buchanan	703
	Buena Vista	703
	Butler	703
	Calhoun	703
	Carroll	703
	Cass	1009
	Cedar	703
	Cerro Gordo	703
	Cherokee	1009
	Chickasaw	703
	Clarke	703
	Clay	1009
	Clayton	703
	Clinton	703
	Crawford	1009
	Dallas	703
	Davis	703
	Decatur	1011
	Delaware	703
	Des Moines	703
	Dickinson	1009
	Dubuque	703
	Emmet	703
	Fayette	703

STATE	COUNTY	AGGREGATED SUB-AREA (ASA) NUMBER
Iowa (continued)	Floyd	703
	Franklin	703
	Fremont	1009
	Greene	703
	Grundy	703
	Guthrie	703
	Hamilton	703
	Hancock	703
	Hardin	703
	Harrison	1009
	Henry	703
	Howard	703
	Humboldt	703
	Ida	1009
	Iowa	703
	Jackson	703
	Jasper	703
	Jefferson	703
	Johnson	703
	Jones	703
	Keokuk	703
	Kossuth	703
	Lee	703
	Linn	703
	Louisa	703
	Lucas	703
	Lyon	1006
	Madison	703
	Mahaska	703
	Marion	703
	Marshall	703
	Mills	1009
	Mitchell	703
	Monona	1009
	Monroe	703
	Montgomery	1009
	Muscatine	703
	O'Brien	1009
	Osceola	1009
	Page	1009
	Palo Alto	703
	Plymouth	1009
	Pocahontas	703
	Polk	703
	Pottawattamie	1009
	Poweshiek	703
	Ringgold	1011

STATE	COUNTY	AGGREGATED SUB-AREA (ASA) NUMBER
Iowa (continued)	Sac	703
	Scott	703
	Shelby	1009
	Sioux	1006
	Story	703
	Tama	703
	Taylor	1009
	Union	1011
	Van Buren	703
	Wapello	703
	Warren	703
	Washington	703
	Wayne	1011
	Webster	703
	Winnebago	703
	Winneshiek	703
	Woodbury	1009
	Worth	703
	Wright	703
Kansas	Allen	1104
	Anderson	1011
	Atchison	1009
	Barber	1103
	Barton	1103
	Bourbon	1011
	Brown	1009
	Butler	1103
	Chase	1104
	Chautauqua	1104
	Cherokee	1104
	Cheyenne	1010
	Clark	1103
	Clay	1010
	Cloud	1010
	Coffey	1104
	Comanche	1103
	Cowley	1103
	Crawford	1104
	Decatur	1010
	Dickinson	1010
	Doniphan	1009
	Douglas	1010
	Edwards	1103
	Elk	1104
	Ellis	1010
	Ellsworth	1010

STATE	COUNTY	AGGREGATED SUB-AREA (ASA) NUMBER
Kansas (continued)	Finney	1103
	Ford	1103
	Franklin	1011
	Geary	1010
	Gove	1010
	Graham	1010
	Grant	1103
	Gray	1103
	Greeley	1103
	Greenwood	1104
	Hamilton	1103
	Harper	1103
	Harvey	1103
	Haskell	1103
	Hodgeman	1103
	Jackson	1010
	Jefferson	1010
	Jewell	1010
	Johnson	1011
	Kearny	1103
	Kingman	1103
	Kiowa	1103
	Labette	1104
	Lane	1103
	Leavenworth	1010
	Lincoln	1010
	Linn	1011
	Logan	1010
	Lyon	1104
	McPherson	1103
	Marion	1104
	Marshall	1010
	Meade	1103
	Miami	1011
	Mitchell	1010
	Montgomery	1104
	Morris	1104
	Morton	1103
	Nemaha	1009
	Neosho	1104
	Ness	1103
	Norton	1010
	Osage	1011
	Osborne	1010
	Ottawa	1010
	Pawnee	1103
	Phillips	1010

STATE	COUNTY	AGGREGATED SUB-AREA (ASA) NUMBER
Kansas (continued)	Pottawatomie	1010
	Pratt	1103
	Rawlins	1010
	Reno	1103
	Republic	1010
	Rice	1103
	Riley	1010
	Rooks	1010
	Rush	1103
	Russell	1010
	Saline	1010
	Scott	1103
	Sedgwick	1103
	Seward	1103
	Shawnee	1010
	Sheridan	1010
	Sherman	1010
	Smith	1010
	Stafford	1103
	Stanton	1103
	Stevens	1103
	Sumner	1103
	Thomas	1010
	Trego	1010
	Wabaunsee	1010
	Wallace	1010
	Washington	1010
	Wichita	1103
	Wilson	1104
	Woodson	1104
	Wyandotte	1011
Kentucky	Adair	505
	Allen	505
	Anderson	505
	Ballard	505
	Barren	505
	Bath	505
	Bell	507
	Boone	502
	Bourbon	505
	Boyd	502
	Boyle	505
	Bracken	505
	Breathitt	505
	Breckinridge	505
	Bullitt	505

STATE	COUNTY	AGGREGATED SUB-AREA (ASA) NUMBER
Kentucky (continued)	Butler	505
	Caldwell	505
	Calloway	602
	Campbell	502
	Carlisle	801
	Carroll	505
	Carter	502
	Casey	505
	Christian	507
	Clark	505
	Clay	505
	Clinton	507
	Crittenden	505
	Cumberland	507
	Daviess	505
	Edmonson	505
	Elliott	502
	Estill	505
	Fayette	505
	Fleming	505
	Floyd	502
	Franklin	505
	Fulton	801
	Gallatin	505
	Garrard	505
	Grant	505
	Graves	801
	Grayson	505
	Green	505
	Greenup	502
	Hancock	505
	Hardin	505
	Harlan	507
	Harrison	505
	Hart	505
	Henderson	505
	Henry	505
	Hickman	801
	Hopkins	505
	Jackson	507
	Jefferson	505
	Jessamine	505
	Johnson	502
	Kenton	502
	Knott	505
	Knox	507
	Larue	505

STATE	COUNTY	AGGREGATED SUB-AREA (ASA) NUMBER
Kentucky (continued)	Laurel	507
	Lawrence	502
	Lee	505
	Leslie	505
	Letcher	505
	Lewis	502
	Lincoln	505
	Livingston	507
	Logan	505
	Lyon	507
	McCracken	505
	McCreary	507
	McLean	505
	Madison	505
	Magoffin	505
	Marion	505
	Marshall	602
	Martin	502
	Mason	505
	Meade	505
	Menifee	505
	Mercer	505
	Metcalfe	505
	Monroe	505
	Montgomery	505
	Morgan	505
	Muhlenberg	505
	Nelson	505
	Nicholas	505
	Ohio	505
	Oldham	505
	Owen	505
	Owsley	505
	Pendleton	505
	Perry	505
	Pike	502
	Powell	505
	Pulaski	507
	Robertson	505
	Rockcastle	507
	Rowan	505
	Russell	507
	Scott	505
	Shelby	505
	Simpson	505
	Spencer	505
	Taylor	505

STATE	COUNTY	AGGREGATED SUB-AREA (ASA) NUMBER
Kentucky (continued)	Todd	507
	Trigg	507
	Trimble	505
	Union	505
	Warren	505
	Washington	505
	Wayne	507
	Webster	505
	Whitley	507
	Wolfe	505
	Woodford	505
Louisiana	Acadia	803
	Allen	803
	Ascension	803
	Assumption	803
	Avoyelles	802
	Beauregard	803
	Bienville	1107
	Bossier	1107
	Caddo	1107
	Calcasieu	803
	Caldwell	802
	Cameron	803
	Catahoula	802
	Claiborne	802
	Concordia	802
	De Soto	1201
	East Baton Rouge	803
	East Carroll	802
	East Feliciana	803
	Evangeline	803
	Franklin	802
	Grant	802
	Iberia	803
	Iberville	803
	Jackson	802
	Jefferson	803
	Jefferson Davis	803
	Lafayette	803
	Lafourche	803
	La Salle	802
	Lincoln	802
	Livingston	803
	Madison	802
	Morehouse	802
	Natchitoches	1107

STATE	COUNTY	AGGREGATED SUB-AREA (ASA) NUMBER
Louisiana (continued)	Orleans	803
	Quachita	802
	Plaquemines	803
	Pointe Coupee	803
	Rapides	802
	Red River	1107
	Richland	802
	Sabine	1201
	St. Bernard	803
	St. Charles	803
	St. Helena	803
	St. James	803
	St. John Baptist	803
	St. Landry	803
	St. Martin	803
	St. Mary	803
	St. Tammany	803
	Tangipahoa	803
	Tensas	802
	Terrebonne	803
	Union	802
	Vermilion	803
	Vernon	803
	Washington	309
	Webster	1107
	West Baton Rouge	803
	West Carroll	802
	West Feliciana	803
	Winn	802
Maine	Androscoggin	101
	Aroostook	101
	Cumberland	102
	Franklin	101
	Hancock	101
	Kennebec	101
	Knox	101
	Lincoln	101
	Oxford	101
	Penobscot	101
	Piscataquis	101
	Sagadahoc	101
	Somerset	101
	Waldo	101
	Washington	101
	York	102

STATE	COUNTY	AGGREGATED SUB-AREA (ASA) NUMBER
Maryland	Allegany	206
	Anne Arundel	205
	Baltimore	205
	Calvert	205
	Caroline	205
	Carroll	205
	Cecil	203
	Charles	206
	Dorchester	205
	Frederick	206
	Garrett	206
	Harford	205
	Howard	205
	Kent	205
	Montgomery	206
	Prince George's	206
	Queen Anne's	205
	St. Mary's	206
	Somerset	205
	Talbot	205
	Washington	206
	Wicomico	205
	Worcester	205
Massachusetts	Barnstable	103
	Berkshire	104
	Bristol	103
	Dukes	103
	Essex	103
	Franklin	105
	Hampden	105
	Hampshire	105
	Middlesex	103
	Nantucket	103
	Norfolk	103
	Plymouth	103
	Suffolk	103
	Worcester	103
Michigan	Alcona	405
	Alger	401
	Allegan	404
	Alpena	405
	Antrim	404
	Arenac	405
	Baraga	401
	Barry	404

STATE	COUNTY	AGGREGATED SUB-AREA (ASA) NUMBER
Michigan (continued)	Bay	405
	Benzie	404
	Berrien	404
	Branch	404
	Calhoun	404
	Cass	404
	Charlevoix	404
	Cheboygan	405
	Chippewa	401
	Clare	405
	Clinton	404
	Crawford	405
	Delta	402
	Dickinson	402
	Eaton	404
	Emmet	404
	Genesee	405
	Gladwin	405
	Gogebic	401
	Grand Traverse	404
	Gratiot	405
	Hillsdale	404
	Houghton	401
	Huron	405
	Ingham	404
	Ionia	404
	Iosco	405
	Iron	402
	Isabella	405
	Jackson	404
	Kalamazoo	404
	Kalkaska	404
	Kent	404
	Keweenaw	401
	Lake	404
	Lapeer	405
	Leelanau	404
	Lenawee	406
	Livingston	406
	Luce	401
	Mackinac	404
	Macomb	406
	Manistee	404
	Marquette	401
	Mason	404
	Mecosta	404
	Menominee	402

STATE	COUNTY	AGGREGATED SUB-AREA (ASA) NUMBER
Michigan (continued)	Midland	405
	Missaukee	404
	Monroe	406
	Montcalm	404
	Montmorency	405
	Muskegon	404
	Newaygo	404
	Oakland	406
	Oceana	404
	Ogemaw	405
	Ontonagon	401
	Osceola	404
	Oscoda	405
	Otsego	405
	Ottawa	404
	Presque Isle	405
	Roscommon	404
	Saginaw	405
	St. Clair	406
	St. Joseph	404
	Sanilac	406
	Schoolcraft	404
	Shiawassee	404
	Tuscola	405
	Van Buren	404
	Washtenaw	406
	Wayne	406
	Wexford	404
Minnesota	Aitkin	701
	Anoka	701
	Becker	901
	Beltrami	901
	Benton	701
	Big Stone	701
	Blue Earth	701
	Brown	701
	Carlton	401
	Carver	701
	Cass	701
	Chippewa	701
	Chisago	701
	Clay	901
	Clearwater	901
	Cook	401
	Cottonwood	701
	Crow Wing	701

STATE	COUNTY	AGGREGATED SUB-AREA (ASA) NUMBER
Minnesota (continued)	Dakota	701
	Dodge	702
	Douglas	701
	Faribault	701
	Fillmore	702
	Freeborn	703
	Goodhue	702
	Grant	901
	Hennepin	701
	Houston	702
	Hubbard	701
	Isanti	701
	Itasca	701
	Jackson	701
	Kanabec	701
	Kandiyohi	701
	Kittson	901
	Koochiching	901
	Lac qui Parle	701
	Lake	401
	Lake of the Wood	901
	Le Sueur	701
	Lincoln	701
	Lyon	701
	McLeod	701
	Mahnomen	901
	Marshall	901
	Martin	701
	Meeker	701
	Mille Lacs	701
	Morrison	701
	Mower	703
	Murray	701
	Nicollet	701
	Nobles	1006
	Norman	901
	Olmsted	702
	Otter Tail	901
	Pennington	901
	Pine	701
	Pipestone	1006
	Polk	901
	Pope	701
	Ramsey	701
	Red Lake	901
	Redwood	701
	Renville	701

STATE	COUNTY	AGGREGATED SUB-AREA (ASA) NUMBER
Minnesota (continued)	Rice	702
	Rock	1006
	Roseau	901
	St. Louis	401
	Scott	701
	Sherburne	701
	Sibley	701
	Stearns	701
	Steele	702
	Stevens	701
	Swift	701
	Todd	701
	Traverse	901
	Wabasha	702
	Wadena	701
	Waseca	701
	Washington	701
	Watonwan	701
	Wilkin	901
	Winona	702
	Wright	701
	Yellow Medicine	701
Mississippi	Adams	802
	Alcorn	801
	Amite	803
	Attala	802
	Benton	802
	Bolivar	802
	Calhoun	802
	Carroll	802
	Chickasaw	308
	Choctaw	802
	Claiborne	802
	Clarke	309
	Clay	308
	Coahoma	802
	Copiah	309
	Covington	309
	De Soto	802
	Forrest	309
	Franklin	802
	George	309
	Greene	309
	Grenada	802
	Hancock	309
	Harrison	309

STATE	COUNTY	AGGREGATED SUB-AREA (ASA) NUMBER
Mississippi (continued)	Hinds	309
	Holmes	802
	Humphreys	802
	Issaquena	802
	Itawamba	308
	Jackson	309
	Jasper	309
	Jefferson	802
	Jefferson Davis	309
	Jones	309
	Kemper	308
	Lafayette	802
	Lamar	309
	Lauderdale	309
	Lawrence	309
	Leake	309
	Lee	308
	Leflore	802
	Lincoln	309
	Lowndes	308
	Madison	802
	Marion	309
	Marshall	802
	Monroe	308
	Montgomery	802
	Neshoba	309
	Newton	309
	Noxubee	308
	Oktibbeha	308
	Panola	802
	Pearl River	309
	Perry	309
	Pike	309
	Pontotoc	802
	Prentiss	308
	Quitman	802
	Rankin	309
	Scott	309
	Sharkey	802
	Simpson	309
	Smith	309
	Stone	309
	Sunflower	802
	Tallahatchie	802
	Tate	802
	Tippah	801
	Tishomingo	602

STATE	COUNTY	AGGREGATED SUB-AREA (ASA) NUMBER
Mississippi (continued)	Tunica	802
	Union	802
	Walthall	309
	Warren	802
	Washington	802
	Wayne	309
	Webster	802
	Wilkinson	802
	Winston	309
	Yalobusha	802
	Yazoo	802
Missouri	Adair	1011
	Andrew	1009
	Atchison	1009
	Audrain	704
	Barry	1101
	Barton	1104
	Bates	1011
	Benton	1011
	Bollinger	705
	Boone	1011
	Buchanan	1009
	Butler	1101
	Caldwell	1011
	Callaway	1011
	Camden	1011
	Cape Girardeau	705
	Carroll	1011
	Carter	1101
	Cass	1011
	Cedar	1011
	Chariton	1011
	Christian	1101
	Clark	704
	Clay	1011
	Clinton	1011
	Cole	1011
	Cooper	1011
	Crawford	705
	Dade	1011
	Dallas	1011
	Daviess	1011
	De Kalb	1011
	Dent	705
	Douglas	1101
	Dunklin	801

STATE	COUNTY	AGGREGATED SUB-AREA (ASA) NUMBER
Missouri (continued)	Franklin	705
	Gasconade	1011
	Gentry	1011
	Greene	1011
	Grundy	1011
	Harrison	1011
	Henry	1011
	Hickory	1011
	Holt	1009
	Howard	1011
	Howell	1101
	Iron	801
	Jackson	1011
	Jasper	1104
	Jefferson	705
	Johnson	1011
	Knox	704
	Laclede	1011
	Lafayette	1011
	Lawrence	1104
	Lewis	704
	Lincoln	704
	Linn	1011
	Livingston	1011
	McDonald	1104
	Macon	1011
	Madison	801
	Maries	1011
	Marion	704
	Mercer	1011
	Miller	1011
	Mississippi	801
	Moniteau	1011
	Monroe	704
	Montgomery	705
	Morgan	1011
	New Madrid	801
	Newton	1104
	Nodaway	1009
	Oregon	1101
	Osage	1011
	Ozark	1101
	Pemiscot	801
	Perry	705
	Pettis	1011
	Phelps	1011
	Pike	704

STATE	COUNTY	AGGREGATED SUB-AREA (ASA) NUMBER
Missouri (continued)	Platte	1011
	Polk	1011
	Pulaski	1011
	Putnam	1011
	Ralls	704
	Randolph	1011
	Ray	1011
	Reynolds	1101
	Ripley	1101
	St. Charles	705
	St. Clair	1011
	St. Francois	705
	St. Louis	705
	Ste. Genevieve	705
	Saline	1011
	Schuyler	704
	Scotland	704
	Scott	801
	Shannon	1101
	Shelby	704
	Stoddard	801
	Stone	1101
	Sullivan	1011
	Taney	1101
	Texas	1011
	Vernon	1011
	Warren	705
	Washington	705
	Wayne	801
	Webster	1011
	Worth	1011
	Wright	1011
Montana	Beaverhead	1002
	Big Horn	1004
	Blaine	1001
	Broadwater	1002
	Carbon	1004
	Carter	1005
	Cascade	1002
	Chouteau	1002
	Custer	1004
	Daniels	1001
	Dawson	1004
	Deer Lodge	1701
	Fallon	1004
	Fergus	1003

STATE	COUNTY	AGGREGATED SUB-AREA (ASA) NUMBER
Montana (continued)	Flathead	1701
	Gallatin	1002
	Garfield	1003
	Glacier	1002
	Golden Valley	1003
	Granite	1701
	Hill	1001
	Jefferson	1002
	Judith Basin	1003
	Lake	1701
	Lewis and Clark	1002
	Liberty	1002
	Lincoln	1701
	McCone	1001
	Madison	1002
	Meagher	1002
	Mineral	1701
	Missoula	1701
	Musselshell	1003
	Park	1004
	Petroleum	1003
	Phillips	1001
	Pondera	1002
	Powder River	1004
	Powell	1701
	Prairie	1004
	Ravalli	1701
	Richland	1004
	Roosevelt	1001
	Rosebud	1004
	Sanders	1701
	Sheridan	1001
	Silver Bowl	1701
	Stillwater	1004
	Sweet Grass	1004
	Teton	1002
	Toole	1002
	Treasure	1004
	Valley	1001
	Wheatland	1003
	Wibaux	1004
	Yellowstone	1004
	Yellowstone Park	1004
Nebraska	Adams	1010
	Antelope	1008
	Arthur	1008

STATE	COUNTY	AGGREGATED SUB-AREA (ASA) NUMBER
Nebraska (continued)	Banner	1007
	Blaine	1008
	Boone	1008
	Box Butte	1008
	Boyd	1008
	Brown	1008
	Buffalo	1008
	Burt	1008
	Butler	1010
	Cass	1009
	Cedar	1006
	Chase	1010
	Cherry	1008
	Cheyenne	1007
	Clay	1010
	Colfax	1008
	Cuming	1008
	Custer	1008
	Dakota	1009
	Dawes	1008
	Dawson	1008
	Deuel	1007
	Dixon	1006
	Dodge	1008
	Douglas	1009
	Dundy	1010
	Fillmore	1010
	Franklin	1010
	Frontier	1010
	Furnas	1010
	Gage	1010
	Garden	1007
	Garfield	1008
	Gosper	1010
	Grant	1008
	Greeley	1008
	Hall	1008
	Hamilton	1010
	Harlan	1010
	Hayes	1010
	Hitchcock	1010
	Holt	1008
	Hooker	1008
	Howard	1008
	Jefferson	1010
	Johnson	1009
	Kearney	1008

STATE	COUNTY	AGGREGATED SUB-AREA (ASA) NUMBER
Nebraska (continued)	Keith	1008
	Keya Paha	1008
	Kimball	1007
	Knox	1006
	Lancaster	1008
	Lincoln	1008
	Logan	1008
	Loup	1008
	McPherson	1008
	Madison	1008
	Merrick	1008
	Morrill	1007
	Nance	1008
	Nemaha	1009
	Nuckolls	1010
	Otoe	1009
	Pawnee	1009
	Perkins	1010
	Phelps	1008
	Pierce	1008
	Platte	1008
	Polk	1010
	Red Willow	1010
	Richardson	1009
	Rock	1008
	Saline	1010
	Sarpy	1009
	Saunders	1008
	Scotts bluff	1007
	Seward	1010
	Sheridan	1008
	Sherman	1008
	Sioux	1008
	Stanton	1008
	Thayer	1010
	Thomas	1008
	Thurston	1009
	Valley	1008
	Washington	1009
	Wayne	1008
	Webster	1010
	Wheeler	1008
	York	1010
Nevada	Carson City	1604
	Churchill	1604
	Clark	1502

STATE	COUNTY	AGGREGATED SUB-AREA (ASA) NUMBER
Nevada (continued)	Douglas	1604
	Elko	1603
	Esmeralda	1603
	Eureka	1603
	Humboldt	1603
	Lander	1603
	Lincoln	1502
	Lyon	1604
	Mineral	1604
	Nye	1603
	Pershing	1603
	Storey	1604
	Washoe	1604
	White Pine	1603
New Hampshire	Belknap	102
	Carroll	102
	Cheshire	105
	Coos	105
	Grafton	105
	Hillsborough	102
	Merrimack	102
	Rockingham	102
	Strafford	102
	Sullivan	105
New Jersey	Atlantic	203
	Bergen	202
	Burlington	203
	Camden	203
	Cape May	203
	Cumberland	203
	Essex	202
	Gloucester	203
	Hudson	202
	Hunterdon	203
	Mercer	203
	Middlesex	202
	Monmouth	202
	Morris	202
	Ocean	203
	Passaic	202
	Salem	203
	Somerset	202
	Sussex	203
	Union	202
	Warren	203

STATE	COUNTY	AGGREGATED SUB-AREA (ASA) NUMBER
New Mexico	Bernalillo	1302
	Catron	1503
	Chaves	1304
	Colfax	1105
	Curry	1203
	De Baca	1304
	Dona Ana	1302
	Eddy	1304
	Grant	1503
	Guadalupe	1304
	Harding	1105
	Hidalgo	1503
	Lea	1204
	Lincoln	1302
	Los Alamos	1302
	Luna	1302
	McKinley	1501
	Mora	1105
	Otero	1302
	Quay	1105
	Rio Arriba	1302
	Roosevelt	1203
	Sandoval	1302
	San Juan	1403
	San Miguel	1304
	Santa Fe	1302
	Sierra	1302
	Socorro	1302
	Taos	1302
	Torrance	1302
	Union	1105
	Valencia	1302
New York	Albany	201
	Allegany	408
	Bronx	202
	Broome	204
	Cattaraugus	407
	Cayuga	408
	Chautauqua	407
	Chemung	204
	Chenango	204
	Clinton	106
	Columbia	201
	Cortland	204
	Delaware	203
	Dutchess	201

STATE	COUNTY	AGGREGATED SUB-AREA (ASA) NUMBER
New York (continued)	Erie	407
	Essex	106
	Franklin	408
	Fulton	201
	Genesee	408
	Greene	201
	Hamilton	201
	Herkimer	201
	Jefferson	408
	Kings	202
	Lewis	408
	Livingston	408
	Madison	408
	Monroe	408
	Montgomery	201
	Nassau	202
	New York	202
	Niagara	407
	Oneida	201
	Onondaga	408
	Ontario	408
	Orange	201
	Orleans	408
	Oswego	408
	Otsego	204
	Putnam	201
	Queens	202
	Rensselaer	201
	Richmond	202
	Rockland	202
	St. Lawrence	408
	Saratoga	201
	Schenectady	201
	Schoharie	201
	Schuyler	408
	Seneca	408
	Steuben	204
	Suffolk	202
	Sullivan	203
	Tioga	204
	Tompkins	408
	Ulster	201
	Warren	201
	Washington	201
	Wayne	408
	Westchester	202

STATE	COUNTY	AGGREGATED SUB-AREA (ASA) NUMBER
New York (continued)	Wyoming	408
	Yates	408
North Carolina	Alamance	301
	Alexander	302
	Alleghany	504
	Anson	302
	Ashe	504
	Avery	601
	Beaufort	301
	Bertie	301
	Bladen	301
	Brunswick	301
	Buncombe	601
	Burke	302
	Cabarrus	302
	Caldwell	302
	Camden	301
	Carteret	301
	Caswell	301
	Catawba	302
	Chatham	301
	Cherokee	601
	Chowan	301
	Clay	601
	Cleveland	302
	Columbus	302
	Craven	301
	Cumberland	301
	Currituck	301
	Dare	301
	Davidson	302
	Davie	302
	Duplin	301
	Durham	301
	Edgecombe	301
	Forsyth	302
	Franklin	301
	Gaston	302
	Gates	301
	Graham	601
	Granville	301
	Greene	301
	Guilford	302
	Halifax	301
	Harnett	301
	Haywood	601

STATE	COUNTY	AGGREGATED SUB-AREA (ASA) NUMBER
North Carolina (continued)	Henderson	601
	Hertford	301
	Hoke	301
	Hyde	301
	Iredell	302
	Jackson	601
	Johnston	301
	Jones	301
	Lee	301
	Lenoir	301
	Lincoln	302
	McDowell	302
	Macon	601
	Madison	601
	Martin	301
	Mecklenburg	302
	Mitchell	601
	Montgomery	302
	Moore	301
	Nash	301
	New Hanover	301
	Northampton	301
	Onslow	301
	Orange	301
	Pamlico	301
	Pasquotank	301
	Pender	301
	Perquimans	301
	Person	301
	Pitt	301
	Polk	302
	Randolph	302
	Richmond	302
	Robeson	302
	Rockingham	301
	Rowan	302
	Rutherford	302
	Sampson	301
	Scotland	302
	Stanly	302
	Stokes	301
	Surry	302
	Swain	601
	Transylvania	601
	Tyrrell	301
	Union	302
	Vance	301

STATE	COUNTY	AGGREGATED SUB-AREA (ASA) NUMBER
North Carolina (continued)	Wake	301
	Warren	301
	Washington	301
	Watauga	504
	Wayne	301
	Wilkes	302
	Wilson	301
	Yadkin	302
	Yancey	601
North Dakota	Adams	1005
	Barnes	901
	Benson	901
	Billings	1005
	Bottineau	901
	Bowman	1005
	Burke	901
	Burleigh	1005
	Cass	901
	Cavalier	901
	Dickey	1006
	Divide	901
	Dunn	1005
	Eddy	1006
	Emmons	1005
	Foster	1006
	Golden Valley	1005
	Grand Forks	901
	Grant	1005
	Griggs	901
	Hettinger	1005
	Kidder	1005
	La Moure	1006
	Logan	1005
	McHenry	901
	McIntosh	1005
	McKenzie	1005
	McLean	1005
	Mercer	1005
	Morton	1005
	Mountrail	1005
	Nelson	901
	Oliver	1005
	Pembina	901
	Pierce	901
	Ramsey	901
	Ransom	901

STATE	COUNTY	AGGREGATED SUB-AREA (ASA) NUMBER
North Dakota (continued)	Renville	901
	Richland	901
	Rolette	901
	Sargent	901
	Sheridan	1005
	Sioux	1005
	Slope	1005
	Stark	1005
	Steele	901
	Stutsman	1006
	Towner	901
	Traill	901
	Walsh	901
	Ward	901
	Wells	1006
	Williams	1005
Ohio	Adams	502
	Allen	406
	Ashland	503
	Ashtabula	407
	Athens	502
	Auglaize	406
	Belmont	502
	Brown	502
	Butler	503
	Carroll	503
	Champaign	503
	Clark	503
	Clermont	502
	Clinton	502
	Columbiana	502
	Coshocton	503
	Crawford	406
	Cuyahoga	407
	Darke	503
	Defiance	406
	Delaware	503
	Erie	406
	Fairfield	502
	Fayette	503
	Franklin	503
	Fulton	406
	Gallia	502
	Geauga	407
	Greene	503
	Guernsey	503

STATE	COUNTY	AGGREGATED SUB-AREA (ASA) NUMBER
Ohio (continued)	Hamilton	502
	Hancock	406
	Hardin	503
	Harrison	503
	Henry	406
	Highland	503
	Hocking	502
	Holmes	503
	Huron	406
	Jackson	502
	Jefferson	502
	Knox	503
	Lake	407
	Lawrence	502
	Licking	503
	Logan	503
	Lorain	407
	Lucas	406
	Madison	503
	Mahoning	502
	Marion	503
	Medina	407
	Meigs	502
	Mercer	406
	Miami	503
	Monroe	502
	Montgomery	503
	Morgan	503
	Morrow	503
	Muskingum	503
	Noble	502
	Ottawa	406
	Paulding	406
	Perry	503
	Pickaway	503
	Pike	503
	Portage	407
	Preble	503
	Putnam	406
	Richland	503
	Ross	503
	Sandusky	406
	Scioto	503
	Seneca	406
	Shelby	503
	Stark	503
	Summit	407

STATE	COUNTY	AGGREGATED SUB-AREA (ASA) NUMBER
Ohio (continued)	Trumbull	502
	Tuscarawas	503
	Union	503
	Van Wert	406
	Vinton	502
	Warren	502
	Washington	503
	Wayne	503
	Williams	406
	Wood	406
	Wyandot	406
Oklahoma	Adair	1104
	Alfalfa	1103
	Atoka	1107
	Beaver	1105
	Beckham	1106
	Blaine	1105
	Bryan	1107
	Caddo	1106
	Canadian	1105
	Carter	1106
	Cherokee	1104
	Choctaw	1107
	Cimarron	1105
	Cleveland	1105
	Coal	1107
	Comanche	1106
	Cotton	1106
	Craig	1104
	Creek	1104
	Custer	1106
	Delaware	1104
	Dewey	1105
	Ellis	1105
	Garfield	1103
	Garvin	1106
	Grady	1106
	Grant	1103
	Greer	1106
	Harmon	1106
	Harper	1103
	Haskell	1104
	Hughes	1105
	Jackson	1106
	Jefferson	1106
	Johnston	1106

STATE	COUNTY	AGGREGATED SUB-AREA (ASA) NUMBER
Oklahoma (continued)	Kay	1103
	Kingfisher	1103
	Kiowa	1106
	Latimer	1104
	Le Flore	1104
	Lincoln	1105
	Logan	1103
	Love	1106
	McClain	1105
	McCurtain	1107
	McIntosh	1105
	Major	1103
	Marshall	1106
	Mayes	1104
	Murray	1106
	Muskogee	1104
	Noble	1103
	Nowata	1104
	Okfuskee	1105
	Oklahoma	1105
	Okmulgee	1105
	Osage	1104
	Ottawa	1104
	Pawnee	1103
	Payne	1103
	Pittsburg	1105
	Pontotoc	1107
	Pottawatomie	1105
	Pushmataha	1107
	Roger Mills	1106
	Rogers	1104
	Seminole	1105
	Sequoyah	1104
	Stephens	1106
	Texas	1105
	Tillman	1106
	Tulsa	1104
	Wagoner	1104
	Washington	1104
	Washita	1106
	Woods	1103
	Woodward	1105
Oregon	Baker	1703
	Benton	1705
	Clackamas	1705
	Clatsop	1705

STATE	COUNTY	AGGREGATED SUB-AREA (ASA) NUMBER
Oregon (continued)	Columbia	1705
	Coos	1705
	Crook	1702
	Curry	1705
	Deschutes	1702
	Douglas	1705
	Gilliam	1702
	Grant	1702
	Harney	1707
	Hood River	1702
	Jackson	1705
	Jefferson	1702
	Josephine	1705
	Klamath	1801
	Lake	1707
	Lane	1705
	Lincoln	1705
	Linn	1705
	Malheur	1703
	Marion	1705
	Morrow	1702
	Multnomah	1705
	Polk	1705
	Sherman	1702
	Tillamook	1705
	Umatilla	1702
	Union	1704
	Wallowa	1704
	Wasco	1702
	Washington	1705
	Wheeler	1702
	Yamhill	1705
Pennsylvania	Adams	204
	Allegheny	502
	Armstrong	501
	Beaver	502
	Bedford	204
	Berks	203
	Blair	204
	Bradford	204
	Bucks	203
	Butler	502
	Cambria	501
	Cameron	204
	Carbon	203
	Centre	204

STATE	COUNTY	AGGREGATED SUB-AREA (ASA) NUMBER
Pennsylvania (continued)	Chester	203
	Clarion	501
	Clearfield	204
	Clinton	204
	Columbia	204
	Crawford	501
	Cumberland	204
	Dauphin	204
	Delaware	203
	Elk	501
	Erie	407
	Fayette	501
	Forest	501
	Franklin	206
	Fulton	206
	Greene	501
	Huntingdon	204
	Indiana	501
	Jefferson	501
	Juniata	204
	Lackawanna	204
	Lancaster	204
	Lawrence	502
	Lebanon	204
	Lehigh	203
	Luzerne	204
	Lycoming	204
	McKean	501
	Mercer	502
	Mifflin	204
	Monroe	203
	Montgomery	203
	Montour	204
	Northampton	203
	Northumberland	204
	Perry	204
	Philadelphia	203
	Pike	203
	Potter	204
	Schuylkill	204
	Snyder	204
	Somerset	501
	Sullivan	204
	Susquehanna	204
	Tioga	204
	Union	204
	Venango	501

STATE	COUNTY	AGGREGATED SUB-AREA (ASA) NUMBER
Pennsylvania (continued)	Warren	501
	Washington	502
	Wayne	203
	Westmoreland	502
	Wyoming	204
	York	204
Rhode Island	Bristol	103
	Kent	103
	Newport	103
	Providence	103
	Washington	103
South Carolina	Abbeville	303
	Aiken	303
	Allendale	302
	Anderson	303
	Bamberg	302
	Barnwell	302
	Beaufort	302
	Berkeley	302
	Calhoun	302
	Charleston	302
	Cherokee	302
	Chester	302
	Chesterfield	302
	Clarendon	302
	Colleton	302
	Darlington	302
	Dillon	302
	Dorchester	302
	Edgefield	303
	Fairfield	302
	Florence	302
	Georgetown	302
	Greenville	302
	Greenwood	302
	Hampton	302
	Horry	302
	Jasper	302
	Kershaw	302
	Lancaster	302
	Laurens	302
	Lee	302
	Lexington	302
	McCormick	303
	Marion	302

STATE	COUNTY	AGGREGATED SUB-AREA (ASA) NUMBER
South Carolina (continued)	Marlboro	302
	Newberry	302
	Oconee	303
	Orangeburg	302
	Pickens	302.
	Richland	302
	Saluda	302
	Spartanburg	302
	Sumter	302
	Union	302
	Williamsburg	302
	York	302
South Dakota	Aurora	1006
	Beadle	1006
	Bennett	1005
	Bon Homme	1006
	Brookings	1006
	Brown	1006
	Brule	1005
	Buffalo	1005
	Butte	1005
	Campbell	1005
	Charles Mix	1005
	Clark	1006
	Clay	1006
	Codington	1006
	Corson	1005
	Custer	1005
	Davison	1006
	Day	1006
	Deuel	1006
	Dewey	1005
	Douglas	1006
	Edmunds	1006
	Fall River	1005
	Faulk	1006
	Grant	701
	Gregory	1005
	Haakon	1005
	Hamlin	1006
	Hand	1006
	Hanson	1006
	Harding	1005
	Hughes	1005
	Hutchinson	1006
	Hyde	1005

STATE	COUNTY	AGGREGATED SUB-AREA (ASA) NUMBER
South Dakota (continued)	Jackson	1005
	Jerauld	1006
	Jones	1005
	Kingsbury	1006
	Lake	1006
	Lawrence	1005
	Lincoln	1006
	Lyman	1005
	McCook	1006
	McPherson	1005
	Marshall	1006
	Meade	1005
	Mcllette	1005
	Miner	1006
	Minnehaha	1006
	Moody	1006
	Pennington	1005
	Perkins	1005
	Potter	1005
	Roberts	701
	Sanborn	1006
	Shannon	1005
	Spink	1006
	Stanley	1005
	Sully	1005
	Todd	1005
	Tripp	1005
	Turner	1006
	Union	1006
	Walworth	1005
	Washabaugh	1005
	Yankton	1006
	Ziebach	1005
Tennessee	Anderson	601
	Bedford	602
	Benton	602
	Bledsoe	601
	Blount	601
	Bradley	601
	Campbell	601
	Cannon	507
	Carroll	801
	Carter	601
	Cheatham	507
	Chester	801
	Claiborne	601

STATE	COUNTY	AGGREGATED SUB-AREA (ASA) NUMBER
Tennessee (continued)	Clay	507
	Cocke	601
	Coffee	602
	Crockett	801
	Cumberland	601
	Davidson	507
	Decatur	602
	De Kalb	507
	Dickson	507
	Dyer	801
	Fayette	801
	Fentress	507
	Franklin	602
	Gibson	801
	Giles	602
	Grainger	601
	Greene	601
	Grundy	601
	Hamblen	601
	Hamilton	601
	Hancock	601
	Hardeman	801
	Hardin	602
	Hawkins	601
	Haywood	801
	Henderson	602
	Henry	602
	Hickman	602
	Houston	507
	Humphreys	602
	Jackson	507
	Jefferson	601
	Johnson	601
	Knox	601
	Lake	801
	Lauderdale	801
	Lawrence	602
	Lewis	602
	Lincoln	602
	Loudon	601
	McMinn	601
	McNairy	801
	Macon	507
	Madison	801
	Marion	601
	Marshall	602
	Maury	602

STATE	COUNTY	AGGREGATED SUB-AREA (ASA) NUMBER
Tennessee (continued)	Meigs	601
	Monroe	601
	Montgomery	507
	Moore	602
	Morgan	601
	Obion	801
	Overton	507
	Perry	602
	Pickett	507
	Polk	601
	Putnam	507
	Rhea	601
	Roane	601
	Robertson	507
	Rutherford	507
	Scott	507
	Sequatchie	601
	Sevier	601
	Shelby	801
	Smith	507
	Stewart	507
	Sullivan	601
	Sumner	507
	Tipton	801
	Trousdale	507
	Unicoi	601
	Union	601
	Van Buren	507
	Warren	507
	Washington	601
	Wayne	602
	Weakley	801
	White	507
	Williamson	507
	Wilson	507
Texas	Anderson	1202
	Andrews	1204
	Angelina	1201
	Aransas	1205
	Archer	1106
	Armstrong	1106
	Atascosa	1205
	Austin	1203
	Bailey	1203
	Bandera	1205
	Bastrop	1204

STATE	COUNTY	AGGREGATED SUB-AREA (ASA) NUMBER
Texas (continued)	Baylor	1106
	Bee	1205
	Bell	1203
	Bexar	1205
	Blanco	1204
	Borden	1204
	Bosque	1203
	Bowie	1107
	Brazoria	1202
	Brazos	1203
	Brewster	1303
	Briscoe	1106
	Brooks	1205
	Brown	1204
	Burleson	1203
	Burnet	1203
	Caldwell	1205
	Calhoun	1205
	Callahan	1204
	Cameron	1305
	Camp	1107
	Carson	1105
	Cass	1107
	Castro	1203
	Chambers	1202
	Cherokee	1201
	Childress	1106
	Clay	1106
	Cochran	1204
	Coke	1204
	Coleman	1204
	Collin	1202
	Collingsworth	1106
	Colorado	1204
	Comal	1205
	Comanche	1203
	Concho	1204
	Cooke	1202
	Coryell	1203
	Cottle	1106
	Crane	1303
	Crockett	1303
	Crosby	1203
	Culberson	1303
	Dallam	1105
	Dallas	1202
	Dawson	1204

STATE	COUNTY	AGGREGATED SUB-AREA (ASA) NUMBER
Texas (continued)	Deaf Smith	1106
	Delta	1107
	Denton	1202
	De Witt	1205
	Dickens	1203
	Dimmit	1205
	Donley	1106
	Duval	1205
	Eastland	1203
	Ector	1204
	Edwards	1205
	Ellis	1202
	El Paso	1302
	Erath	1203
	Falls	1203
	Fannin	1107
	Fayette	1204
	Fisher	1203
	Floyd	1203
	Foard	1106
	Fort Bend	1202
	Franklin	1107
	Freestone	1202
	Frio	1205
	Gaines	1204
	Galveston	1202
	Garza	1203
	Gillespie	1204
	Glasscock	1204
	Goliad	1205
	Gonzales	1205
	Gray	1106
	Grayson	1106
	Gregg	1201
	Grimes	1203
	Guadalupe	1205
	Hale	1203
	Hall	1106
	Hamilton	1203
	Hansford	1105
	Hardeman	1106
	Hardin	1201
	Harris	1202
	Harrison	1107
	Hartley	1105
	Haskell	1203
	Hays	1204

STATE	COUNTY	AGGREGATED SUB-AREA (ASA) NUMBER
Texas (continued)	Hemphill	1105
	Henderson	1202
	Hidalgo	1305
	Hill	1203
	Hockley	1203
	Hood	1203
	Hopkins	1107
	Houston	1202
	Howard	1204
	Hudspeth	1302
	Hunt	1201
	Hutchinson	1105
	Irion	1204
	Jack	1202
	Jackson	1205
	Jasper	1201
	Jeff Davis	1303
	Jefferson	1201
	Jim Hogg	1205
	Jim Wells	1205
	Johnson	1202
	Jones	1203
	Karnes	1205
	Kaufman	1202
	Kendall	1205
	Kenedy	1205
	Kent	1203
	Kerr	1205
	Kimble	1204
	King	1106
	Kinney	1305
	Kleberg	1205
	Knox	1106
	Lamar	1107
	Lamb	1203
	Lampasas	1203
	La Salle	1205
	Lavaca	1205
	Lee	1203
	Leon	1202
	Liberty	1202
	Limestone	1203
	Lipscomb	1105
	Live Oak	1205
	Llano	1204
	Loving	1303
	Lubbock	1203

STATE	COUNTY	AGGREGATED SUB-AREA (ASA) NUMBER
Texas (continued)	Lynn	1203
	McCulloch	1204
	McLennan	1203
	McMullen	1205
	Madison	1202
	Marion	1107
	Martin	1204
	Mason	1204
	Matagorda	1204
	Maverick	1305
	Medina	1205
	Menard	1204
	Midland	1204
	Milam	1203
	Mills	1204
	Mitchell	1204
	Montague	1106
	Montgomery	1202
	Moore	1105
	Morris	1107
	Motley	1106
	Nacogdoches	1201
	Navarro	1202
	Newton	1201
	Nolan	1204
	Nueces	1205
	Ochiltree	1105
	Oldham	1105
	Orange	1201
	Palo Pinto	1203
	Panola	1201
	Parker	1203
	Parmer	1203
	Pecos	1303
	Polk	1202
	Potter	1106
	Presidio	1303
	Rains	1201
	Randall	1106
	Reagan	1204
	Real	1205
	Red River	1107
	Reeves	1303
	Refugio	1205
	Roberts	1105
	Robertson	1203
	Rockwall	1202

STATE	COUNTY	AGGREGATED SUB-AREA (ASA) NUMBER
Texas (continued)	Runnels	1204
	Rusk	1201
	Sabine	1201
	San Augustine	1201
	San Jacinto	1202
	San Patricio	1205
	San Saba	1204
	Schleicher	1204
	Scurry	1204
	Shackelford	1203
	Shelby	1201
	Sherman	1105
	Smith	1201
	Somervell	1203
	Starr	1305
	Stephens	1203
	Sterling	1204
	Stonewall	1203
	Sutton	1305
	Swisher	1106
	Tarrant	1202
	Taylor	1203
	Terrell	1303
	Terry	1204
	Throckmorton	1203
	Titus	1107
	Tom Green	1204
	Travis	1204
	Trinity	1202
	Tyler	1201
	Upshur	1107
	Upton	1303
	Uvalde	1205
	Val Verde	1305
	Van Zandt	1201
	Victoria	1205
	Walker	1202
	Waller	1203
	Ward	1303
	Washington	1203
	Webb	1305
	Wharton	1204
	Wheeler	1106
	Wichita	1106
	Wilbarger	1106
	Willacy	1305
	Williamson	1203

STATE	COUNTY	AGGREGATED SUB-AREA (ASA) NUMBER
Texas (continued)	Wilson	1205
	Winkler	1303
	Wise	1202
	Wood	1201
	Yoakum	1204
	Young	1203
	Zapata	1305
	Zavala	1205
Utah	Beaver	1602
	Box Elder	1601
	Cache	1601
	Carbon	1401
	Daggett	1401
	Davis	1601
	Duchesne	1401
	Emery	1401
	Garfield	1403
	Grand	1402
	Iron	1602
	Juab	1602
	Kane	1403
	Millard	1602
	Morgan	1601
	Piute	1602
	Rich	1601
	Salt Lake	1601
	San Juan	1403
	Sanpete	1602
	Sevier	1602
	Summit	1601
	Tooele	1601
	Uintah	1401
	Utah	1601
	Wasatch	1601
	Washington	1502
	Wayne	1403
	Weber	1601
Vermont	Addison	106
	Bennington	201
	Caledonia	105
	Chittenden	106
	Essex	105
	Franklin	106
	Grand Isle	106
	Lamoille	106

STATE	COUNTY	AGGREGATED SUB-AREA (ASA) NUMBER
Vermont (continued)	Orange	105
	Orleans	106
	Rutland	106
	Washington	105
	Windham	105
	Windsor	105
Virginia	Accomack	205
	Albemarle	205
	Alexandria	206
	Alleghany	205
	Amelia	205
	Amherst	205
	Appomattox	205
	Arlington	206
	Augusta	206
	Bath	205
	Bedford	301
	Bland	504
	Botetourt	205
	Bristol	601
	Brunswick	301
	Buchanan	502
	Buckingham	205
	Buena Vista	205
	Campbell	205
	Caroline	205
	Carroll	504
	Charles City	205
	Charlotte	301
	Charlottesville	205
	Chesapeake	205
	Chesterfield	205
	Clarke	206
	Clifton Forge	205
	Colonial Heights	205
	Covington	205
	Craig	205
	Culpeper	205
	Cumberland	205
	Danville	301
	Dickenson	502
	Dinwiddie	301
	Emporia	301
	Essex	205
	Fairfax	206
	Falls Church	206

STATE	COUNTY	AGGREGATED SUB-AREA (ASA) NUMBER
Virginia (continued)	Fauquier	206
	Floyd	504
	Fluvanna	205
	Franklin	301
	Frederick	206
	Fredericksburg	205
	Galax	504
	Giles	504
	Gloucester	205
	Goochland	205
	Grayson	504
	Greene	205
	Greenville	301
	Halifax	301
	Hampton	205
	Hanover	205
	Harrisonburg	206
	Henrico	205
	Henry	301
	Highland	205
	Hopewell	205
	Isle of Wight	205
	James City	205
	King and Queen	205
	King George	206
	King William	205
	Lancaster	205
	Lee	601
	Lexington	205
	Loudoun	206
	Louisa	205
	Lunenburg	301
	Lynchburg	205
	Madison	205
	Manassas	206
	Manassas Park	206
	Martinsville	301
	Mathews	205
	Mecklenburg	301
	Middlesex	205
	Montgomery	504
	Nansemond	205
	Nelson	205
	New Kent	205
	Newport News	205
	Norfolk	205
	Northampton	205

STATE	COUNTY	AGGREGATED SUB-AREA (ASA) NUMBER
Virginia (continued)	Northumberland	205
	Norton	601
	Nottoway	205
	Orange	205
	Page	206
	Patrick	301
	Petersburg	301
	Pittsylvania	301
	Poquoson	205
	Portsmouth	205
	Powhatan	205
	Prince Edward	205
	Prince George	205
	Prince William	206
	Pulaski	504
	Radford	504
	Rapahannock	205
	Richmond	205
	Roanoke	301
	Rockbridge	205
	Rockingham	206
	Russell	601
	Salem	301
	Scott	601
	Shenandoah	206
	Smyth	601
	Southampton	301
	South Boston	301
	Spotsylvania	205
	Stafford	206
	Staunton	206
	Suffolk	205
	Surry	205
	Sussex	301
	Tazewell	601
	Virginia Beach	205
	Warren	206
	Washington	601
	Waynesboro	206
	Westmoreland	206
	Williamsburg	205
	Winchester	206
	Wise	601
	Wythe	504
	York	205
Washington	Adams	1702

STATE	COUNTY	AGGREGATED SUB-AREA (ASA) NUMBER
Washington (continued)	Asotin	1704
	Benton	1702
	Chelan	1702
	Clallam	1706
	Clark	1705
	Columbia	1702
	Cowlitz	1705
	Douglas	1702
	Ferry	1702
	Franklin	1702
	Garfield	1704
	Grant	1702
	Grays Harbor	1705
	Island	1706
	Jefferson	1706
	King	1706
	Kitsap	1706
	Kittitas	1702
	Klickitat	1702
	Lewis	1705
	Lincoln	1702
	Mason	1706
	Okanogan	1702
	Pacific	1705
	Pend Oreille	1701
	Pierce	1706
	San Juan	1706
	Skagit	1706
	Skamania	1705
	Snohomish	1706
	Spokane	1701
	Stevens	1702
	Thurston	1706
	Wahkiakum	1705
	Walla Walla	1702
	Whatcom	1706
	Whitman	1704
	Yakima	1702
West Virginia	Barbour	501
	Berkeley	206
	Boone	504
	Braxton	504
	Brooke	502
	Cabell	502
	Calhoun	502
	Clay	504

STATE	COUNTY	AGGREGATED SUB-AREA (ASA) NUMBER
West Virginia (continued)	Doddridge	502
	Fayette	504
	Gilmer	502
	Grant	206
	Greenbrier	504
	Hampshire	206
	Hancock	502
	Hardy	206
	Harrison	501
	Jackson	502
	Jefferson	206
	Kanawha	504
	Lewis	501
	Lincoln	502
	Logan	502
	McDowell	502
	Marion	501
	Marshall	502
	Mason	504
	Mercer	504
	Mineral	206
	Mingo	502
	Monongalia	501
	Monroe	504
	Morgan	206
	Nicholas	504
	Ohio	502
	Pendleton	206
	Pleasants	502
	Pocahontas	504
	Preston	501
	Putnam	504
	Raleigh	504
	Randolph	501
	Ritchie	502
	Roane	502
	Summers	504
	Taylor	501
	Tucker	501
	Tyler	502
	Upshur	501
	Wayne	502
	Webster	504
	Wetzel	502
	Wirt	502
	Wood	502
	Wyoming	502

STATE	COUNTY	AGGREGATED SUB-AREA (ASA) NUMBER
Wisconsin	Adams	702
	Ashland	401
	Barron	702
	Bayfield	401
	Brown	402
	Buffalo	702
	Burnett	701
	Calumet	402
	Chippewa	702
	Clark	702
	Columbia	702
	Crawford	702
	Dane	703
	Dodge	703
	Door	402
	Douglas	401
	Dunn	702
	Eau Claire	702
	Florence	402
	Fond du Lac	402
	Forest	402
	Grant	703
	Green	703
	Green Lake	402
	Iowa	703
	Iron	401
	Jackson	702
	Jefferson	703
	Juneau	702
	Kenosha	403
	Kewaunee	402
	La Crosse	702
	Lafayette	703
	Langlade	402
	Lincoln	702
	Manitowoc	402
	Marathon	702
	Marinette	402
	Marquette	402
	Menominee	402
	Milwaukee	403
	Monroe	702
	Oconto	402
	Oneida	702
	Outagamie	402
	Ozaukee	403
	Pepin	702

STATE	COUNTY	AGGREGATED SUB-AREA (ASA) NUMBER
Wisconsin (continued)	Pierce	702
	Polk	701
	Portage	702
	Price	702
	Racine	403
	Richland	702
	Rock	703
	Rusk	702
	St. Croix	701
	Sauk	702
	Sawyer	702
	Shawano	402
	Sheboygan	402
	Taylor	702
	Trempealeau	702
	Vernon	702
	Vilas	702
	Walworth	403
	Washburn	701
	Washington	403
	Waukesha	403
	Waupaca	402
	Waushara	402
	Winnebago	402
	Wood	702
Wyoming	Albany	1007
	Big Horn	1004
	Campbell	1004
	Carbon	1007
	Converse	1007
	Crook	1005
	Fremont	1004
	Goshen	1007
	Hot Springs	1004
	Johnson	1004
	Laramie	1007
	Lincoln	1401
	Natrona	1007
	Niobrara	1005
	Park	1004
	Platte	1007
	Sheridan	1004
	Sublette	1401
	Sweetwater	1401
	Teton	1703
	Uinta	1401

STATE	COUNTY	AGGREGATED SUB-AREA (ASA) NUMBER
Wyoming (continued)	Washakie	1004
	Weston	1005

APPENDIX C

NEEDS AND POTENTIAL FOR ARTIFICIAL RECHARGE

		NEEDS							POTENTIAL				
ASA	State	Year 1985 Quantity Problem Indicated	*1000 Acres Irrigated Lands (ASA)	Sum	Year 2000 Δ_1	Sum	Year 2030 Δ_2	Sum	Heath Region No.	H,M,L	Field Studies (Published)	Verbal Information	Sum
101	ME	X	3	P	0	N	0	N	9	L			L
102			11	P		N		N	9	L			L
	ME	X			0		0		9	L			
	NH	X			0		0		9	L			
103			22	P		N		N	7 9	M L	X	X	MSV
	MA	X			0		0		7 9	M L			
	RI	X			0		0		9	L			

		NEEDS							POTENTIAL			
ASA	State	Year 1985 Quantity Problem Indicated	*1000 Acres Irrigated Lands (ASA)	Sum	Year 2000 Δ_1	Sum	Year 2030 Δ_2	Sum	Heath Region No. H,M,L	Field Studies (Published)	Verbal Information	Sum
104			7	P		N		N	7 M 9 L			M
	CT	X			0		0		7 M 9 L			
	MA	-			0		0		9 L			
105			9	X		N		N	9 L			L
	CT	-			0		0		9 L			
	MA	-			0		0		9 L			
	NH	-			0		0		9 L			

		NEEDS							POTENTIAL				
ASA	State	Year 1985 Quantity Problem Indicated	*1000 Acres Irrigated Lands (ASA)	Sum	Year 2000 Δ_1	Sum	Year 2030 Δ_2	Sum	Heath Region No.	H,M,L	Field Studies (Published)	Verbal Information	Sum
105 (continued)													
	VT	-			0		0		9	L			
106			1	X		N		N	7 9	M L			M
	VT	-			0		0		7 9	M L			
	NY	-			0		0		7 9	M L			
201			4	P		N		N	7 9	M L			M
	NY	X			0		0		7 9	M L			

ASA	State	NEEDS: Year 1985 Quantity Problem Indicated	*1000 Acres Irrigated Lands (ASA)	Sum	Year 2000 Δ_1	Sum	Year 2030 Δ_2	Sum	POTENTIAL: Heath Region No.	H,M,L	Field Studies (Published)	Verbal Information	Sum
201 (continued)	VT	-			0		0		9	L			
202			40	P		I		I	6 9	M L	X	X	MSV
	NJ	X			7		19		6	M			
	NY	X			0		0		9	L			
203			127	P		I		I	6 7 8 10	M M L L	X	X	MSV
	MD	-			63		189		10	M			

ASA	State	NEEDS Year 1985 Quantity Problem Indicated	*1000 Acres Irrigated Lands (ASA)	Sum	Year 2000 Δ_1	Sum	Year 2030 Δ_2	Sum	POTENTIAL Heath Region No.	H,M,L	Field Studies (Published)	Verbal Information	Sum
203 (continued)	DE	X			65		195		10	M			
	NJ	X			8		19		6 10	M			
	NY	X			0		0		7	M			
	PA	-			0		0		6 8	M L			
204			18	X		N		N	6 7 8	M M L			M
	NY	-			0		0		7	M			

ASA	State	NEEDS: Year 1985 Quantity Problem Indicated	*1000 Acres Irrigated Lands (ASA)	Sum	Year 2000 Δ_1	Sum	Year 2030 Δ_2	Sum	POTENTIAL: Heath Region No. H,M,L	Field Studies (Published)	Verbal Information	Sum
204 (continued)												
	PA	-			0		0		6 M 7 M 8 L			
205			100	P		I		I	6 M 8 L 10 M	X		MS
	DE	-			65		195		10 M			
	MD	-			63		189		10 M			
	VA	X			19		50		6 M 8 L			

		NEEDS							POTENTIAL				
ASA	State	Year 1985 Quantity Problem Indicated	*1000 Acres Irrigated Lands (ASA)	Sum	Year 2000 Δ_1	Sum	Year 2030 Δ_2	Sum	Heath Region No.	H,M,L	Field Studies (Published)	Verbal Information	Sum
206			19	P		I		I	6 8	M L			M
	DC	-							6	M			
	MD	X			63		189		6 8	M L			
	PA	-			0		0		6	M			
	VA	-			19		50		6 8	M L			
	WV	X			0		0		6	M			

ASA	State	NEEDS: Year 1985 Quantity Problem Indicated	*1000 Acres Irrigated Lands (ASA)	Sum	Year 2000 Δ_1	Sum	Year 2030 Δ_2	Sum	POTENTIAL: Heath Region No.	H,M,L	Field Studies (Published)	Verbal Information	Sum
301			306	P		N		N	6 8 10	M L M	X		MS
	VA	X			19		50		6 8 10	M L M			
	NC	X			0		0		6 8 10	M L M			
302			96	P		I		I	8 10 11	L M H			H
	NC	-			0		0		8 10	L M			H

		NEEDS							POTENTIAL				
ASA	State	Year 1985 Quantity Problem Indicated	*1000 Acres Irrigated Lands (ASA)	Sum	Year 2000 Δ_1	Sum	Year 2030 Δ_2	Sum	Heath Region No.	H,M,L	Field Studies (Published)	Verbal Information	Sum
302 (continued)													
	SC	X			129		388		8 10 11	L M H			
303			304	P		I		I	8 10 11	L M H			H
	FL	X			0		0		11	H			
	GA	X			75		224		8 11 10	L H M			
	SC	X			129		388		8 (10)	L (M)			

		NEEDS							POTENTIAL				
ASA	State	Year 1985 Quantity Problem Indicated	*1000 Acres Irrigated Lands (ASA)	Sum	Year 2000 Δ_1	Sum	Year 2030 Δ_2	Sum	Heath Region No.	H,M,L	Field Studies (Published)	Verbal Information	Sum
304			1061	P		I		I	11	H	X	X	HSV
	FL	X			0		0		11	H			
	GA	X			75		224		11	H			
305			1502	P		N		N	11	H	X		HS
	FL	X			0		0		11	H			
306			620	P		I		I	8 10 11	L M H			H

ASA	State	NEEDS: Year 1985 Quantity Problem Indicated	*1000 Acres Irrigated Lands (ASA)	Sum	Year 2000 Δ_1	Sum	Year 2030 Δ_2	Sum	POTENTIAL: Heath Region No.	H,M,L	Field Studies (Published)	Verbal Information	Sum
306 (continued)													
	AL	X			65		196		11 8 10	H L M			
	FL	X			0		0		11	H			
	GA	X			75		224		8 10 11	L M H			
307			76	P		I		I	6 10 11 (8)	M M H (L)	X		HS

ASA	State	NEEDS: Year 1985 Quantity Problem Indicated	*1000 Acres Irrigated Lands (ASA)	Sum	Year 2000 Δ_1	Sum	Year 2030 Δ_2	Sum	POTENTIAL: Heath Region No.	H,M,L	Field Studies (Published)	Verbal Information	Sum
307 (continued)													
	AL	X			65		196		6 10 11 (8)	M M H (L)			
	FL	X			0		0		10 11	M H			
	GA	-			75		224		6 (8)	M (L)			
308			12	P		I		I	10 6	M			M
	AL	-			65		196		10 6	M			

ASA	State	NEEDS Year 1985 Quantity Problem Indicated	*1000 Acres Irrigated Lands (ASA)	Sum	Year 2000 Δ_1	Sum	Year 2030 Δ_2	Sum	POTENTIAL Heath Region No.	H,M,L	Field Studies (Published)	Verbal Information	Sum
308 (continued)													
	MS	X			77		231		10	M			
309	(LA)												
	MS	X	0	P	77	I	231	I	10	M			M
401			5	P		I		I	9 (7)	L (M)			L
	MN	-			69		208		9	L			
	MI	X			72		217		7 9	M L			
	WI	-			51		152		9	L			

		NEEDS							POTENTIAL				
ASA	State	Year 1985 Quantity Problem Indicated	*1000 Acres Irrigated Lands (ASA)	Sum	Year 2000 Δ_1	Sum	Year 2030 Δ_2	Sum	Heath Region No.	H,M,L	Field Studies (Published)	Verbal Information	Sum
402			119	P		I		I	7 9	M L			M
	MI	X			72		217		7 9	M L			
	WI	X			51		152		7 9	M L			
403			59	P		I		I	7	M	X	X	MSV
	WI	X			51		152		7	M			
	IL	X			69		206		7	M			
	IN	X			101		304		7	M			

ASA	State	NEEDS: Year 1985 Quantity Problem Indicated	*1000 Acres Irrigated Lands (ASA)	Sum	Year 2000 Δ_1	Sum	Year 2030 Δ_2	Sum	POTENTIAL: Heath Region No.	H,M,L	Field Studies (Published)	Verbal Information	Sum
404			390	P		I		I	7	M	X		MS
	MI	-			72		217		7	M			
	IN	X			101		304		7	M			
405	MI	-	36	X	72	N	217	N	7	M	X		MS
406			25	P		I			7	M			M
	IN	X			101		304		7	M			
	OH	-			18		48		7	M			
	MI	-			72		217		7	M			

		NEEDS							POTENTIAL				
ASA	State	Year 1985 Quantity Problem Indicated	*1000 Acres Irrigated Lands (ASA)	Sum	Year 2000 Δ_1	Sum	Year 2030 Δ_2	Sum	Heath Region No.	H,M,L	Field Studies (Published)	Verbal Information	Sum
407			6	P		N		N	7	M			M
	NY	X			0		0		7	M			
	PA	-			0		0		7	M			
	OH	-			18		48		7	M			
408	NY	-	14	X	0	N	0	N	7 (9)	M (L)			M
501			0	P		N		N	6 (7)	M			M
	PA	-			0		0		6 (7)	M			

		NEEDS							POTENTIAL				
ASA	State	Year 1985 Quantity Problem Indicated	*1000 Acres Irrigated Lands (ASA)	Sum	Year 2000 Δ_1	Sum	Year 2030 Δ_2	Sum	Heath Region No.	H,M,L	Field Studies (Published)	Verbal Information	Sum
501 (continued)	WV	X			0		0		6	M			
502			5	P		I		I	6 (7)	M			M
	KY	X			84		251		6	M			
	IN	-			101		304		7	M			
	PA	-			0		0		6 7	M			
	OH	-			18		48		6 7	M			
	VA	-			19		50		6	M			
	WV	X			0		0		6	M			

		NEEDS							POTENTIAL				
ASA	State	Year 1985 Quantity Problem Indicated	*1000 Acres Irrigated Lands (ASA)	Sum	Year 2000 Δ_1	Sum	Year 2030 Δ_2	Sum	Heath Region No.	H,M,L	Field Studies (Published)	Verbal Information	Sum
503			18	P		I		I	6 7	M	X		MS
	IN	X			101		304		7	M			
	OH	-			18		48		7 6	M			
504			0	P		I		I	6 (8)	M (L)	X		MS
	VA	-			19		50		6	M			
	WV	X			0		0		6	M			
	NC	-			0		0		8	L			

		NEEDS							POTENTIAL				
ASA	State	Year 1985 Quantity Problem Indicated	*1000 Acres Irrigated Lands (ASA)	Sum	Year 2000 Δ_1	Sum	Year 2030 Δ_2	Sum	Heath Region No.	H,M,L	Field Studies (Published)	Verbal Information	Sum
505			20	P		I		I	6 (10)	M		X	MV
	KY	X			84		251		6 (10)	M			
	IL	-			69		206		6 10	M			
	IN	-			101		304		6	M			
506			65	P		I		I	7 (6)	M			M
	IL	-			69		206		7 (6)	M			

		NEEDS							POTENTIAL				
ASA	State	Year 1985 Quantity Problem Indicated	*1000 Acres Irrigated Lands (ASA)	Sum	Year 2000 Δ_1	Sum	Year 2030 Δ_2	Sum	Heath Region No.	H,M,L	Field Studies (Published)	Verbal Information	Sum
506 (continued)													
	IN	X			101		304		7 6	M			
507			5	P		I		I	6	M			M
	KY	X			84		251		6	M			
	IN	X			65		195		6	M			
601			11	P		I		I	6 8	M L			M
	GA	-			75		224		6 8	M L			
	NC	X			0		0		8	L			

		NEEDS							POTENTIAL				
ASA	State	Year 1985 Quantity Problem Indicated	*1000 Acres Irrigated Lands (ASA)	Sum	Year 2000 Δ_1	Sum	Year 2030 Δ_2	Sum	Heath Region No.	H,M,L	Field Studies (Published)	Verbal Information	Sum
601 (continued)													
	TN	X			65		195		6	M			
	VA	-			19		50		6	M			
602			9	P		I		I	6 10	M			M
	KY	-			84		251		10	M			
	AL	-			65		196		6 10	M			
	TN	X			65		195		6 10	M			

		NEEDS							POTENTIAL			
ASA	State	Year 1985 Quantity Problem Indicated	*1000 Acres Irrigated Lands (ASA)	Sum	Year 2000 Δ_1	Sum	Year 2030 Δ_2	Sum	Heath Region No. H,M,L	Field Studies (Published)	Verbal Information	Sum
701			354	P		I		I	7 M 9 L	X		MS
	SD	X			89		198		7 M			
	MN	X			69		208		7 M 9 L			
	WI	-			51		152		7 M 9 L			
702			185	P		I		I	7 M 9 L	X		MS
	WI	X			51		152		7 M 9 L			

ASA	State	NEEDS Year 1985 Quantity Problem Indicated	*1000 Acres Irrigated Lands (ASA)	Sum	Year 2000 Δ_1	Sum	Year 2030 Δ_2	Sum	POTENTIAL Heath Region No.	H,M,L	Field Studies (Published)	Verbal Information	Sum
702 (continued)													
	MN	X			69		208		7 9	M L			
703			111	P		I		I	6 7	M		X	MV
	IL	-			69		206		7	M			
	IA	X			41		123		6 7	M			
	MN	-			69		208		6 7	M			
	WI	-			51		152		6 7	M			

		NEEDS							POTENTIAL				
ASA	State	Year 1985 Quantity Problem Indicated	*1000 Acres Irrigated Lands (ASA)	Sum	Year 2000 Δ_1	Sum	Year 2030 Δ_2	Sum	Heath Region No.	H,M,L	Field Studies (Published)	Verbal Information	Sum
704													
			138	P		I		I	6 7	M	X	X	MSV
	IL	-			69		206		6 7	M			
	IN	X			101		304		7	M			
	MO	X			64		191		7	M			
705													
			16	X		I		I	6 7	M			M
	IL	-			69		206		6 7	M			

		NEEDS							POTENTIAL				
ASA	State	Year 1985 Quantity Problem Indicated	*1000 Acres Irrigated Lands (ASA)	Sum	Year 2000 Δ_1	Sum	Year 2030 Δ_2	Sum	Heath Region No.	H,M,L	Field Studies (Published)	Verbal Information	Sum
705 (continued)													
	MO	-			64		191		6 7	M			
801			2856	P		I		I	6 10	M	X		MS
	AR	X			51		153		6 10	M			
	MO	-			64		191		6 10	M			
	TN	-			65		195		10	M			
	MS	-			77		231		10	M			

		NEEDS							POTENTIAL				
ASA	State	Year 1985 Quantity Problem Indicated	*1000 Acres Irrigated Lands (ASA)	Sum	Year 2000 $\Delta 1$	Sum	Year 2030 $\Delta 2$	Sum	Heath Region No.	H,M,L	Field Studies (Published)	Verbal Information	Sum
801 (continued)	KY	-			84		251		10	M			
802			1422	P		I		I	6 10	M			M
	MS	X			77		231		10	M			
	LA	X			9		22		10	M			
	AR	X			51		153		6 10	M			
803			1053	P		I		I	10	M	X		MS
	LA	X			9		22		10	M			

		NEEDS							POTENTIAL				
ASA	State	Year 1985 Quantity Problem Indicated	*1000 Acres Irrigated Lands (ASA)	Sum	Year 2000 Δ_1	Sum	Year 2030 Δ_2	Sum	Heath Region No.	H,M,L	Field Studies (Published)	Verbal Information	Sum
803 (continued)													
	MS	-			77		231		10	M			
901			95	P		I		I	7 9	M (L)	X	X	MSV
	ND	X			164		269		7	M			
	MN	X			69		208		7 9	M L			
1001													
	MT	X	206	P	8	I	9	I	6 7	M			M

		NEEDS							POTENTIAL			
ASA	State	Year 1985 Quantity Problem Indicated	*1000 Acres Irrigated Lands (ASA)	Sum	Year 2000 Δ_1	Sum	Year 2030 Δ_2	Sum	Heath Region No. H,M,L	Field Studies (Published)	Verbal Information	Sum
1002												
	MT	-	998	X	8	I	9	I	1 L 6 M 7 M			M
1003												
	MT	-	193	X	8	I	9	I	6 M			M
1004												
			1287	P		I		I	1 L 4 L 6 M	X		MS
	MT	X			8		9		1 L 6 M			
	WY	-			0		0		1 L 4 L 6 M			

ASA	State	NEEDS: Year 1985 Quantity Problem Indicated	*1000 Acres Irrigated Lands (ASA)	Sum	Year 2000 Δ_1	Sum	Year 2030 Δ_2	Sum	POTENTIAL: Heath Region No.	H,M,L	Field Studies (Published)	Verbal Information	Sum
1005			399	P		I		I	5 6 7	H M M	X		HS
	MT	X			8		9		6	M			
	WY	-			0		0		6	M			
	ND	X			164		269		6 7	M			
	SD	X			89		198		5 6 7	H M M			
1006			390	P		I		I	6	M			MS

		NEEDS							POTENTIAL				
ASA	State	Year 1985 Quantity Problem Indicated	*1000 Acres Irrigated Lands (ASA)	Sum	Year 2000 Δ_1	Sum	Year 2030 Δ_2	Sum	Heath Region No.	H,M,L	Field Studies (Published)	Verbal Information	Sum
1006 (continued)													
	ND	-			164		269		6	M			
	NE	-			0		0		6	M			
	SD	X			89		198		6	M			
	MN	X			69		208		6	M			
	IA	X			41		123		6	M			
1007			2079	P		N		N	1 4 5 6	L L H M	X	X	HSV

		NEEDS							POTENTIAL			
ASA	State	Year 1985 Quantity Problem Indicated	*1000 Acres Irrigated Lands (ASA)	Sum	Year 2000 Δ_1	Sum	Year 2030 Δ_2	Sum	Heath Region No. H,M,L	Field Studies (Published)	Verbal Information	Sum
1007 (continued)												
	CO	X			0		0		1 L 5 H 6 M			
	WY	-			0		0		1 L 4 L 5 H 6 M			
	NE	X			0		0		5 H 6 M			
1008												
	NE		3892	P	0	N	0	N	5 H 1 L 7 M	X	X	HSV

		NEEDS							POTENTIAL			
ASA	State	Year 1985 Quantity Problem Indicated	*1000 Acres Irrigated Lands (ASA)	Sum	Year 2000 Δ_1	Sum	Year 2030 Δ_2	Sum	Heath Region No. H,M,L	Field Studies (Published)	Verbal Information	Sum
1009			200	P		I		I	7 M		X	MV
	NE	-			0		0		7 M			
	MO	-			64		191		7 M			
	KS	-			32		95		7 M			
	IA	X			41		123		7 M			
1010			4003	P		I		I	5 H 6 M 7 M	X	X	HSV
	CO	X			0		0		5 H (6) (M)			

ASA	State	Year 1985 Quantity Problem Indicated	*1000 Acres Irrigated Lands (ASA)	Sum	Year 2000 Δ_1	Sum	Year 2030 Δ_2	Sum	Heath Region No. H,M,L	Field Studies (Published)	Verbal Information	Sum
		NEEDS							POTENTIAL			
1010 (continued)												
	KS	X			32		95		5 H 6 M 7 M			
	NE	X			0		0		5 H 6 M (7)			
1011			120	P		I		I	6 7 M		X	MV
	KS	-			31		95		6 (7) M			
	IA	X			41		123		7 M			

ASA	State	NEEDS: Year 1985 Quantity Problem Indicated	*1000 Acres Irrigated Lands (ASA)	Sum	Year 2000 Δ_1	Sum	Year 2030 Δ_2	Sum	POTENTIAL: Heath Region No.	H,M,L	Field Studies (Published)	Verbal Information	Sum
1011 (continued)													
	MO	X			64		191		6 7	M			
1101			282	X		I		I	6	M	X		MS
	AR	-			51		153		6	M			
	MO	-			64		191		6	M			
1102			519	P		N		N	1 6 5	L M H	X	X	HSV
	CO	X			0		0		1 6 5	L M H			

		NEEDS							POTENTIAL				
ASA	State	Year 1985 Quantity Problem Indicated	*1000 Acres Irrigated Lands (ASA)	Sum	Year 2000 Δ_1	Sum	Year 2030 Δ_2	Sum	Heath Region No.	H,M,L	Field Studies (Published)	Verbal Information	Sum
1103			2837	P		I		I	5 6	H M	X	X	HSV
	CO	-			0		0		5 (6)	H (M)			
	OK	X			1		1		6 (5)	M (H)			
	KS	X			32		95		5 6	H M			
1104			139	P		I		I	6	M	X		MS
	KS	-			32		95		6	M			
	AR	-			51		153		6	M			

		NEEDS							POTENTIAL				
ASA	State	Year 1985 Quantity Problem Indicated	*1000 Acres Irrigated Lands (ASA)	Sum	Year 2000 Δ_1	Sum	Year 2030 Δ_2	Sum	Heath Region No.	H,M,L	Field Studies (Published)	Verbal Information	Sum
1104 (continued)													
	MO	X			64		191		6	M			
	OK	-			1		1		6	M			
1105			2051	P		I		I	1 5 6	L H M	X	X	HSV
	NM	X			104		104		1 5 6	L H M			
	OK	X			1		1		5 6	H M			
	TX	X			0.06		0.06		5 6	H M			

ASA	State	NEEDS: Year 1985 Quantity Problem Indicated	*1000 Acres Irrigated Lands (ASA)	Sum	Year 2000 Δ_1	Sum	Year 2030 Δ_2	Sum	POTENTIAL: Heath Region No.	H,M,L	Field Studies (Published)	Verbal Information	Sum
1106			1367	P		N		N	6 5	M H	X	X	HSV
	TX	X			0.06		0.06		6 5	M H			
	OK	X			1		1		6	M			
1107			81	P		I		I	6 10	M			M
	LA	X			9		22		10	M			
	AR	X			51		153		6 10	M			
	OK	–			1		1		6	M			

ASA	State	NEEDS: Year 1985 Quantity Problem Indicated	*1000 Acres Irrigated Lands (ASA)	Sum	Year 2000 Δ_1	Sum	Year 2030 Δ_2	Sum	POTENTIAL: Heath Region No.	H,M,L	Field Studies (Published)	Verbal Information	Sum
1107 (continued)													
	TX	-			0.06		0.06		6 10	M			
1201			155	P		I		I	10 (6)	M			M
	LA	-			9		22		10	M			
	TX	X			0.06		0.06		10 (6)	M			
1202													
	TX	X	629	P	0.06	N	0.06	N	6 10	M	X		MS

ASA	State	NEEDS							POTENTIAL			
		Year 1985 Quantity Problem Indicated	*1000 Acres Irrigated Lands (ASA)	Sum	Year 2000 Δ_1	Sum	Year 2030 Δ_2	Sum	Heath Region No. H,M,L	Field Studies (Published)	Verbal Information	Sum
1203			3536	P		I		I	6 M 5 H 10 M (2) (H)	X	X	HSV
	TX	X			0.06		0.06		6 M 5 H 10 M			
	NM	X			104		104		5 H (2)			
1204			1629	P		I		I	6 M 5 H 10 M (2) (H)	X	X	HSV

ASA	State	NEEDS: Year 1985 Quantity Problem Indicated	*1000 Acres Irrigated Lands (ASA)	Sum	Year 2000 Δ_1	Sum	Year 2030 Δ_2	Sum	POTENTIAL: Heath Region No.	H,M,L	Field Studies (Published)	Verbal Information	Sum
1204 (continued)													
	TX	X			0.06		0.06		6 5 10 (2)	M H M (H)			
	NM	X			104		104		5 (2)	H			
1205			654	P		N		N	6 10	M	X		MS
	TX	X			0.06		0.06		6 10	M			
1301	CO		619	P	0	N	0	N	1 2	L H	X		HS

ASA	State	NEEDS: Year 1985 Quantity Problem Indicated	NEEDS: *1000 Acres Irrigated Lands (ASA)	NEEDS: Sum	NEEDS: Year 2000 Δ_1	NEEDS: Sum	NEEDS: Year 2030 Δ_2	NEEDS: Sum	POTENTIAL: Heath Region No.	POTENTIAL: Heath Region H,M,L	POTENTIAL: Field Studies (Published)	POTENTIAL: Verbal Information	POTENTIAL: Sum
1302			669	P		I		I	2 1 4	H L L	X		HS
	NM	X			104		104		2 1 4	H L L			
	TX	X			0.06		0.06		2	H			
1303	TX	X	281	P	0.06	N	0.06	N	2 6	H M	X		HS
1304	NM	-	253	X	104	I	104	I	2 (1) (5) (6)	H (L) (H) (M)	X		HS

ASA	State	NEEDS: Year 1985 Quantity Problem Indicated	*1000 Acres Irrigated Lands (ASA)	Sum	Year 2000 Δ_1	Sum	Year 2030 Δ_2	Sum	POTENTIAL: Heath Region No.	H,M,L	Field Studies (Published)	Verbal Information	Sum
1305	TX	X	755	P	0.06	N	0.06	N	6 10	M			M
1401			857	X		N		N	5 1	H L	X		HS
	UT	-			1		1		5	H			
	WY	-			0		0		5 1	H L			
	CO	-			0		0		5 (1)	H (L)			
1402			503	X		N		N	4 1	L			L

		NEEDS							POTENTIAL				
ASA	State	Year 1985 Quantity Problem Indicated	*1000 Acres Irrigated Lands (ASA)	Sum	Year 2000 Δ_1	Sum	Year 2030 Δ_2	Sum	Heath Region No.	H,M,L	Field Studies (Published)	Verbal Information	Sum
1402 (continued)													
	UT	-			1		1		4	L			
	CO	-			0		0		4 1	L			
1403			313	P		I		I	4 1 (2)	L L (H)			L
	UT	-			1		1		4 (2)	L (H)			
	NM	X			104		104		4	L			
	CO	-			0		0		4 1	L			

		NEEDS							POTENTIAL				
ASA	State	Year 1985 Quantity Problem Indicated	*1000 Acres Irrigated Lands (ASA)	Sum	Year 2000 Δ_1	Sum	Year 2030 Δ_2	Sum	Heath Region No.	H,M,L	Field Studies (Published)	Verbal Information	Sum
1501			114	P		I		I	4 2	L H			H
	AZ	X			3		3		4 2	L H			
	NM	X			104		104		4	L			
1502			162	P		N		N	2 4	H L			H
	UT	X			1		1		2 4	H L			
	AZ	X			3		3		2 4	H L			

		NEEDS							POTENTIAL				
ASA	State	Year 1985 Quantity Problem Indicated	*1000 Acres Irrigated Lands (ASA)	Sum	Year 2000 Δ_1	Sum	Year 2030 Δ_2	Sum	Heath Region No.	H,M,L	Field Studies (Published)	Verbal Information	Sum
1502 (continued)													
	NV	X			0		0		2	H			
1503			155	P		I		I	2	H	X		HS
	AZ	X			3		3		2	H			
	NM	X			104		104		2	H			
1601			902	P		I		I	1 2 (4)	L H L	X	X	HSV

		NEEDS							POTENTIAL				
ASA	State	Year 1985 Quantity Problem Indicated	*1000 Acres Irrigated Lands (ASA)	Sum	Year 2000 Δ_1	Sum	Year 2030 Δ_2	Sum	Heath Region No.	H,M,L	Field Studies (Published)	Verbal Information	Sum
1601 (continued)													
	ID	-			17		51		1 2	L H			
	UT	X			1		1		2 1 (4)	H L L			
1602													
	UT	X	436	P	1	N	1	N	2 (4) (1)	H L L			H
1603													
	NV	X	805	P	0	N	0	N	2 3	H M			H
1604													
	NV	X	265	P	0	N	0	N	2 3	H M			H

		NEEDS							POTENTIAL				
ASA	State	Year 1985 Quantity Problem Indicated	*1000 Acres Irrigated Lands (ASA)	Sum	Year 2000 Δ_1	Sum	Year 2030 Δ_2	Sum	Heath Region No.	H,M,L	Field Studies (Published)	Verbal Information	Sum
1701			708	X		I		I	1 (3)	L M			M
	ID	-			17		51		1 (3)	L M			
	MT	-			8		9		1	L			
	WA	-			31		84		1 3	L M			
1702			2250	P		I		I	3 1	M L	X		MS
	OR	X			9		27		3 (1)	M L			

		NEEDS							POTENTIAL				
ASA	State	Year 1985 Quantity Problem Indicated	*1000 Acres Irrigated Lands (ASA)	Sum	Year 2000 Δ_1	Sum	Year 2030 Δ_2	Sum	Heath Region No.	H,M,L	Field Studies (Published)	Verbal Information	Sum
1702 (continued)													
	WA	X			31		84		3 1	M L			
1703			4101	P		I		I	3 1 2	M L H	X		HS
	WY	-			0		0		1	L			
	OR	X			9		27		3	M			
	ID	-			17		51		3 1 2	M L H			

ASA	State	NEEDS: Year 1985 Quantity Problem Indicated	*1000 Acres Irrigated Lands (ASA)	Sum	Year 2000 Δ1	Sum	Year 2030 Δ2	Sum	POTENTIAL: Heath Region No.	H,M,L	Field Studies (Published)	Verbal Information	Sum
1704			308	X		I		I	1 3	L M			M
	ID	-			17		51		1 (3)	L M			
	OR	-			9		27		3	M			
	WA	-			31		84		3	M			
1705			525	P		I		I	1 2	L H	X	X	HSV
	WA	-			31		84		1 2	L H			

ASA	State	Year 1985 Quantity Problem Indicated	*1000 Acres Irrigated Lands (ASA)	Sum	Year 2000 Δ_1	Sum	Year 2030 Δ_2	Sum	Heath Region No.	H,M,L	Field Studies (Published)	Verbal Information	Sum
1705 (continued)													
	OR	X			9		27		1 2	L H			
1706													
	WA	X	74	P	31	I	84	I	1 2	L H			H
1707													
	OR	-	492	X	9	I	27	I	3	M			M
1801													
			482	X		I		I	1 2 3	L H M	X		HS
	CA	-			0		0		1 2 3	L H M			

		NEEDS							POTENTIAL				
ASA	State	Year 1985 Quantity Problem Indicated	*1000 Acres Irrigated Lands (ASA)	Sum	Year 2000 Δ_1	Sum	Year 2030 Δ_2	Sum	Heath Region No.	H,M,L	Field Studies (Published)	Verbal Information	Sum
1801 (continued)													
	OR	-			9		27		3 1	M L			
1802	CA	X	2298	P	0	N	0	N	1 2 3	L H M			H
1803	CA	X	5608	P	0	N	0	N	1 2	L H	X	X	HSV
1804	CA	X	302	P	0	N	0	N	2	H	X		HS
1805	CA	X	472	P	0	N	0	N	2	H	X		HS

		NEEDS							POTENTIAL				
ASA	State	Year 1985 Quantity Problem Indicated	*1000 Acres Irrigated Lands (ASA)	Sum	Year 2000 Δ_1	Sum	Year 2030 Δ_2	Sum	Heath Region No.	H,M,L	Field Studies (Published)	Verbal Information	Sum
1806	CA	X	808	P	0	N	0	N	2	H	X	X	HSV
1807	CA	X	62	P	0	N	0	N	2 (1)	H L			H

Sources:

Published Case Studies
Verbal Information
National Water Summary 1983
$\Delta_1 = [(IWU_{2000}-IWU_{1985}) \div IWU_{1985}] \times 100$
$\Delta_2 = [(IWU_{2030}-IWU_{1985}) \div IWU_{1985}] \times 100$
Table 7
Phone calls (summary)
Heath Regions

CODE:

P: quantity problem exists
X: no quantity problem exists for first Sum column under needs; X under field studies and verbal information denotes availability of information; X under year 1985 quantity problem indicated denotes a problem
I: future increase in irrigated water usage
N: no future increase in irrigated water usage
S: published paper or reports (studies, demonstrations, projects)
V: telephone contact (verbal communication)
H: potential for artificial recharge is high
M: potential for artificial recharge is medium
L: potential for artificial recharge is low

INDEX